IET CONTROL ENGINEERING SERIES 96

Cyber-Physical System Design with Sensor Networking Technologies

Other volumes in this series:

Cyber-Physical System Design with Sensor Networking Technologies

Edited by Sherali Zeadally
and Nafaâ Jabeur

The Institution of Engineering and Technology

Published by The Institution of Engineering and Technology, London, United Kingdom

The Institution of Engineering and Technology is registered as a Charity in England & Wales (no. 211014) and Scotland (no. SC038698).

First published 2016

The Institution of Engineering and Technology
Michael Faraday House
Six Hills Way, Stevenage
Herts, SG1 2AY, United Kingdom

www.theiet.org

British Library Cataloguing in Publication Data
A catalogue record for this product is available from the British Library

ISBN 978-1-84919-824-0 (hardback)
ISBN 978-1-84919-825-7 (PDF)

Typeset in India by MPS Limited
Printed in the UK by CPI Group (UK) Ltd, Croydon

All chapters in this book have been rigorously peer-reviewed by the following international referees.

International Editorial Review Board:

To my wife Borrara, my daughters Zobia and Zofia, and my parents – Sherali Zeadally

To my beloved mother, my sons Jihed and Hedi, my wife Ines, and my brothers and sisters – Nafaâ Jabeur

Contents

Cyber-physical systems – List of acronyms

AC	alternating current
ADC	analog-to-digital converter
AHS	Automated Highway System
AMI	advanced metering infrastructure
AP	access point
API	application programming interface
APTEEN	adaptive threshold-sensitive energy-efficient sensor network
AQI	air quality index
ARQ	automatic repeat request
ASHRAE	American Society of Heating, Refrigerating and Air-Conditioning Engineers
BAN	bicycle-area network/body-area network
BOSS	beaconless on-demand strategy for geographic routing
BSN	body sensor network
CDMA	code division multiple access
CEP	complex event processing
CNT	carbon nanotube
CPS	cyber-physical system
CPSS	cyber-physical social system
CPU	central processing unit
CRN	cognitive radio network
DAHT	database active, human passive
DAST	dynamic application security testing
DBMS	database management system
DMM	data management module
DoS	denial of service
DPM	dynamic probabilistic model
DSMS	data stream management system
EBGRES	energy-efficient beaconless geographic routing

ECA	event chasing agent
EHR	electronic health record
EM	electromagnetic
ETOA	elapsed time of arrival
FCFS	first-come, first-served
FDMA	frequency division multiple access
FDS	fairness on dominant shares
FFD	full-function device
FPGA	field-programmable gate array
GAF	geographic adaptive fidelity
GEAR	geography- and energy-aware routing
GFJ	generalized fairness on jobs
GNR	graphene nanoribbon
GPRS	general packet radio service
GPS	global positioning system
GREES	geographic routing with environmental energy supply
GSM	global system for mobile communications
HCPS	human-centric cyber-physical system
HMM	hidden Markov model
HVAC	heating, ventilation and air-conditioning
IaaS	infrastructure as a service
ICB	intelligent control box
ICT	information and communication technology
ID	identifier
IEFT	Internet Engineering Task Force
IFP	information flow processing
IoT	Internet of Things
IP	internet protocol
IR	infrared
IS	intelligent system
ITS	intelligent transportation system
ITU-T	Telecommunication Standardization Sector of the International Telecommunication Union
iWSN	intelligent wireless sensor network
LEACH	low-energy adaptive clustering hierarchy
LMP	locational marginal price

LP	local point
LPOI	live point of interest
M2M	machine to machine
MAC	medium-access control
MANET	mobile ad-hoc network
MAP	memory authentication protocol
MCN	mobile cellular network
MCPS	mobile cyber-physical system/medical cyber-physical system
ME	mobile element
MEMS	micro-electromechanical systems
MOS	mean opinion score
MR	mobile relay
MRT	mass rapid transport
MSN	mobile wireless sensor network
NAHSC	National Automated Highway System Consortium
NASS	network-aware supervisory system
NBC	nuclear, biological and chemical
NoC	network-on-chip
NSF	National Science Foundation
OCI	on-chip interconnect
OS	operating system
P2P	point-to-point
PaaS	platform as a service
PCAST	President's Council of Advisors on Science and Technology
PDA	personal digital assistant
PE	processing element
PEGASIS	power-efficient gathering in sensor information systems
PIR	passive infrared
PMU	phasor measurement unit
POI	point of interest
PSNR	peak signal to noise ratio
QoS	quality of service
RCA	resource chasing agent
RF	radio frequency
RFD	reduced function device
RFID	radio frequency identification

RSS	received signal strength
RTO	regional transmission organization
SaaS	software as a service
SDN	software-defined networking
SIG	self-adapting intelligent gateway
SIS	self-adapting intelligent sensor
S-MAC	sensor-MAC
SNR	signal-to-noise ratio
SOA	service-oriented architecture
SoC	system-on-chip
SoS	system of systems
SPeC	smart pest control
SPIN	sensor protocols for information via negotiation
SSID	service set identifier
TDMA	time division multiple access
TEEN	threshold-sensitive energy-efficient network
UAV	unmanned aerial vehicle
V2I	vehicle to infrastructure
V2V	vehicle to vehicle
VANET	vehicular ad-hoc network
VCPS	vehicular cyber-physical system
WEAR	balanced, fault-tolerant and energy-aware routing
WLAN	wireless area network
WoT	Web of Things
WNSN	wireless nanosensor network
WSN	wireless sensor network
WUSN	wireless underground sensor network

Preface

Overview

We have witnessed significant technological advances in the fields of computing and communications over the past few decades. These advances have led to the emergence of a wide range of technologies (e.g. wireless embedded sensors and actuators, wired/wireless networks, machine-to-machine (M2M) communications) which currently underpin both cyber and physical systems. There is growing interest in academia and industry in a move towards seamless and tight integration (at all levels) of the networking, communication and computing capabilities of cyber systems with the control and management of entities (sensors, actuators, communication devices, etc.) that are located in the physical world, resulting in the development of cyber-physical systems (CPSs). Many CPS applications are currently being developed for a wide range of application domains such as healthcare, transportation, energy and manufacturing. These CPS applications will play a pivotal role in ensuring the safety, security and privacy of the many underlying physical systems which they control, as many people depend on these systems. To meet these goals, the design, implementation, deployment, management and maintenance of CPSs and their applications require a range of innovative systems and engineering solutions.

The main aim of this book is to present state-of-the-art research achievements and results in the field of CPS. The book will represent a valuable and authoritative point of reference for designers, developers, engineers, students, faculty members and researchers working in the area of CPS. We have organized this book into four parts: (1) CPS basics and fundamentals; (2) architecture, security, routing and data management in CPS; (3) resource management and reliability in CPS; and (4) CPS case study applications.

Cyber-Physical System Design with Sensor Networking Technologies is a collection of outstanding contributions from recognized experts around the world who are active in CPS research. We present below a summary of the various chapters this book covers.

Wireless sensor networks: basics and fundamentals

Chapter 1 presents the basic and fundamental concepts related to wireless sensor networks (WSNs) with a focus on the special case of wireless nanosensor networks (WNSNs). This chapter presents an overview of some types of wireless ad-hoc networks including mobile ad-hoc networks (MANETs), WSNs, vehicular ad-hoc networks (VANETs) and WNSNs' communication mechanisms and applications. The chapter further investigates the integration of WNSNs into embedded system designs and highlights their benefits. An example of WNSN as a communication medium for a system-on-chip design is also presented.

Cyber-physical systems: basics and fundamentals

Chapter 2 describes the basic and fundamental concepts of CPSs while providing the reader with an overview of the main characteristics and challenges related to the design and implementation of these systems. The chapter outlines some existing and emerging applications of CPSs including smart grid, intelligent vehicular management and agricultural applications. The chapter also discusses several research and development opportunities in the area of CPS.

Integrating wireless sensor networks and cyber-physical systems: challenges and opportunities

Chapter 3 focuses on the integration of WSNs and CPSs. It outlines the challenges posed by CPSs, which rely on inputs collected from large numbers of wireless sensors about the physical systems and events of interest. The chapter also discusses benefits brought about by leveraging the use of WSN capabilities when designing and deploying CPS applications. Architectural issues related to the integration of WSN and CPS, along with additional scientific, technical, social and institutional barriers for this integration, are also discussed. Examples of such barriers include power consumption, communication costs, dependability, security, quality of services, Big Data, and data analysis and modelling environments. Furthermore, this chapter highlights opportunities offered by the development of new integrated WSN-CPS concepts, models and applications.

Enabling cyber-physical system architectural design with wireless sensor network technologies

Chapter 4 highlights the main features of MANET, M2M, WSN and CPS technologies and compares them. The idea is to clarify the position of CPS with respect to some related technologies. The chapter also outlines the challenges related to CPSs' design and focuses on the main components of these systems by presenting some high-level architectures described in the literature. Since a CPS is better understood within its context of use, the chapter describes several examples of CPS architectures proposed for various domains, such as healthcare and intelligent transportation. To meet the requirements of these application domains, the role of WSN technologies and their contributions to CPS are highlighted. Finally, the chapter proposes a new architecture where WSN technologies, the multi-agent system paradigm and natural ecosystems are all integrated to enable CPS representation and operations.

Cyber security in cyber-physical systems: on false data injection attacks in the smart Grid

In Chapter 5, the authors review the coexistence of both CPS and WSN within the same system while highlighting their differences. They investigate various cyber threats in CPSs. To systematically study the effects of these threats, the authors take the example of the smart grid and present various false data injection attacks along with their formal models. Through a combination of theoretical analysis and performance evaluation, they discuss the effects of these attacks on electricity market operations.

Data management in cyber-physical systems with wireless sensor networks

In Chapter 6, the authors look to the integration of both WSN and CPS technologies from a data management perspective. They highlight the differences between WSNs and integrated WSN-CPSs. They describe some activities of data management when both technologies are integrated, with a particular focus on data collection in a mobile context. The authors investigate how data management in WSNs can drive the progress of CPS towards more intelligent applications. Within this context, they highlight important factors to add intelligence to the data management process, which may lead to more successful CPS applications. Furthermore, the authors identify the main opportunities and challenges that integrated WSN-CPS solutions have to deal with in the future.

Data routing in cyber-physical Systems

Sensors have limited computing resources and they face various challenges when routing data within the sensor network. When it comes to the extended context of CPSs, additional challenges must be addressed to route data, not only between sensors but also involving physical systems that may use heterogeneous communication technologies, data formats, quality of services and safety measures. To address these issues, Chapter 7 focuses on designing appropriate routing protocols for WSNs in the context of CPS applications. A comprehensive survey of routing techniques in WSNs for CPS from networks addressing, protocol operation/design, and network hierarchy perspectives, is therefore presented. The chapter also discusses future research directions related to routing protocols in an integrated WSN-CPS context.

Resource management in cyber-physical systems

A CPS strongly relies on the capabilities of WSNs to remotely control the spatially distributed physical systems. The limited capabilities of sensor nodes require efficient resource allocations to be implemented in order to maintain effective data collection and communication from the physical world. This issue is addressed in Chapter 8 from different aspects, including a single sensor perspective (to manage the limited power and bandwidth), network perspective (to find the optimum scheduling access to the shared wireless channel in order to reduce the interference) and design objectives related to the envisioned applications. The chapter addresses this multifaceted optimization problem of resource allocation using a game theoretical approach. It reviews the different aspects of resource allocation in WSN in the context of CPS. The chapter also discusses the implementation and the results of a multi-hop communication and cooperative relaying approach in WSNs. Finally, it highlights some future research and development directions particularly related to the issues of scalability, latency, throughput and fairness.

Mobile sensors in cyber-physical systems

The introduction of mobility in WSNs helps to overcome the problem of coverage, implement efficient data management and track objects and events of interest. In addition, mobility, in the context of CPS, enables sophisticated and ubiquitous mobile sensing applications, where mobile sensors are capable of acquiring data on mobile physical facilities, shifting the processing load to the areas that currently need extensive data collection and reporting, and maintains the reachability of physical systems. However, in order to adequately benefit from the mobility of sensors, several challenges must be addressed. Chapter 9 addresses these challenges through a survey of mobile WSN approaches and applications within a CPS context. These applications are then classified into three main groups and are compared within the context of every group. Moreover, the chapter also presents some future research challenges and opportunities for mobile WSN-CPS which require further investigation by researchers in industry and academia.

Intelligent wireless sensor networks (iWSNs) in cyber-physical Systems

Starting from the vision that CPS aims to bring smartness everywhere and in everything, Chapter 10 introduces the fundamentals of an intelligent wireless sensor network (iWSN) to meet this objective. The chapter presents the requirements, characteristics and applications of iWSN in the context of CPS. It highlights the differences between a commonly used WSN and an iWSN while focusing on the different levels of smartness needed by several applications. The chapter explores two particular applications: the smart grid and a smart field monitoring system for pest control. It then compares these applications in terms of the intelligence needed at different functionality levels, including energy efficiency, congestion control, adaptive routing, cross-layer design, programmability, data processing and reliability.

Resilient wireless sensor networks for cyber-physical systems

Resilience in WSNs is a critical issue as these networks are a key enabling technology in CPS. Resilience is particularly important because WSNs pose substantial security challenges, thereby affecting the closed control loop of CPS. This happens, for instance, when control decisions are being manipulated by attackers who have gained unauthorized access to the WSN. Chapter 11 provides an overview of key topics related to the resilience of WSN in CPS. It discusses potential applications of WSN in CPS followed by an overview of possible attacks against WSN. The chapter then formalizes the notion of attack resilience in WSN and discusses the main challenges to attack-resilient design. Furthermore, the chapter presents various approaches to achieve attack resilience along with detailed examples. Finally, it outlines interesting directions for future research and development works in the area of resilient WSN for CPS.

Case studies of integrated wireless sensor networks and cyber-physical systems

CPSs are enabling many opportunities that can enhance the interaction between the cyber world and the physical world. Chapter 12 systematically reviews WSN-CPS applications from static to dynamic networks, from small-scale to large-scale service coverage and from simple to more complex system interactions. More precisely, the chapter starts by investigating three types of smart space systems to control the usage of household appliances and utilities intelligently. These types of smart systems are fixed monitoring systems, energy-efficient smart buildings and preference-based smart home. The chapter examines two types of healthcare systems: sensor-assisted sleep-facilitating systems and body motion monitoring that focus on the identification of body motion patterns. Furthermore, the chapter studies the integration of real-time monitoring and intelligent decision-making in emergency response systems.

This chapter also presents a comprehensive study on how to infer everyday human activities automatically and reviews five types of applications, including conversation behaviour analyzer, mood detection, road safety helpers, social activity inference and virtual–physical games. In addition, interesting applications in a smart city, including urban-scale environmental monitoring, urban mobility and activity diaries, and intelligent transportation systems, are also reviewed. A fine-grained classification based on technical features and requirements of WSN-CPS systems is also discussed along with highlights of some important challenges that need to be addressed in the future.

Medical cyber-physical system

Chapter 13 addresses the issue of medical cyber-physical systems (MCPSs). It outlines the background and related works which include model-based developments, medical guidelines and user-centric design. It also describes functional active and passive data acquisition methods as well as functional sensor interpretation methods. Additionally, the chapter discusses the challenges and opportunities of MCPSs, including real-time clinical decision support. Throughout the chapter, special attention is given to data semantics.

Acknowledgments

We express our deepest gratitude to the authors who contributed their chapters to make this book possible. We are thankful to all the reviewers who spent their precious time providing comments and valuable feedback that greatly helped the authors improve the quality and presentation of their chapters.

Finally, we also thank the following staff at IET: Paul Deards, Jennifer Grace and Joanna Hughes for their patience, constant support and advice throughout the preparation of this book.

We hope you will enjoy reading this book as much as we enjoyed editing it.

Sherali Zeadally and Nafaâ Jabeur

Chapter 1

Wireless sensor networks: basics and fundamentals

O. Yalgashev[1], M. Bakhouya[2], A. Nait-Sidi-Moh[3] and J. Gaber[1]

Abstract

A wireless ad-hoc network is a decentralized type of wireless network, where communication does not rely on a pre-existing infrastructure, such as routers in wired networks or access points in infrastructure-based wireless networks. Ad-hoc communications are present in wireless sensor networks (WSNs), mobile ad-hoc networks (MANETs) and vehicular ad-hoc networks (VANETs). Recently, wireless nano sensor networks (WNSNs) have emerged as a typical subclass of WSN, but at the nanometer scale. The vision of WNSN could achieve the functionality and performance of today's WSN with the exception that node size is measured in nanometers and up to hundreds or thousands of nanometers physically separate channels. In addition, nodes are assumed to be mobile and rapidly deployable. In this chapter, we survey existing wireless ad-hoc communication mechanisms and their applications. We then focus on WNSN and its applications in system-on-chip together with some preliminary results.

1.1 Introduction

A Wireless Ad hoc networks (WSNs, VANETs, MANETs) are self-organizing networks that are created from distributed autonomous nodes without wired infrastructure. These nodes operate cooperatively to achieve specific tasks and monitor physical or environmental conditions. The development of these networks is motivated by unlimited applications such as health care, biomedical, environmental,

[1]University of Technology of Belfort-Montbéliard, 90010 Belfort, France, e-mail: olimjon.yalgashev@utbm.fr, gaber@utbm.fr
[2]International University of Rabat, Parc Technopolis, Salé, Morocco: e-mail: mohamed.bakhouya@uir.ac.ma
[3]University of Picardie Jules Verne, INSSET, 02100 Saint Quentin, France, e-mail: ahmed.nait-sidi-moh@u-picardie.fr

industrial and military applications. The emergence and development of wireless communications, micro-electromechanical systems technology and digital electronics has allowed these networks to be widely deployed and explored. The contribution of these systems to the improvement of users' daily activities has been strongly observed. Factors contributing to this evolution include the development of wireless communication systems, data processing and related technologies such as routing techniques and media-access control (MAC) protocols and the understanding and exploitation of the potential of the impressive growth of wireless networks. Furthermore, wireless ad-hoc applications occupy an important place in our daily activities and in various fields (e.g. security, health, business, environment). For example, sensor nodes are able to monitor a wide variety of states and ambient conditions including vehicular movements, detection of certain obstacles on the road, vehicular speed, temperature, humidity, pressure, etc.

Recently, WNSNs (wireless nanosensor networks) have emerged as a typical subclass of WSNs, but at nanoscale (1–100 nanometers, 1 nm $= 10^{-9}$ m). The vision of WNSN could achieve the functionality and performance of today's WSN with the exception that nodes are very tiny devices able to perform very simple computation, sensing and/or actuation tasks. In addition, nodes are assumed to be mobile and rapidly deployable. However, in order to develop WNSN applications and services, many research issues still have to be investigated. In this chapter, we survey existing wireless ad-hoc types and measure the performance of WNSN as a communication fabric for system-on-chip.

The remainder of this chapter is structured as follows. An introduction to wireless ad-hoc network types will first be presented in Section 1.2. In Section 1.3, an overview of wireless nanosensor networks together with existing communication mechanisms will be presented. Our preliminary investigations in using WNSN as a communication medium for system-on-chip design will be presented in Section 1.4. A summary and discussion will be presented in Section 1.5.

1.2   Wireless ad-hoc networks

A wireless ad-hoc network is a decentralized type of wireless network. In this type of network, each node participates in the routing process by forwarding data to other nodes according to the network connectivity [1]. There are three main categories of wireless ad-hoc networks: WSNs, MANETs and VANETs. Table 1.1 illustrates the difference and common characteristics of WSN/WNSN, MANET and VANET, in terms of node storage, communication carrier, energy consumption, etc. These types of ad-hoc networks are described in the rest of this section.

In WSN, sensors are hardware devices that produce a measurable response to a change in a physical condition such as temperature or pressure. Sensors measure the physical data of the parameter to be monitored. These devices also should operate as sensors of WSN. Molecules, or hand-made tiny devices that are called nanomachines (or nanodevices, nanosensors), are components of a WNSN. The components of MANET are personal digital assistants (PDAs), mobile phones and smartphones that

Table 1.1   Requirements and common characteristics of wireless ad-hoc networks (*depending on the scenario)

Requirements/ characteristics	WSN/WNSN	MANET	VANET
Nodes	Sensors, nanosensors	Devices (PDA, smartphones)	Vehicles, roadside units
Communication infrastructure	Infrastructure-less	Infrastructure-less	Infrastructure-less
Power consumption	High	Medium	Low
Mobility	Limited*	Low	High
Communication mechanism	Wireless, molecular*	Wireless	Wireless
Data storage	Low	Medium	High
Heterogeneity	Low	High	High
Security	Low	High	High
Scalability	High	High	High

move slowly and randomly. Vehicles and roadside units are nodes of vehicular communication system that provide different warnings and road information [2].

WSN/WNSN mostly use a broadcast communication paradigm and most ad-hoc networks are based on point-to-point communications. It is stated that WNSN is a subclass of WSN; despite the tiny size of the nanomachines, these machines, like WSN devices, are equipped with batteries according to their size. This means that WSN/WNSN devices are limited in power. These networks also use broadcasting protocols and, consequently, power consumption of WSN/WNSN is a major concern.

All the above-mentioned networks use wireless communication mechanisms. The communication mechanisms of WNSN can be classified into four categories: molecular (or chemical), electromagnetic, nano-mechanical and acoustic. These communication mechanisms are described in Section 1.3.1. WSN and WNSN devices are designed to be micro-sized, therefore the data storage of these devices is low-volume. As mentioned earlier, in terms of MANET and VANET, these devices have higher computational and data storage capabilities.

Security in WSN, WNSN, MANET and VANET networks is an ongoing research issue. When it comes to MANET and VANET, devices move in open areas, and exchanged information might be accessed. WNSNs are initially designed to be deployed in closed areas, such as a human or animal body, and data security is not a major concern.

In addition to security, scalability is another important issue that needs to be addressed when designing these networks. For example, in some ad-hoc network scenarios, the network can be composed of hundreds or thousands of nodes. This means that protocols for ad-hoc networking must be able to operate efficiently in the presence of a very large number of nodes also. WSNs are typically composed of hundreds or even several thousands of nodes. Thus, the scalability of protocols for WSNs must be explicitly considered at the design stage.

1.2.1   Wireless sensor networks

The wireless sensor network (WSN) is a particular type of ad-hoc network, in which nodes are 'smart sensors/devices'. These devices are equipped with advanced sensing functionalities (thermal, pressure, acoustic), a small processor and a short-range wireless transceiver. In this type of network, sensors exchange information in order to build a global view of the monitored region, which is made accessible to the external user through one or more gateway node(s). Sensor networks are expected to herald a breakthrough in the way natural phenomena are observed; the accuracy of the observation will be considerably improved, leading to a better understanding and forecasting of such phenomena. The expected benefits to society will be considerable [3].

WSNs are expected to have significant impact on the efficiency of many fields including civil and military applications. A sample application scenario could be a situation in which a WSN is used to monitor a vast and remote geographical region; in such a way, abnormal events (e.g. a forest fire) can be quickly detected. In this scenario, smart sensors, each equipped with a battery and significant processing and wireless communication capabilities, are placed in strategic positions – for example, on the top of a hill or in locations with a panoramic view. Each sensor covers an area of a few hectares and can communicate with other sensors in the vicinity. The sensor node gathers atmospheric data (e.g. temperature, pressure, humidity, wind velocity and direction) and analyzes the make-up of the atmosphere to detect particular particles (e.g. ash). Furthermore, each sensor node is equipped with an infrared camera, which is able to detect thermal variations. Every sensor knows its geographic position via localization algorithms [4], expressed in terms of degree of latitude and longitude. This can be accomplished either by equipping every node with a GPS receiver, or, since in this scenario sensor position is fixed, by setting the position in a sensor register at the time of deployment. Periodically, sensors exchange data with neighboring nodes in order to detect unusual situations that could be caused, for instance, by the start of a fire (e.g. temperature at a sensor is much higher than that of its neighbors).

These routine data are aggregated and propagated throughout the network and can be gathered by the external operator to collect atmospheric data (e.g. to check the air quality). When a potentially dangerous situation is detected (e.g. the infrared camera detects a rapid thermal increase in a certain zone), an emergency procedure is initiated. In fact, the sensor node that has detected the abnormal condition communicates with its neighbors in order to verify whether the same condition has been detected by other sensors; then, it tries to accurately determine the geographic position of the hazard. If the same abnormal situation has been detected by other sensors, this can be accomplished using triangulation techniques. Furthermore, the information on the wind velocity and direction can be useful both in the localization of the fire and in forecasting the direction of its propagation. Once the position of the fire has been determined, an alarm message containing the fire's geographic coordinates and (possibly) its propagation direction is disseminated with the maximum priority. The external operator (e.g. a park ranger equipped with a portable device) is promptly alerted of the presence of fire, its position and the forecasted propagation direction of the fire, and he or she can quickly intervene [3].

In addition, WSNs are deployed widely in various applications such as structural health monitoring [5] and monitoring of the agricultural environment [6]. For example, Hwang *et al.* [6] discuss how WSN-based sensors are used to collect environmental and soil information on the outdoors and to collect location information using GPS modules for producers. WSN technology has also been used for habitat monitoring [7]. In fact, habitat and environmental monitoring represent an important class of sensor network applications. Furthermore, WSNs offer great promise for information capture and processing for military and defense applications. The military is using this concept rigorously for their advancement in warfare and in the field [8]. A WSN is typically composed of nodes with the same features. Typically, the number of nodes composing a WSN is quite large, ranging from a few tens to thousands of nodes that are dispersed in a relatively large geographical region, so that 1-hop communication between nodes is, in general, not possible.

As mentioned previously, WNSNs have emerged as a typical subclass of WSN, but at the nanometer scale. There are a number of applications that might be designed only in nanoscale. For example, in military applications, chemical and biological nanosensors can be used to detect harmful chemical and biological weapons in a distributed manner. However, taking into account that these sensors need direct contact with the molecules, having a network with a very large number of nanosensor nodes is necessary [9]. Moreover, WNSN might operate for monitoring blood composition of the human body for detecting an infection or a cancer. Further details regarding technologies and applications based on WNSN are presented later in this chapter.

1.2.2 Mobile ad-hoc networks

MANET is a special form of ad-hoc network that differs from WSNs in respect of its node mobility features. It is a self-configuring network of autonomous and mobile nodes (devices) that do not rely on a predefined architecture. In a MANET, each node is free to move independently from other nodes and in any direction and at any moment. The nodes communicate with each other via wireless links while maintaining their connectivity in a decentralized way. A node frequently changes its links with other nodes without notice, thus routing in such networks is a challenging issue. The main role of each node in a MANET is to continuously maintain the information required to its own usage as a host and be a router for neighbor nodes enabling the information transmission.

Combining WSN and MANET technologies provides powerful, mobile and headline applications in various domains. It has been shown in the literature that most of the cooperative networks developed are based on the coordination of different spatially distributed nodes within a wireless network, such as MANET and WSN. In Samanta and Pal [10], a conceptual framework of cooperative network was proposed for coordination among different wireless networks to achieve perfect overall network performance with anytime and anywhere connectivity. Several network designs and technologies are available in the wireless domain. Each one of these technologies and designs arose to fulfill specific technological needs and requirements.

Much research work has been conducted and has targeted the development of applications and prototypes. Among these applications, the MANET principle is applied for military communication and operations; emergency services (disaster recovery, policing and firefighting, rescue operations, etc.); commercial and civilian environments (e-commerce, vehicular services, sports stadiums, etc.); home and enterprise networking (home/office wireless networking, meeting rooms, personal area networks, etc.); education and learning (virtual classrooms, ad-hoc communication during meeting, etc.); context-aware services (information services, follow-on service, etc.); and sensor networks (data tracking, movement detection, smart sensors, etc.) [11].

MANET resources are highly crucial for the successful communication between mobile nodes. In situations where both sending and receiving nodes are placed within the transmission range of each other, communication is possible through a single-hop connection. In all other scenarios where nodes are distanced, the exchange of packets is possible as long as a multi-hop path is available between them. Despite the unique characteristics of MANETs, they share many attributes and operations with other traditional networks [12].

1.2.3   Vehicular ad-hoc networks

Vehicular ad-hoc networks are a special subclass of MANETs for inter-vehicle communication and have a relatively more dynamic nature compared to MANETs due to the rapid network topology changes. In VANETs, nodes are vehicles equipped with embedded calculators, sensors and wireless communication technologies. Unlike MANETs, where nodes can freely move in a certain area, the movement of vehicles in VANETs is always at high speed and can be predicted, because it is dependent on streets, traffic and specific rules. Communication between nodes in VANETs is less reliable due to the high mobility, dynamic topology and different traffic patterns compared to MANETs [13].

Due to the large requirement of intelligent transportations systems (ITSs), VANET constitutes an emerging branch of wireless ad-hoc network where communication can be conducted between vehicles, vehicle to vehicle (V2V) and/or between vehicle and infrastructure (V2I) [14–16]. It is seen as hopeful technology for the efficiency, safety, comfort and assistance of users in ITS [2]. In a VANET, routing issues and maintaining the communication between vehicles – considering the highly dynamic topology of the network, which is frequently disconnected, and mobility constraints – are important challenges. In recent decades, with the aim of increasing the levels of intelligence and safety in transportation systems, VANET technology has become an attractive research area for many research communities [17], [18].

As has been the case in many research works, as well as under real-life driving conditions, VANET technology may fail to maintain communication connectivity and ensure timely detection of dangerous road conditions when the network density is low. Since vehicles have high velocity in a VANET, they could be disconnected from each other. Therefore, the density of VANET might be low and not all nodes in the network might receive messages. Combining VANET technology with other technologies is one of the options to address this problem. The integration of WSN and VANET overcomes the inherent limitations of using VANET alone [15]. In Hua *et al.*'s research work, the authors have proposed a VANET–WSN system for timely detection of road conditions and to help connect partitioned segments of VANET. This system has been developed to collect, process, share, and deliver real-time information about road conditions in order to make roads safer, cleaner and smarter. As detailed in references [16] and [19–21], sensor nodes can be deployed along the infrastructure with higher density than current roadside stations, enabling a strong connectivity with the VANET. The sensor nodes can sense the road conditions, collect and process the sensing data to find out useful information for safe driving and deliver the information to vehicles. Some examples showing the effectiveness of integrating WSNs with VANETs in preventing road accidents are given in reference [15]. These include: deploying WSN along rural roads in order to prevent vehicle–animal collision; and detecting, relaying and propagating the information about bad infrastructure and roads from connected sensors to any vehicle approaching the dangerous area. Furthermore, the VANET–WSN concept was used for collecting state messages from roadside sensors for the routing protocol with the aim of assisting drivers and for driving safety. WSN was used also as a solution for smart transportation [22]. In fact, with their low power consumption of node sensors, their wireless distribution and their flexibility without cable restrictions, WSNs have been used for transportation information collection and communication. These wireless networks have proved their effectiveness in the field of ITS.

1.3   Wireless nanosensor networks (WNSNs)

In 1959, Richard Feynman, the Nobel Laureate physicist, in his famous lecture entitled 'There's plenty of room at the bottom', described for the first time how the manipulation of individual atoms and molecules would give rise to more functional and powerful devices [23]. This was the vision of many researchers who were willing to explore matter at the nanoscale. For example, Akyildiz *et al.* [23] have stated that nanomachines could be classified into two categories: those that mimic existing electro-mechanical machines and those that mimic the nanomachines of nature (e.g. molecular motors) as illustrated in Figure 1.1.

The aim of nanotechnology is to create nanomachines with new functionalities stemming from these unique characteristics, not just developing miniaturized classical machines [24]. The vision of WNSN could achieve the functionality and performance of today's WSN with the exception that device size is measured in nanometers and up to hundreds or thousands of physically separate nanometer channels. In addition, nanomachines are assumed to be mobile and rapidly deployable. Therefore, ad-hoc communication mechanisms are required to allow these entities to exchange sensed information between each other and with external units.

In this research direction, nanotechnology has emerged as a field of manipulating individual units at the atomic or molecular level. According to the US

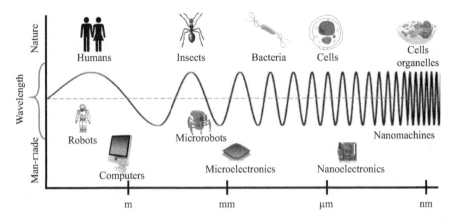

Figure 1.1   The vision of shifting from top-down to bottom-up machine design

National Nanotechnology Initiative, nanotechnology is a research field for manipulating matter with at least one dimension sized from 1 to 100 nanometers. At this scale, a nanomachine can be considered as the most basic functional unit. Nanomachines are tiny components consisting of an arranged set of molecules, which are able to perform very simple computation, sensing and/or actuation tasks [25]. Nanomachines can be further used as building blocks for the development of more complex systems such as nanorobots and computing devices such as nano-processors, nano-memory or nano-clocks [23]. More than half a century later, nanotechnology is providing a new set of tools to the engineering community to control entities at the atomic and molecular scale. Foremost among these new capabilities are nanomachines, integrated functional devices consisting of nanoscale components. Nanomachines used in applications today typically operate independently and accomplish tasks ranging from computing and data storage to sensing and actuation [26].

Nanomachines should have a sensing ability that is capable of sensing events within the environment and propagating these to other members of the WNSN. To perform these functions, it must have units that act as sensor, power supplier, processor, storage unit and communicator [24].

1.3.1   Nano-communication mechanisms

Nano-communication networks are communication networks that exist at the nanometer scale. Classical communication paradigms need to undergo a profound rethinking and redesign in order to meet the requirements (e.g. size, power consumption, etc.) of new WNSN applications. Existing networking architectures and communication protocols/software have to be completely rethought in light of these new communication paradigms [26]. Nanomachines can be interconnected to execute collaborative tasks in a distributed manner. The resulting WNSNs are envisaged to expand the capabilities and applications of single nanomachines in the following ways. For example, nanomachines such as chemical sensors, nano-valves, nano-switches, or molecular elevators [27] cannot execute complex tasks by themselves. The exchange of information and commands between networked nanomachines will allow them to work in a cooperative and synchronous manner to perform more complex tasks such as in-body drug delivery or disease treatments. WNSN will allow dense deployments of interconnected nanomachines. Thus, larger application scenarios will be enabled, such as the monitoring and control of chemical agents in ambient air. In some application scenarios, nanomachines will be deployed over large areas, ranging from meters to kilometers. In these scenarios, the control of a specific nanomachine is extremely difficult due to its small size. WNSN will thus enable interaction with a remote nanomachine by means of nano-communication mechanisms.

Communication between nanomachines might be designed through electro-magnetic, molecular (chemical), nanomechanical or acoustic communication mechanisms [27]. In acoustic communication, the transmitted message is encoded using acoustic energy, i.e. pressure variations. At the nano-level, acoustic commu-nication is mainly based on the transmission of ultrasonic waves. Similar to the com-munication based on electromagnetic waves, the acoustic communication implies the integration of ultrasonic transducers in the nanomachines. These transducers should be capable of sensing the rapid variations of pressure produced by ultrasonic waves and emitting acoustic signals accordingly. Again, the size of these transducers represents the major barrier in their integration into the nanomachines. The propagation speed of signals used in traditional communication networks, such as electromagnetic or acoustic waves, is much faster than that of molecular messages [28], [29].

Nanomechanical communications is defined as the transmission of information through mechanical contact between the transmitter and the receiver, i.e. infor-mation flows through the mechanical connection of nanoscale devices. Molecular communication-based WNSNs use molecules and their components, such as gap junctions, receptors and nucleus [29]. Molecular communication between nanos-cale entities occurs in nature; it explains why using molecular communication is particularly useful. Moreover, the WNSN built upon such naturally occurring phenomena with appropriate instrumentation offers a faster engineering pathway to viable solutions. The most important reason is that several applications require biocompatibility and hence require properties that are readily offered by WNSNs that use molecular communication [27].

Electromagnetic (EM) radiation-based communication is rapidly gaining the attention of the scientific community. In fact, it is already possible to exploit some features of nano-materials for performing, at the nanoscale, the transmission and processing of electromagnetic signals. Research on WNSN is still at an early stage and a number of critical issues have to be addressed before the widespread adoption of such technology in the biomedical, industrial and military fields [24].

Recent advancements in molecular and carbon electronics have opened the door to a new generation of electronic nano-components, such as nano-batteries, nano-memories, logical circuitry at the nanoscale and nano-antennas. From a commu-nication perspective, the unique properties observed in novel nano-materials will determine the specific bandwidths for emission of electromagnetic radiation, the time lag of the emission or the magnitude of the emitted power for a given input energy. All these require a fundamental change in the current state of the art of analytical channel models, network architectures and communication protocols [24].

It is worth noting that terahertz band (0.1–10 THz) communication might allow for the designing of EM radiation-based WNSN. The terahertz band is between infrared and microwave radiation in the electromagnetic spectrum. Therefore, this band might present some characteristics of neighboring radiations. The terahertz band communication channel depends on the molecular composition of the medium and transmission distance. In the very short range, i.e. for a transmission distance of the order of several tens of millimeters, the terahertz band can be considered as a single transmission window almost 10 THz wide. This is the main difference with existing terahertz communications systems, which are focused on using a single transmission window below 350 GHz [30].

In WNSN, nano-antennas perform radiation and reception processes at the nanoscale; they might be several nanometers wide and a few micrometers long. Moreover, nano-antennas should be easily integrated into nanomachines. In the development of nano-antennas, the use of carbon nanotubes (CNTs) and graphene nanoribbons (GNRs) could prove effective. In the modeling of nano-antennas based on nanotubes and nanoribbons, the specific operating range, the bandwidth radiation and the efficiency of radiation should be taken into consideration. All these will determine the communication capabilities of nanosensor devices [24].

1.3.2   Applications

The potential applications of WNSN are unlimited. We classify them into four groups: biomedical, environmental, industrial and military applications. However, since nanotechnologies have a key role in the manufacturing process of several devices, WNSN could be used extensively in many other fields such as consumer electronics, lifestyle and home appliances, among others. The most direct applications of nanomachines and WNSN are in the biomedical field. Biological models inspire and encourage the use of nanotechnologies to interact with organs and tissues. The advantages provided by WNSN are clearly in terms of size, biocompatibility and biostability, enabled by the control of nanomachines at the molecular level. More precisely, since WNSN applications play an important role in the healthcare domain, in designing nanodevices, compatibility of these devices with the biological environment is very important, and this is called the biocompatibility issue. The main factor is related to the rejection of implants and drugs by the host organism. Accordingly, the solution for this biocompatibility problem lies in using nanotechnology to design nanodevices that are more suited to biological organisms. When it comes to the use of WNSN in the biological environment, biostability is a critical issue. Biostability is the ability of nanomachines to maintain their integrity following implantation in the biologic environment.

Here are some of the envisaged applications: immune system support, bio-hybrid implants, drug delivery systems, health monitoring and genetic engineering. WNSN will be used not only in intra-body but also in industries. WNSN can help with the development of new materials, manufacturing processes and quality control procedures. More specifically, these applications have already been proposed for food and water quality control, functionalized materials and fabrics.

Nanotechnologies can also have several applications in the military field. While in the applications mentioned earlier, the range covered by WNSN is short, in the military field the deployment range of WNSN can be widely variable depending on the application. Battlefield monitoring and actuation demand a dense deployment of WNSN over large areas, while systems aimed at monitoring soldier performance are deployed in smaller areas, i.e. the human body. Among military applications, we can include nuclear, biological and chemical (NBC) defenses and nano-functionalized equipment. Since WNSNs are inspired from biological systems found in nature, they can also be applied in environmental fields to achieve several goals that could not be solved with current technologies. Some environmental applications are biodegradation, animals and biodiversity control and air pollution control [24].

Several issues need to be addressed to deploy WNSN applications. For example, the physical and practical limitations of nanomachines should be taken into consideration in the development of communication techniques between nanomachines. In the implementation process of WNSN, an interface that performs interconnection between the nano and micro world should be realized. For example, selecting an effective transmission range is a critical issue in the design and performance of WNSN applications. Increasing the transmission range increases the carrying capacity of the network. Choosing too high a transmission range increases the number of forwarding nodes needed to reach the intended destination. Conversely, choosing a lower transmission range reduces the interference seen by potential transmitters but packets require more forwarding nodes to reach their intended destination. Obviously, increasing the number of forwarder nodes increases the energy requirement of any network. Our preliminary investigation of the EM-based broadcasting mechanism for nano-communication shows that in order to limit packet flooding while increasing throughput and decreasing latency according to a given network density, the nodes' transmission range needs to be tuned. In other words, in forming well-connected WNSN, the transmission range of nanonodes should be carefully tuned according to the network density. For example, a larger transmission range will increase the number of neighbors, which could guarantee a low latency with high packet delivery ratio. For example, increasing the transmission range too high might increase the probability of packet collision, and shortening it too much might disconnect nodes from each other. In this case, a small number of packets will be delivered with high latency, since the number of hops among nanonodes will be high.

Taking into account the above-mentioned problems associated with the transmission range, approaches are required to allow adaptation of the transmission range of nanonodes based on network density. An adaptive transmission range of an EM-based communication mechanism has been introduced by Yalgashev *et al.* [31]. We consider that nanonodes are replaced in a cuboid and they move according to the Gauss–Markov mobility model. In our first evaluation of WNSN, several issues have been noticed. For example, when a node should send or rebroadcast a packet, it checks first for its neighbors using information from the channel layer.

If there are no neighbor nodes in its coverage area, packets will be destroyed. There is a possibility that the packet could be got/returned back/from to the transmitter of other neighbor nodes after several hops or directly after transmission. In this case, packets will be also destroyed. In order to avoid this issue, we should consider the following cases: (i) in sparse WNSN, the probability of finding neighbors is very low. In this case, the transmission range of nanonodes needs to be adapted to retain network connectivity. (ii) In dense WNSN, throughput could be low because of collisions. In this case, the transmission range needs to be adapted to minimize the collisions.

It is worth noting that when we assign a short/long transmission range in a dense/sparse network, results for throughput and average latency are better compared to a fixed transmission range. For adapting the node's transmission range, the two cases described above have been considered. In the first case, when an intermediate transmitter has no neighbors, we assign a higher transmission range in the channel layer, which might increase the number of neighbors. In the second case, the transmission range takes the value of the average distances to all its neighbors [31].

The performance evaluation study was conducted using a simulation tool, named Nano-Sim, for EM-based WNSN. In this work, we have used the flooding algorithm already implemented in Nano-Sim. In this algorithm, packets travel in the WNSN hop by hop, until they reach the gateway node. Nanonodes generate packets according to a certain time interval and broadcast them to neighbors. Each packet is identified with a unique packet ID and a source node ID.

Simulations have been conducted using several parameters as follows: 100 to 500 nanonodes are randomly generated. The position of each nanonode may change during simulation time and in accordance with the mobility model. Moreover, the simulation area is finite and nanonodes can move only inside this area. Nanonodes collect information about the surrounding environment and periodically broadcast them to neighboring nodes. Thus, packets are broadcast from one to others until they reach the gateway. The gateway node is an interface between the WNSN and the external network, e.g. the internet [31].

For example, Figure 1.2 presents the comparison of average latency when we adapt and fix transmission ranges in nano-networks with dense, medium and sparse networks. Simulation results were better with adaptive transmission range in terms of average latency compared with fixed transmission range. When we fix a 0.002 value for transmission range, average latency is the lowest, but throughput is almost near to zero. In case of fixing 0.006 for transmission range, we observed the highest latency compared with others. This could be explained by the high number of hops required to deliver packets among nanonodes. When we fix 0.008 for the transmission range, it can be seen that latency is lower and throughput is higher. When we use an adaptive transmission range, the average packet delivery time gets shorter for all nano-networks. It can be observed that when we use an adaptive transmission range, the average latency is low.

The obtained results show that adapting the transmission range of nanonodes gives good results for throughput and latency. Our ongoing work focuses on

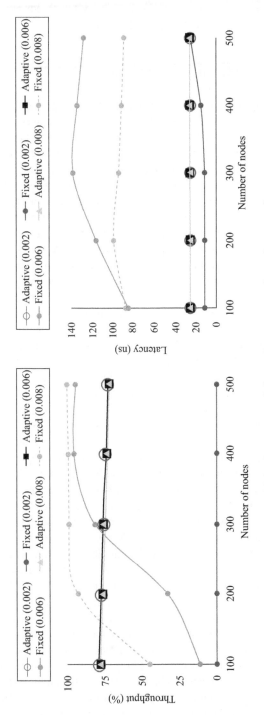

Figure 1.2 Latency and throughput vs. number of nanonodes

developing adaptive broadcasting approaches in order to further minimize the broadcast messages while achieving good throughput.

1.4   WNSN for embedded system design

In parallel to this progress, industry is reaching limits regarding the speed of processors that can be placed in an on-chip system with low power consumption and heat dissipation. In other words, embedded computing systems have become widespread in various application domains, as evident from their use in products such as PDAs, household appliances, telecommunication and automotive systems. This has been further accelerated by the advances in silicon technology, which has led to the design of complex and large systems-on-chip (SoCs). More precisely, rapid advances in technology and design tools have enabled engineers to design SoC with hundreds of cores. These systems are composed of several processing elements (PEs) and dedicated hardware and software components that are interconnected by a network-on-chip (NoC). According to Moore's law, the number of cores on-chip will double every 18 months, and therefore thousands of cores on-chip will be integrated in the next 20 years in order to meet the power and performance requirements of applications. The current and future tendency will be the integration of more and more cores in a single chip. A key element in the performance of, and energy consumption in, SoCs is the NoC infrastructure [32]. More precisely, while microchip technology miniaturizes and gates become faster and more energy efficient, wires used for communication between cores are performing slowly and are more energy hungry. Therefore, NoC infrastructure represents one of most important components that determine the overall performance (e.g. latency and throughput), reliability and cost (e.g. energy consumption and area overhead) of future SoC [33].

Recently, radio frequency (RF)/wireless NoC, 3D NoC and photonic NoC have been introduced to alleviate these problems by, for example, replacing some of the multi-hop wire links with high-bandwidth single-hop long-range wireless channels [34]. Several studies have shown that wireless NoCs are promising alternatives to conventional OCIs (on-chip interconnects), which mitigate latency and energy dissipation in communication between remote nodes. Other studies show that nano-photonic communication is expected to reach levels of performance-per-watt scaling by supporting energy-efficient high-bandwidth data transfers among processing cores. Carloni *et al.* [35] have stated clearly that 3D NoC, nano-photonic communication and on-chip wireless links are all promising alternatives to traditional interconnects for building on-chip interconnects of future multi-core SoC.

The advantages of nano-NoC-based architectures include high scalability, low latency, high bandwidth, distributed routing decisions and low energy consumption. In our recent work, the performance of a nano-NoC is evaluated using throughput, average latency and energy consumption. Figure 1.3 shows an example of nano-on-chip topology in which every nanonode is coupled with a nano-router.

Preliminary results using traffic models (Uniform, Transpose, Bit-Reversal and Shuffle) most widely used for the analysis of energy consumption and latency in

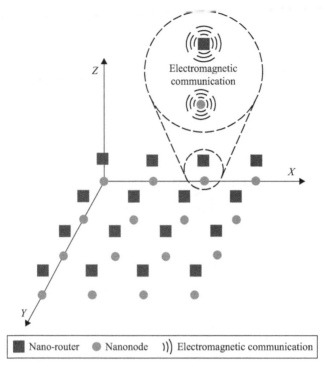

Figure 1.3 Nano-NoC-Like-Mesh topology

interconnection networks are shown in Figure 1.4. It can be seen that throughput is higher with the growing number of nodes when we use Bit-Reversal and Uniform traffic patterns. When we use Shuffle and Transpose traffic patterns, all packets arrive at destination nodes. Regarding latency, we can observe that the Shuffle traffic pattern gives the best results, and Bit-Reversal traffic pattern gives the worst. This is due to the selection rules of source and destination nodes in each traffic pattern. Furthermore, less energy consumption is achieved by using the Shuffle traffic pattern.

1.5 Summary and discussion

Nanotechnology is one of the fields of fundamental sciences and applied technologies that enables the design and engineering of miniaturized nanomachines at the scale of a few nanometers. These nanomachines are very tiny devices able to perform very simple computation, sensing and/or actuation tasks. Like other wireless ad-hoc networks, these machines could be interconnected in order to increase their capabilities and then execute collaborative tasks in a distributed manner. Several applications, ranging from biomedical, environmental, industrial, to military services, could benefit from this new communication paradigm. For example, the largest number of applications is in the biomedical field.

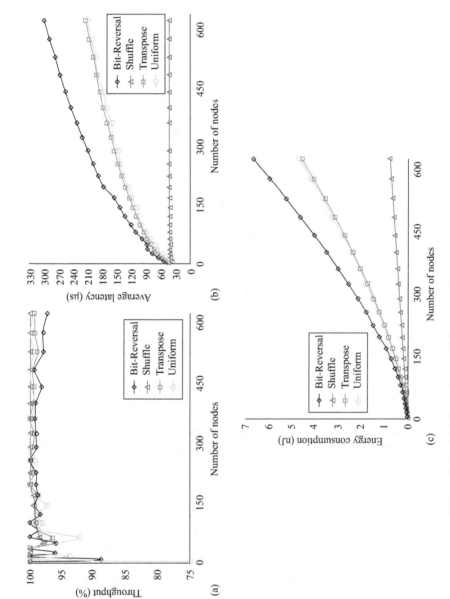

Figure 1.4 (a) Throughput, (b) latency, (c) energy vs. number of nanonodes

Nano-networks could also be deployed in industries to help develop new quality control procedures and manufacturing processes. These applications have been proposed to improve systems of controlling quality of food and water, materials and fabrics. However, research into nano-networks is not yet mature. Specifically, analytical performance evaluation and simulation tools are required to study new protocols and standards. Furthermore, new cost-effective solutions are required for communication among thousands of distributed nanomachines. For example, adaptive transmitting range for connectivity and energy consumption is one of the most important problems in nano-networking. Several adaptive techniques have been proposed in the context of ad-hoc networks (e.g. MANET, WSN, VANET) that could also be adapted in nano-networks by considering the limited capabilities of nanomachines.

In summary, these networks should scale and reach an acceptable level of performance in terms of throughput and latency with little overhead (e.g. energy consumption) even when large numbers of nodes have to be deployed. Furthermore, since these networks do not rely on any fixed infrastructure, and nodes should relay messages, energy-efficient protocols are required. Due to the dynamic nature of these networks, adaptive approaches and protocols are required and are still an ongoing research issue.

In addition, approaches are required to aim at investigating nano-networking mechanisms as an on-chip communication fabric for high-performance SoC design. In particular, new techniques in modeling, design and simulation of NoC at the nanoscale are of greatest importance. While several challenges have to be addressed, efficient design of nano-on-chip communication techniques may offer a unique solution for high-performance SoC design. This could provide new opportunities for lowering computing power which can be deployed in portable embedded devices and sensors.

References

[1]   Z. J. Haas, J. Deng, B. Liang, P. Papadimitratos and S. Sajama. *Wireless ad hoc Networks*. Wiley Online Library, 2003.

[2]   M. Bakhouya, K. Dar, J. Gaber, M. Wack and P. Lorenz. Wireless communication technologies for ITS applications. *Communications Magazine, IEEE*, 48(5): 156–162, May 2010.

[3]   P. Santi. Topology control in wireless ad hoc and sensor networks. *ACM Computing Surveys (CSUR)*, 37(2): 164–194, June 2005.

[4]   A. Hadir, K. Zine-Dine, M. Bakhouya and J. El Kafi. An improved DV-Hop localization algorithm for wireless sensor networks. In *Proceedings of Next Generation Networks and Services (NGNS), 2014 Fifth International Conference*, Casablanca, 28–30 May 2014.

[5]   G. Anastasi, G. Lo Re and M. Ortolani. WSNs for structural health monitoring of historical buildings. In *Proceedings of Human System Interactions, 2009, HSI '09, 2nd Conference*, Catania, 2009.

[6]   J. Hwang, C. Shin and H. Yoe. Study on an agricultural environment monitoring server system using wireless sensor networks. *Sensors*, 10(12): 11189–11211, 2010.

[7]   A. Mainwaring, D. Culler, J. Polastre, R. Szewczyk and J. Anderson. Wireless sensor networks for habitat monitoring. In *WSNA '02 Proceedings of the 1st ACM International Workshop on Wireless Sensor Networks and Applications*, New York, 2002.

[8]   M. Hussain, P. Khan and K. K. Sup. WSN research activities for military application. In *Proceedings of Advanced Communication Technology, 2009, ICACT 2009, 11th International Conference*, February 2009.

[9]   N. Kumar, A. Kumar and D. Chaudhry. A novel approach to use nano-sensor in WSN applications. *International Journal of Computer Applications (0975 – 8887)*, 14(2): 31–34, January 2011.

[10] R. Samanta and A. Pal. An approach for the integration of the existing wireless networks. *International Journal of Computer, Information, Systems and Control Engineering*, 8(9): 1588–1595, 2014.

[11] J. Hoebeke, I. Moerman, B. Dhoedt and P. Demeester. An overview of mobile ad hoc networks: applications and challenges. *Journal of the Communications Network*, 3(3): 60–66, 2004.

[12] D. Liarokapis and A. Shahrabi. An adaptive broadcasting scheme in mobile ad hoc networks. In Xin Wang (ed.) *Mobile Ad-Hoc Networks: Protocol Design*. InTech, Rijeka, Croatia, January 2011, pp. 243–261.

[13] M. Bakhouya, S. Cai, M. Becherif, J. Gaber and M. Wack. An in-vehicle embedded system for CAN-bus events monitoring. *Mobile Multimedia*, 10(1–2): 128–140, May 2014.

[14] W. Ait-Cheik-Bihi, A. Chariette, M. Bakhouya, A. Nait-Sidi-Moh, J. Gaber and M. Wack. An in-vehicle emergency call platform for efficient road safety. Inn *Proceedings of the 8th ITS European Congress*, Lyon, France, June 2011.

[15] Q. Hua, Z. Li, Y. Wang, X. Lu, W. Zhang and G. Wang. An integrated network of roadside sensors and vehicles for driving safety: Concept, design and experiments. In *Proceedings of the 2010 IEEE International Conference on Pervasive Computing and Communications (PerCom)*, Mannheim, 2010.

[16] A. Festag, A. Hessler, R. Baldessari, L. Le, W. Zhang and D. Westhoff. Vehicle-to-vehicle and road-side sensor communication for enhanced road safety. In *Proceedings of the 15th World Congress on Intelligent Transport Systems and ITS America's 2008 Annual Meeting*, New York, 2008.

[17] S. Medetov, M. Bakhouya, J. Gaber, K. Zinedine, M. Wack and P. Lorenz. A decentralized approach for information dissemination in vehicular ad hoc networks. *Journal of Network and Computer Applications*, 46: 154–165, November 2014.

[18] M. Bakhouya, J. Gaber and P. Lorenz. An adaptive approach for information dissemination in vehicular ad hoc networks. *Journal of Network and Computer Applications*, 34(6): 1971–1978, November 2011.

[19] E. Weingärtner and F. Kargl. *A Prototype Study on Hybrid Sensor–Vehicular Networks*. 6. KuVS Fachgespräch Sensornetzwerke, RWTH-Aachen Technical Report, Aachen, Germany, 2007.

[20] J.-M. Bohli, A. Hessler, O. Ugus and D. Westhoff. A secure and resilient WSN roadside architecture for intelligent transport systems. In *WiSec '08 Proceedings of the 1st ACM Conference on Wireless Network Security*, New York, 2008.

[21] B. Z. Uichin Lee, M. Gerla, E. Magistretti, P. Bellavista and A. Corradi. Mobeyes: smart mobs for urban monitoring with a vehicular sensor network. *Wireless Communications, IEEE*, 13(5): 52–57, 2006.

[22] A. Ali Khan, S. Ahmed Shah, M. Abdul Aleem, Z. Ali Bhutto, A. Ali Shaikh and M. Aslam Kumbhar. Wireless sensor networks: a solution for smart transportation. *Emerging Trends in Computing and Information Sciences*, 3(4): 566–571, 2012.

[23] I. F. Akyildiz, F. Brunetti and C. Blázquez. Nanonetworks: A new communication paradigm. *Computer Networks: The International Journal of Computer and Telecommunications Networking*, 52(12): 2260–2279, 2008.

[24] I. F. Akyildiz and J. M. Jornet. Electromagnetic wireless nanosensor networks. *Nano Communication Networks*, 1(1): 3–19, 2010.

[25] L. Parcerisa Giné and I. F. Akyildiz. Molecular communication options for long range nanonetworks. *Computer Networks: The International Journal of Computer and Telecommunications Networking*, 53(16): 2753–2766, November 2009.

[26] MoNaCo: Molecular Nano-Communication Networks [online]. Available from: http://www.ece.gatech.edu/research/labs/bwn/monaco/projectdescription.html [Accessed 08 Jan 2016].

[27] S. G. Glisic. *Advanced Wireless Communications & Internet: Future Evolving Technologies*, Third Edition. Wiley, Chichester, 2011.

[28] K. Bengston and M. Dunbabin. Design & performance of a networked ad-hoc acoustic communications system using inexpensive commercial CDMA modems. In *Proceedings of OCEANS 2007*, Aberdeen, UK, June 2007.

[29] R. Diamant and L. Lampe. A hybrid spatial reuse MAC protocol for ad-hoc underwater acoustic communication networks. In *Proceedings of Communications Workshops (ICC), 2010 IEEE International Conference*, Cape Town, 2010.

[30] I. Akyildiz and J. Jornet. Channel modeling and capacity analysis for electromagnetic wireless nanonetworks in the terahertz band. *IEEE Transactions on Wireless Communications*, 10(10): 3211–3221, 2011.

[31] O. Yalgashev, M. Bakhouya and J. Gaber. An adaptive transmission range for electromagnetic-based broadcasting in nanonetworks. In *Proceedings of the 5th International Conference on Ambient Systems, Networks and Technologies (ANT-2015)*, London, UK, 2015.

[32] A. Chariete, M. Bakhouya, J. Gaber and M. Wack. A design space exploration methodology for customizing on-chip communication architectures: towards fractal NoCs. *Integration, VLSI*, 158–172, June 2014.

[33]   O. Yalgashev, M. Bakhouya, A. Chariete and J. Gaber. Performance eva-luation of nano-on-chip interconnect for SoCs. In *Proceedings of High Performance Computing & Simulation (HPCS), 2014 International Con-ference*, Bologna, Italy, 2014.

[34]   P. Pande, C. Grecu, A. Ivanov, R. Saleh and G. De Micheli. Design, synth-esis, and test of networks on chips. *Design & Test of Computers*, 22(5): 404–413, 2005.

[35]   L. Carloni, P. Pande and X. Yuan. Networks-on-chip in emerging inter-connect paradigms: Advantages and challenges. In *Proceedings of Net-works-on-Chip, 2009 (NoCS 2009), 3rd ACM/IEEE International Symposium*, San Diego, CA, 2009.

Chapter 2

Cyber-physical systems: basics and fundamentals

Syed Hassan Ahmed[1], Safdar Hussain Bouk[2], Dongkyun Kim[3] and Mahasweta Sarkar[4]

Abstract

In the first decade of the 21st century, the cyber world and the physical world were considered as two different entities. However, in the literature we can easily find that these two entities are closely correlated with each other after integration of sensor and actuators in the cyber systems. Cyber systems became responsive to the physical world by enabling real-time control emanating from conventional embedded systems, thus giving birth to a new research paradigm named the cyber-physical system (CPS). In this chapter, we investigate the major challenges in integrating the cyber world with the physical world and its applications, followed by the basics and fundamentals of CPS. In addition, we discuss the CPS requirements for building its architectures, which should contain several modules supporting the CPS. The motivation of this chapter is to provide an overview of the CPS and the prerequisite knowledge for modeling and simulations.

2.1 Introduction

Computer systems provide a way of working, organizing or performing one or many tasks according to a fixed set of rules, program or plan. In other words, the main objective of computer(s) and the software(s) operating on them is to process information to perform better decisions. Apart from information processing, computing devices have also paved the way forward in various other systems, ranging from

[1]School of Computer Science and Engineering, Kyungpook National University, Daegu, Republic of Korea, e-mail: hassan@knu.ac.kr
[2]School of Computer Science and Engineering, Kyungpook National University, Daegu, Republic of Korea, e-mail: bouk@knu.ac.kr
[3]School of Computer Science and Engineering, Kyungpook National University, Daegu, Republic of Korea, e-mail: dongkyun@knu.ac.kr
[4]Department of Electrical and Computer Engineering, San Diego State University, California, USA, e-mail: msarkar2@mail.sdsu.edu

vehicles, manufacturing plants, cellphones, to home appliances. Following the emergence of computational capability, these systems became more interactive and intelligent. For example, vehicle breaks are applied if the vehicle comes into close proximity of another vehicle or object. Manufacturing plants work continuously without any human intervention and in the event of any fault in the production line, it will pause automatically. Home appliances, such as heating or cooling systems, react according to the home environment, the tenant's preferences and his/her presence in the home. This type of computing system is termed an embedded system and the software designed to run on that system is called embedded software.

An embedded system interacts with the physical process or environment and reacts accordingly. It is designed for dedicated application(s), a specific subsection of an application or product or part of a larger system. Successful applications include aircraft control systems, automotive electronics, home appliances, weapons systems, games, toys and so on. However, most of these embedded systems are closed 'boxes' that do not expose the computing capability to the outside. A simplified version of an embedded system is shown in Figure 2.1. The information about the physical environment is gathered by the system through sensors, either continuously or periodically. Sensed information is provided through the analog-to-digital (A/D) converters. The embedded software residing in the memory processes

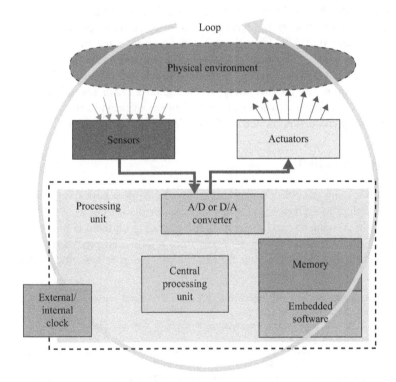

Figure 2.1   Embedded system

this information and sends instructions to the actuator circuit through the A/D converter. The actuators are mainly the electromechanical devices that perform actions to control the properties of the physical environment.

Initially, computing and embedded systems worked as standalone components. For example, embedded systems were largely an industrial problem and they used computing component(s) to enhance the performance or functionality of the system. In this earlier context, embedded software differed from other software only in its resource limitations (small memory, small data word sizes and relatively slow clocks). In this context, the 'embedded software problem' is an optimization problem.

Solutions rely on efficiency; engineers write software at a very low level (in assembly code or C), avoid operating systems with a rich suite of services and use specialized computer architectures such as programmable Digital Signal Processors (DSPs) and network processors that provide hardware support for common operations. These solutions have defined the practice of embedded software design and development for the past 30 years or so. In an analysis that remains as valid today as a couple of decades ago, Rajkumar *et al.* [1] lament the resulting misconceptions that real-time computing 'is equivalent to fast computing' or 'is performance engineering' (most embedded computing is real-time computing). But the resource limitations of 30 years ago are surely not resource limitations today. Indeed, the technical challenges have centered more on predictability and robustness than on efficiency.

The characteristics of an embedded system are outlined as follows.

The embedded system, either as a whole or subsystem, must exhibit the following characteristics:

1. The embedded system must be reactive to the physical environment. This can be achieved by continuously monitoring the environment and later on performing actions within the physical environment via sensors and actuators respectively.
2. The reaction of the system should be within the time constraints, which means that the embedded system must respond to the changes in physical environment within the dedicated time interval. Any late, but correct response from the system is considered as an incorrect response.
3. Along with the restiveness, the system should unveil maximum robustness with minimum resources (cost, energy, minimum code size, etc.).
4. The system should be safe, reliable, secure, extendable and available.

Soon, it was realized that there should be a mechanism so that these standalone systems can interact with each other and, consequently, the communication networks emerged. The communication system enabled all standalone systems to ubiquitously communicate information (such as digital data, text, video, audio) with each other through wires or wirelessly. The research community has intensively explored the networks in past decades due to their wide variety of applications. Enormous improvements have been made in both wired and wireless communication technology that lead us to the new communication paradigms such as ubiquitous and pervasive computing, wireless sensor networks (WSNs), underwater sensor networks (UWSNs) and machine-to-machine (M2M) communication systems. More recently, embedded systems have dominated the computing market because of their usage trends, i.e. anytime, anywhere information access, wearable devices, intelligent appliances and buildings, the Internet of Things (IoT), etc.

The radical transformation that we envision today solely came from the networking of these devices. Such networking also posed considerable technical challenges. For example, a usual practice in embedded systems is that they rely on bench testing for concurrency and timing properties. This has worked reasonably well, because programs are small, and because software gets encased in a box with no outside connectivity that can alter the behavior. However, the applications that we envision today demand that embedded systems be feature-rich and networked, so bench testing and encasing become inadequate. In a networked environment, it becomes impossible to test the software under all possible conditions. Moreover, general-purpose networking techniques themselves make program behavior much more unpredictable. A major technical challenge is to achieve predictable timing in the face of such openness.

2.1.1   Cyber-physical system

The integration of networked computational devices with physical processes has found the new system paradigm called CPSs, as shown in Figure 2.2. It shows that the communication-enabled embedded systems monitor and control the physical processes and the computerized system is closely and seamlessly integrated with the dynamics of the physical processes. Put simply, the CPS provides connectivity between computer, network, control and physical systems. Therefore, the CPS is not merely a simple control system but a real-time, intelligent, predictive, adaptive, distributed and networked system with little human involvement [2]. As CPSs are

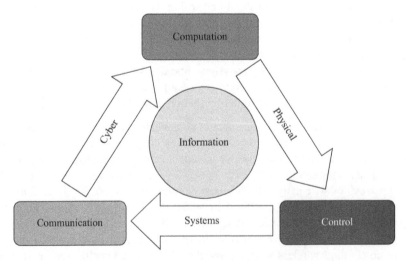

Figure 2.2   A general cyber-physical system

enabling the upcoming new networking model, industry and academic stakeholders are investing in the exploration of this future network.

As stated earlier, CPS seamlessly integrates physical processes with computation and communication and provides the abstractions, design, modeling and analysis techniques for the integrated whole [3]. Therefore, the networking and computation technologies in CPS need to embrace information and the physical dynamics as well. CPS as an integrated whole necessitates new design technologies because CPS merges multiple disciplines, such as computing science, control theories and communication engineering. In addition, software is embedded in devices, the principal mission of which is not computation alone. CPSs range from relatively small systems, such as aircrafts and automobiles, to large systems such a national power grid [4].

Before proceeding further, we need to discuss the behavior of the current networking techniques and the embedded systems. The most widely used networking techniques present today introduce a great deal of timing variability and random behavior. As we know that the embedded systems are real-time and often use specialized networking technologies (such as CAN bus in manufacturing systems and FlexRay in automotive applications), therefore as far as concurrency is concerned, the physical processes are intrinsically concurrent and their interaction with computing requires, at a minimum, concurrent composition of the computing with the physical processes. The current embedded systems must simultaneously react to multiple real-time streams from sensors and control multiple actuators. The interaction mechanism between sensors and actuator hardware is done by the notion of interrupts that lies in the area of the operating system. Unfortunately, the programming languages cannot achieve the concurrency because the interactions with the hardware are exposed to the programmers through abstraction of threads. It is stated in Tang *et al.* [5] that the threads are by nature problematic. The cyber-physical systems that are extensively networked will alleviate the concurrency problem because of their stochastic behavior. Hence, before realizing the future networks and CPS that are envisioned to be concurrent by nature, the following questions must be answered:

- What aspects of those networking technologies should or could be important in large-scale networks?
- Are they compatible with global networking? To be specific, recent advances in time synchronization across networks promise networked platforms that share a common notion of time to a known precision [6].
- How are distributed cyber-physical applications to be developed?
- What are the implications for security measures?
- Can we mitigate security risks created by the possibility of disrupting the shared notion of time?
- Can security techniques effectively exploit a shared notion of time to improve robustness?

Recently, it has been observed that CPS is basically the emergence of wireless networks (such as WSNs) and the IoTs. CPSs have gained much attention from many vendors, researchers and manufacturers [7]. This emergence of multiple disciplines has accelerated the new series of applications and thus brings new challenges. Following this, in the recent past, some significant achievements have been made in the said emerging domains. Later on, those achievements endorse the expansion of CPS. However, it is also true that instead of a rapid development, we continuously face new challenges and implementation issues. Added to this are security issues, since emerging different domains compromise security, privacy and data integrity.

The main objective of research in the CPS domain is to tightly amalgamate the cyber and physical world in such a way that designing of computing, control and communication becomes possible. It is also true that CPS differs from traditional desktop computers, and similarly from conventional real-time/embedded systems, and therefore WSN; however, they do have some essential properties as listed below [8]:

- CPS expects to provide cyber capabilities to every component in the physical world that is supposed to be a resource constraint. We must understand that the software is a part of every embedded system and physical module, and so is counted as the system resource. But such software brings several challenges as well.
- CPS is an intensely integrated computation with physical processes.
- CPS is usually designed and networked at multiple and extremely wide scales. Therefore, it is considered as networks, which include both wired and wireless domains. Moreover, highly varied system scales and device category needs are supported by CPS.
- In CPS, the different components likely have unequal granularity of time and spatiality. CPSs are strictly constrained by spatiality and real-time capacity.
- Dynamically reorganizing/reconfiguring. CPSs, as very complicated and large-scale systems, must have adaptive capabilities.
- Closed-loop control and high degrees of automation. CPSs favor convenient man–machine interaction, and advanced feedback control technologies widely apply to these systems.
- The CPS operations must be dependable and certified in some cases. On the other hand, reliability and security are also necessary for CPSs because of their extreme scales and complexities.

2.2   CPS concept and requirements

In the literature, the term 'CPS' has recently been introduced where it leads toward the development of a modern and new vision for services to society. Those services may extend the dimensions of communications to a breakthrough performance level in the future. For instance, CPS is defined as an integration of physical processes with computation. Here it is worth mentioning that CPS is not about combining physical and cyber characteristics, but it is about their intersection. From the recent literature we can obtain a few definitions of CPS; for example, one complex

CPS definition was made by Shankar Sastry from the University of California, Berkeley, in 2008: 'A cyber-physical system (CPS) integrates computing, communication and storage capabilities with monitoring and/or control of entities in the physical world, and must do so dependably, safely, securely, efficiently and in real-time' [9]. It is quite clear that CPS is definitely not similar to the traditional or existing embedded systems, including real-time systems. More precisely, CPSs differ from the current wireless or wired sensor networks and so-called desktop applications, and they have certain properties that define them as cyber-physical integration, which are as follows:

1. Cyber/communication capabilities in every physical component available.
2. Networked at various multiple and huge scale.
3. Reconfiguring and reorganizing with respect to the dynamics of the environment.
4. The control loops must be close with high degrees of computation and automation.
5. The certified and dependable operations are key elements in some cases.
6. The physical and cyber components are integrated for initially learning and later on adaptation purposes.
7. CPSs are expected to provide high-performance, self-organized, auto-assembly operations.

As for other communication and information systems, CPSs are also distinguished by the following basic properties:

1. Operability/functionality
2. Performance metrics
3. Security and dependability
4. Cost effectiveness

Additional characteristics affecting the overall system security and dependability are not limited to CPS usability, its management and system adaptability. The main expected features of CPS are as follows:

1. CPS will support the input from the physical environment and then feed back to the physical environment in the presence of the secured channels for communication.
2. Furthermore, CPS will guarantee a combined management approach and distributed controlling mechanism.
3. CPS will also meet the real-time communication and performance requirements.
4. Different from current embedded systems, a wide geographical coverage without the physical security components will be supported by CPS.
5. Forthcoming technology, known as system of systems (SoS), will be as a result of CPS.

Due to the higher reliability and predictability standard, the existing embedded systems have always been considered more reliable than other generic computing systems. For example, customers do not expect their TV/other electronic devices to crash and restart. People have started counting on secure and reliable vehicles, where the use of computer controllers has effectively improved both the reliability and increased efficiency of cars and so on. Indeed, if the improved reliability and predictability function is skipped, the CPS will not be able to perform as expected for various applications such as automotive safety, health care and traffic control. On the other hand, we must remember that the physical world is not completely predictable.

2.3   CPS architectures

To date there is no identified or standard concept of a CPS. However, in general, the CPS is referred to as the combination of the dynamic physical world and the cyber world. In the case of CPS, it makes a distinction between the physical and cyber world and processes the data by computers. Later on, that data effectively changes the physical world. From the literature, we can find some examples, such as He Jifeng who proposed the concepts of Computation, Communication and Control and named this 3C [33]. The main objective was to fuse computation, communications and controlling with 'information' at the center, thus achieving dynamic control, real-time sensing and information services in a large-scale system. CPS is closely related to embedded systems, wireless networks and ad-hoc sensor networks, but has its own characteristics or properties as well. Those properties include the dynamics and complexity of an environment that together bring a big space problem and demand a highly reliable system, i.e. CPS. Since CPS is in its early stages, few researchers have divided CPS into a simple two-tier structure, which inherits the physical part and computing part from the above-mentioned definitions. More specifically, the physical part includes sensors or receivers that sense the physical environment and collect data. Later on, the computing part executes the decision made by itself on behalf of the data collected by the physical part. Here it is worth mentioning that the computing part analyzes and processes the data from the physical part, and then makes a decision. We can collectively refer to it as a feedback control relation of the two parts. On the other hand, Hu *et al.* [10] proposed a three-tier architecture of CPS. This architecture consists of the Environmental Tier that is made up of various physical devices and a target environment with end users having the devices and being a part of their associated physical environment. The second tier, named the Service Tier, is composed of a classic computing environment with several services and cloud computing. Similarly, the Control Tier is designed to receive the monitored data, which are collected via sensors, to make further control decisions. Also, control tiers help the CPS to find the right services by a service-consulting framework and invoke the services on the physical devices (Figure 2.3).

It is expected that the architecture of a CPS will provide homogeneous treatment of the cyber and physical elements together. Likewise, a new architecture for software proves to be a good initiative; however, the concept should be extended to CPS by using new terminologies in vocabulary for physical and cyber-physical elements,

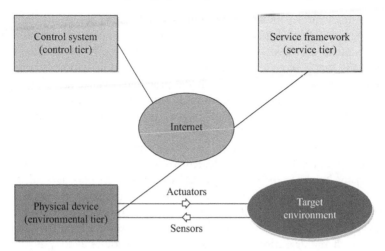

Figure 2.3   Tire-based architecture of cyber-physical systems

which are necessary to analyze the system behavior. To achieve this, one effort has been made in Hoang *et al.* [11], where the authors have proposed a prototype of the CPS concept, in which events or information of the cyber world are highlighted. In addition, also an abstraction of the real physical world ruled by semantic laws is illustrated which is evolving the current architecture of the embedded systems and aligning that abstract to the current CPS technology requirements. More precisely, we list the outcomes of the proposed architecture as follows:

1. The proposed architecture provides the Global Reference Time with the help of the next-generation networks that are expected by all the CPS components.
2. Remember that the events are 'raw data/facts' noted down or collected by a set of sensor nodes and so on. Similarly, some actuator modules or humans make the 'actions'. Here it is worth explaining that system control units or humans through event processing are the providers of information set for a CPS following the proposed architecture.
3. Quantified Confidence – a standard method to calculate the confidence of the events/information at any point in time.
4. The proposed work also enables a CPS control unit to subscribe for specific data and once the provider (i.e. sensors) collects that data it will transfer that data to the subscriber autonomously.
5. Any CPS architecture is expected to support the Semantic Control Laws, such as possessing the event–condition–action form, the precise laws defining the behaviors of a control system which are relative to the environment aligned with the user-defined scenarios or conditions.

Similarly, the paper entitled 'A software architecture for next-generation cyber-physical systems' by West and Parmar in 2006 [34] proposes to develop a CPS on their software architecture, which is in actual fact a collection of specific-application services or domains. According to the authors, a CPS organizes itself with the more suitable methods of communication and continuously switches to an isolation mode between various services. Additionally, the proposed software architecture also considers the autonomous composition of different services to satisfy the given constraints of an application. Moreover, the proposed architecture also takes into account the underlined limitations of hardware and its heterogeneity in the generation from which the hardware belongs. Furthermore, the verification of a software system for a given application is also valued.

In the past decade, architectural evolution has shifted into systems integration, ensuring that SOA (Service Oriented Architecture) will have a major role to play in many branches of technology. SOA basically enables a rapid, low-cost composition of interoperable and scalable systems based on reusable services exposed by these systems. In Tan *et al.* [12], the authors proposed a simplified middleware CPS architecture, integrating with web services named 'WebMed', through which an interaction with physical devices becomes as easy as invoking a computation service. Emphasizing the basics of service-oriented guidelines, the authors built a loosely coupled infrastructure that exposes the functionality of physical devices to the web for application development. However, Sanislav *et al.* [13] addressed the architecture of a CPS to precisely provide a uniform treatment of cyber and physical elements. The software architecture provides a good starting point, but the concept should be extended to CPS by using a new vocabulary for physical and cyber-physical elements necessary to analyze the system behavior. Tan *et al.* [14] presented representative prototype architecture of the CPS concept. They highlighted the cyber world represented by events/information as an abstraction of the real physical world governed by semantic laws, evolving the typical architecture of the embedded systems and aligned it to current technological requirements. In Wan *et al.* [15], the authors analyzed the features of M2M, WSNs, CPS and IoT, and highlighted the correlations among them. Then, home M2M networks were reviewed. The authors gave a CPS scenario, called human-centric cyber-physical system (HCPS), which takes into account human activities to design and develop the CPS-based social system. This demonstrates how M2M systems with the capabilities of decision-making and autonomous control can be upgraded to CPS and sketches the important research proposals and challenges related to CPS designs.

Recently a new architecture has been designed to meet the basic requirements of future CPS systems and their applications [16]. This is depicted in Figure 2.4, where we have five main modules for CPS architecture. For data collection of the physical world through sensors, the main function of this module works for environmental awareness, which is achieved by preliminary data pre-processing. The data are provided to the data management module (DMM). The Sensing module supports multiple networks. It depends on the nature of networks that are deployed. For example, in a WSN, each sensor node is equipped with a sensing module for real-time sensing. Other network nodes can also operate with a part of this module in different scenarios. In the case of the vehicular cyber-physical system (VCPS), VANETS nodes (i.e. cars) can be equipped with a sensing

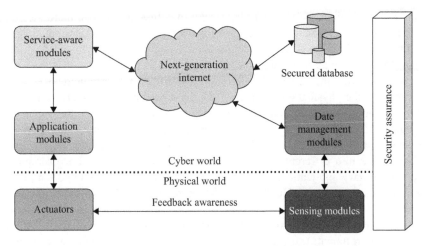

Figure 2.4   Recently proposed architecture of cyber-physical systems

module to sensor data from the physical world. In the case of HCPS, using BAN, sensors attached to patients are equipped with sensing module nodes to enable real-time control.

DMM consists of the computational devices and storage media. This provides heterogeneous data processing such as normalization, noise reduction, data storage and other similar functions. DMM is considered as the bridge between the dynamic environment and services as it is collecting the sensed data from sensors and forwards the data to service-aware modules (SAM) using next-generation internet (NGI).

A common feature of the emerging NGI is the ability for applications to select the path, or paths, that their packets take between the source and destination. This dynamic nature of internet service is required for designing CPSs. Unlike the current internet architecture where routing protocols find a single (the best) path between a source and destination, future internet routing protocols will need to present applications with a choice of paths. To achieve this, research is still ongoing to find quality of service (QoS) routing. While QoS routing provides applications with a path that best meets the application's needs, it does not scale to the size of the current internet, let alone the NGI. IPv6 and exploiting 802.16n and 802.21 are ongoing projects and expected to be included in NGI services trial. On the other hand, the SAM provides the typical functions of the whole system, including decision-making, task analysis, task schedule and so on. After receiving sensed data, this module recognizes and sends data to the services available.

In the Application Module, a number of services are deployed and interact with the NGI. Simultaneously, information is saved in a secure database for QoS support. The database is stored locally and simultaneously on cloud platforms in order to keep data safe. We can use a concept of NoSQL for saving data [17].

Although the NoSQL systems have a variety of different features, there are some common ones. First, many NoSQL systems manage data that is distributed across multiple sites. This saved data over the cloud system can be accessed from anywhere followed by authenticated access.

Actuators and the Sensing Modules are two different electronic devices which interact with the physical environment [18]; the actuator may be a physical device, a car, a lamp or a watering pump. It receives the commands from the Application Module, and executes. The security assurance part is inherently important in a whole system, from the access security, data security to device security. We divide CPS security into different requirements in different scenarios. For example, in terms of military applications, the confidentiality feature is more important, but in the smart home system or HCPS, the real-time requirements are more emphasized. Security of CPS can be divided into the following three phases: awareness security, which is to ensure the security and accuracy of the information collected from the physical environment; transport security, which is to prevent the data from being destroyed during the transmission processes; physical security, such as safety procedures in servers or workstations. Feedback Awareness is one of the advanced level services to minimize the data processing by communication between sensor and actuator for executing required actions directly.

So far, we have seen that CPS presents a set of advantages and advancements in the field of wireless communications and so on. It is also expected that CPS is a collection or combination of a secured and efficient systems. Those systems enable individual entities to work together in order to form various complex systems with a new set of applications and capabilities. As previously mentioned, cyber-physical technology may be applied in a wide range of domains, offering numerous opportunities such as critical infrastructure control, providing security and efficient transportation systems, alternative energy, environmental control, tele-presence, medical devices and integrated systems, telemedicine, assisted living, social networking and gaming, manufacturing, agriculture. Similarly, we have critical infrastructure, and assets that are essential for the functioning of a society and the economy for any region. Those assets include but are not limited to the facilities for water supply (storage, treatment, transport and distribution, waste water), generating or producing electricity, transmission and distribution of gas and its production, the supply of various oil products and the oil itself, and last but not the least telecommunication. With this economic impact in mind, Wan *et al.* [35] proposed some requirements of CPS that the authors believed every CPS should meet according to the business sectors in which they would be used. For example, we can consider CPS enabling intelligent automotive, environment monitoring/ protection, aviation and defense, critical infrastructure, healthcare (see Table 2.1 later). In addition, the physical platforms and support for CPS must satisfy five essential capabilities: (a) computing, (b) communication, (c) precise control, (d) remote cooperation and (e) autonomy. Different from traditional embedded systems, CPSs interact directly with the physical world where they detect environmental changes; therefore, adaptation of system behavior must be considered as one of the key challenges in the design of CPS.

Table 2.1 CPS applications and requirements

Applications	CPS requirements
Vehicular CPS	Due to the complex algorithms for traffic control, for example the best route calculation in given traffic situations, the automotive industry requires more computing power for enabling CPS.
Environmental CPS	For monitoring the environment in a wide geographical area such as mountains, rivers and forests, the CPS system must avoid human physical intervention for prolonged durations, thus guaranteeing low energy consumption. For such a scenario, enabling low-powered ad-hoc networks to collect time-sensitive data using a precise and secured CPS is still an ongoing research issue.
Air CPS	An aviation and defense CPS requires highly secured and precise controlling with low power computing. Therefore, the designing and developing of security protocols is an ongoing research issue.
Smart grid CPS	Water resource management and energy control are the key requirements of smart grid CPSs. Furthermore, application software methodologies are still missing from the literature to guarantee QoS in current software.
Health CPS	Health care and medical equipment demand a complete new series of synthesis, analysis and hypothesis. In addition, emerging technologies are still needed for medical or health CPSs. Therefore, we are still lacking the design and development of such applications for inter-operable algorithms.

2.4 CPS applications

CPS can be used in a wide range of applications, including intelligent transportation, precision agriculture, health CPS, water and mine monitoring, aerospace and so on. Figure 2.5 illustrates the range of possible forthcoming applications to be supported by CPS.

2.4.1 *CPS for vehicular environments*

The rapid increase in the number of the vehicles on the roads brings several challenges, for example an increased volume of air pollution, traffic congestion, safety issues. Therefore, much effort is required to improve the driver's experience and road safety. It is expected that in the near future, roads will be significantly aided by the next-generation intelligent vehicle with the support of an Intelligent Transportation System (ITS), inter-vehicle sensing and advanced computing. Those technologies will be widely used in order to provide protected air traffic, a secured railway traveling experience and automatic car control for safety throughput.

Similarly, other applications for CPS include the VCPS, which is not a new concept. For instance, VCPS may refer to a wide range of transportation management systems that are highly integrated and should be highly accurate, real-time and efficient. Due to modern computing technologies such as electronic circuits,

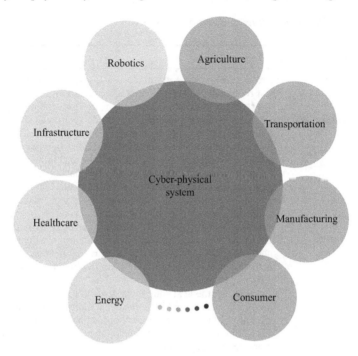

Figure 2.5   Applications of cyber-physical systems

sensors, computers and wireless networks, traditional transport modules are becoming more intelligent and effective. For that reason, the NAHSC (the National Automated Highway System Consortium) is studying and developing an independent high-speed system known as AHS (Automated Highway System), whose aim is to achieve more intelligent and secure traffic. Furthermore, one of the many American National Science Foundation (NSF) projects is CarTel, originally developed by MIT [19]. The main objective of the CarTel project is to combine sensing and mobile computing via wireless networking. To achieve this, data-intensive algorithms are under construction to be evaluated on servers located in the cloud to address challenges such as the VCPS. Moreover, the CarTel project supports applications designed for easy data collection, data processing, delivery, data analyzing and its visualization from sensors located on/in vehicle units. To date, the main contributions of the CarTel project include wireless traffic mitigation, monitoring the road surface and hazard detection to avoid accidents, vehicular ad-hoc networking and so on.

2.4.2   CPS for agriculture

During the first decade of the 21st century, a lot of attention has been given to the agricultural domain in order to improve crop reproduction and quality. In addition, human input has reduced and currently numerous experimental studies are

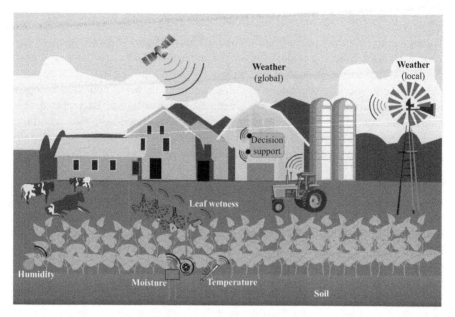

Figure 2.6   Agricultural CPS

illustrating the benefits of using WSNs in the agriculture sector. However, great accuracy in agriculture is demanded in order to meet the global demands of food consumption. As we know, the main goal of achieving accuracy in agriculture is to reduce wastage during the entire agricultural cycle and also to increase crop yield (i.e. from preparation of the crop field through to harvesting). In order to achieve this, we continuously evaluate the field crop with respect to macro- and micro-level weather and climate information. Also, we need to consider the effect of natural components in the soil, the fundamental geographic of farmland and so on, throughout the crop production cycle. See Figure 2.6 for a possible scenario of agricultural CPS.

One of the many efforts to bring CPS into the agriculture sector was undertaken by the University of Nebraska–Lincoln. Agnelo R. Silva and Mehmet C. Vuran, under the project entitled 'Underground Wireless Sensor Network', developed a novel CPS, where the integration of center pivot systems with wireless underground sensor networks was deployed and the main objective was to make a CPS for an accurate agriculture monitoring system [20]. This system was proposed by the Cyber-Physical Networking Lab where the wireless underground sensor networks (WUSNs) consisted of wirelessly connected underground sensor nodes. These nodes were capable of subsoil communications. Moreover, the experimental results illustrated that the concept of the CPS is possible and can be utilized to make a highly reliable agricultural CPS. This combination of CPS and precision agriculture is one of the typical applications of CPS, with good prospects.

2.4.3   CPS for health and medical sciences

Through embracing the potential of embedded network and software connectivity, a rapid transformation is under way in the medical industry. In the near future, distributed systems will control and monitor the multiple aspects of a patient's physiology; hence, the standalone systems designed, certified and used independently for different purposes for the treatment of patients will be replaced. The modern medical device systems, which combine embedded software controlled devices, networking capabilities and complex physical dynamics revealed by patient bodies, can be classified as a distinct class of CPSs termed medical cyber-physical systems (MCPSs).

MCPSs face numerous developmental challenges due to their increased size and complexity relative to conventional medical systems. However, new developments in design, verification, composition and validation techniques can address the challenges faced by MCPSs. These developments provide new opportunities for research into embedded systems, CPS and specifically MCPSs. In the case of MCPSs, new protocols need to be approved for their use in the treatment of patients. Furthermore, the conventional approval method used by the US Food and Drug Administration (FDA) – a process-based regulatory regime for the approval of medical devices – is not feasible as it is too lengthy and expensive for the increased complexity of MCPSs. Possible solutions to ease this process are presented here. In this chapter, we describe some of the research directions that we are taking toward addressing some of the challenges involved in building MCPSs. The ultimate goal is to develop foundations and techniques for building safe and effective MCPSs.

Overall, we advocate a systematic analysis and design of MCPS for handling its inherent complexity. Consequently, MCPS design is largely affected by model-based design techniques. These models should include the devices, the communication between them, and the relationship between these and the patients and caregivers. These models help the developers to build systems confidently, by accessing the system properties earlier in the development process prior to manufacturing. At the modeling level the analysis of system effectiveness and safety is further complemented by using generative implementing techniques, preserving the properties of the model in the implementation stage. The basis of evidence-based regulatory approval is formed through the results of model analysis and the guarantees of the generation process. MCPSs are safety-critical, interconnected, intelligent systems of medical devices. Conventional clinical scenarios are depicted as closed-loop systems controlled through the caregivers, with sensing achieved through medical devices and actuators and the plants being the patients. The introduction of MCPSs gives additional computational procedures to the caregivers, thus aiding in the controlling of the plant.

In Figure 2.7, a conceptual overview of the MCPS is shown. Two large groups of devices that are used in MCPSs can be categorized on the basis of their primary functions: monitoring devices, which provide the patient's physiologic state information, such as the oxygen-level monitors, heart rate and sensors; and delivery

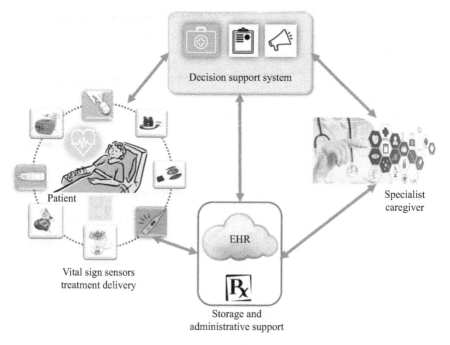

Figure 2.7   CPS for medical overview

devices, which target changing the patient's physiologic state through therapy, such as ventilators and infusion pumps. In MCPSs, the decision support or administrative support entities are directly fed the data collected through the monitoring devices, each of which serves a complementary and different purpose.

When we look at a medical set-up in a hospital, we come across different administrative units, including electronic health records (EHRs) and various pharmacies. The main objective of any medical care unit is to manage patients' health and treatment information which is collected periodically. Assuming the access of those medical units to provide the very best outcome from personalized information, they do have the potential to provide better combined treatment actuation based on a more comprehensive view of the particular patient's health, such as considering potential drug interactions or by considering the long-term progress of a patient's physiological limit. In such cases, those medical units can assist in fulfilling the demand for the continuous care of any specific patient. The periodic data collection and its management is important for today's health challenges and issues such as dealing with the older population and the rapid rise in the number of people with chronic symptoms such as diabetes and asthma. Therefore, decision-making entities can process the collected data and can generate alarms for various situations that may require medical attention. Such alarms are essential to let clinicians know when the patient's state has deteriorated and what information is relevant to treat them properly. Thus, we can say that it is a present-day necessity to develop a smart alarm system that will go beyond current state-of-the-art time-based methodologies and provide more accurate and targeted alarms, along with related information. Medical caregivers can then analyze the information provided and use it properly to deliver a clear message to devices to initiate any necessary treatment immediately. In this way, we bring the caregiver into the controlled loop around the patient. Alternatively, decision-making entities can utilize a smart controller that analyzes the received data from the monitoring modules and can estimate the state of the patient's condition, and it is expected that it may start the treatment autonomously. Those treatments may include drug infusion by releasing commands to drug delivery smart devices, thereby closing the controlled loop. However, building such MCPSs brings a significant number of challenges and issues which have to be addressed. These include the following:

- Software plays a vital and essential role in designing medical devices. Many functions traditionally implemented in hardware, including safety interlocks, are now being implemented in software. Thus, high-confidence software development is absolutely essential to ensure the safety and effectiveness of medical CPS.
- Since medical devices should include communication interfaces, it is also essential to ensure that integrated medical devices are effectively safe, secure, and can ultimately be certified.
- The patient information exchanged while the device is interoperating may not only ensure a better understanding of the general health conditions of the patient but can also enable early detection of infirmities. As a result, the generation of effective and smart alarms is imperative in case of emergency situations. One must remember the complexity of the human body and thus the huge variations of physiological parameters over the number of patients in one medical unit; therefore, developing such computational intelligence is not an easy task.
- Computing intelligence should be a property of MCPS. Going forward, that property can be used for increasing the authority of the system by enabling actuation of therapies based on the patient's current health status. Therefore, closing the controlled loop in this manner must be done safely and effectively.
- MCPS collects very critical medical data and also aims to manage that data. Therefore, any unauthorized access or tampering with this information can have severe consequences for the patient's health and can breach or break privacy laws, resulting in attempted abuse, discrimination and physical harm. Preserving the security of MCPS is therefore crucial and would be a compulsory prerequisite.
- The demonstration of dependability and cost-effectiveness for medical device software is the backbone of complex and safety-critical MCPSs. Certification of such medical devices provides a way of achieving this goal. Verification is therefore an essential requirement for the eventual viability of MCPS devices and is, therefore, an important challenge to be addressed.

Similarly, health CPSs (HCPSs) will soon replace traditional health devices which work independently, and will become part of our lives in the future.

Currently, along with sensors and other networks, various health devices work together to detect patients' physical condition in real time, especially for critical patients, such as patients with heart disease. These tasks are achieved by using body area network (BAN) technology. There are also portable terminal devices on the market which are carried by the patient and which can detect the patient's condition at any time and send timely warnings or predictions. Furthermore, the merging of health equipment and real-time data delivery would be much more advantageous for patients. To this end Lee and Sokolsky [21] reviewed the issues regarding the development of a highly trustworthy health CPS system. Their study suggested that HCPS would depend on software for new function development, and the need for network connections together with the continuous monitoring of patients, plus further research into the future development of health CPS. For such health devices, we are presented with interoperability challenges as well. Similarly, Kim *et al.* [22] proposed a generic framework named the network-aware supervisory systems (NASS), whose main purpose would be to integrate health devices into such a clinical interoperability system that uses real networks. It provides a development environment, in which health-device supervisory logic can be developed based on the assumptions of an ideal, robust network. In view of the complexity of these health applications, features such as higher security, real time and reduced network delay will need to be considered in the design of health CPSs.

2.4.4 CPS for smart grid

The development and value of smart applications are bound up in the interactions between people, their businesses and general entities. Wireless networks need to operate by their own rules over an infrastructure to ensure that these interactions occur. One example is the electricity grid infrastructure, which is connected to limited interactive entities. Nowadays, consumers' energy requirements are met through a few central power stations, following the typical centralized approach. However, due to the development of renewable energy resources, in the future we can expect users to also produce energy – hence the term 'prosumers'. Thus consumers can interchange the consumer/producer roles. Karim and Phoba [23] propose the concept of a smart grid to provide a next-generation electricity network boasting advanced reactiveness, configurability and self-management. This is achieved through a complex infrastructure, SoS, taking care of a number of characteristics [24] such as geographical distribution, independent operational and managerial elements, interdisciplinary nature, highly heterogeneous networked systems as well as their behavior and evolutionary development. This technology is expected to be the key part of the worldwide ecosystem of interacting entities, thus innovating the future of cross-industry services. An efficient management and energy utilization requirement is the key driving force of these efforts, such as to obtain granularity in the monitoring and management in the utilization of both local and global available resources (see Figure 2.8).

CPSs [25] describe the combination of computer and physical properties; it is found in multiple domains, particularly the electricity grid. Networked embedded

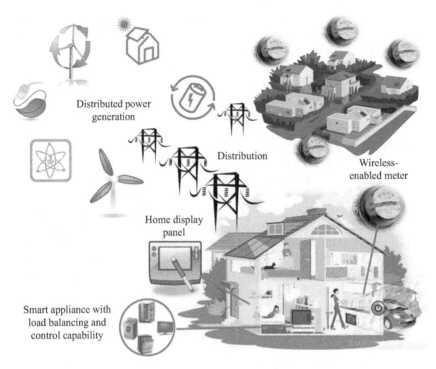

Distributed power
generation

Distribution

Wireless-
enabled meter

Home display
panel

Smart appliance with
load balancing and
control capability

Figure 2.8   CPS smart home applications

devices that offer real-time information exchange between the real world and the virtual world have, over the past decade, reached such an advanced stage that these can no longer be distinguished. With the increase in computation and communication capabilities and the decrease in size, these devices provide control and monitoring of real-world processes to an exceptional scale. CPS is relied upon by modern businesses in the synchronization of their real-world status on backend systems and processes. At its core, the combination of the physical and virtual world helps gather useful information about physical objects seamlessly which in turn is used in various applications throughout the objects' whole life span. Information collection, such as the objects' and goods' source, geographic location, mobility, usage history, physical properties and context, can help business owners create and improve current intracompany and intercompany business processes. This direct information from the real-world scenarios of the objects can be used to enhance decision-making algorithms, leading to more accurate business processes. The need to evolve embedded and ubiquitous computing technologies, by decreasing costs and increasing capabilities, has led businesses to gain access to the network and network edge, i.e. the CPS, which in turn has simplified a number of limitations faced by centralized approaches. The impact of CPS on current business is evident from industry predictions and visions for the future. The network industry, especially mobile and the all-pervasive CPS, has high stakes in, and expectations from, this technology. According to Hakan Djuphammar, VP of systems architecture at Ericsson, '[In 10 years' time], everything will be connected. We're talking about 50 billion connections, all devices will have connectivity...'. Hans Vestberg, President and CEO of Ericsson, who stated that 50 billion devices would be connected to the web by 2020, further endorsed this belief. Furthermore, the late John Woodget, Intel's global director of the Telecom sector, made a more moderate prediction, forecasting approximately 20 billion connected devices by 2020 [26]. Marie Hattar, VP of marketing at Cisco's network systems solutions group, specifically targeting smart grid networks, estimated that these networks would be '100 or 1000 times larger than the Internet' [27]. Vishal Sikka, CTO of SAP, expressed similar views that 'the next billion SAP users will be smart meters' [28].

One example of smart grid CPS is smart meters in homes and the installation cost of this system is estimated at $4.8 billion according to ABI Research [29]. According to another study by Pike Research, by 2020 the energy management systems market will be worth an estimated $6.8 billion per year. This includes lighting controls, WSNs, cooling and heating management in buildings, and it is estimated that a total investment of $67.6 billion will have been generated between 2010 and 2020 [30]. Also, currently (2015) an estimated $4.3 billion will have been spent on the installation, management and maintenance service for smart grids [31].

According to an estimation by the Canada Electronics Research Network, the smart home WSN market will rise from $470 million in 2007 to $2.8 billion in 2018 [32]. The massive shipping market and WSN-based home monitoring services provided by AT&T and SK Telecom have made the smart home (Figure 2.8) a reality. Another breakthrough is the latest CPS-enabled smartphones and tablets which promise powerful communication and computation along with various on-board sensors. As per Gartner's report [36], worldwide shipment of devices will reach 2.5 billion units in 2015 and will exceed 2.228 billion in 2017. These estimations further enhance the fact that the CPS-dominated era has begun. Currently, a number of research and deployment projects are under way worldwide in the field of smart grids and their CPS nature.

2.4.5 Overview

Table 2.1 gives a tabulated overview and summary of the major applications and requirements of CPS as discussed in the preceding subsections.

2.5 Future aspects of CPS

For CPS, research areas are distributed across isolated sub-domains, including communications and networks, mathematical modeling, systems theory, software engineering, wireless networks (i.e. sensor and ad-hoc networks) and computer science. Therefore, designing and developing digital systems is achieved by using various models, formalisms and tools. Those systems illustrate a set of features, and on the other hand can skip a complete system. More often, any formalism depicts either physical processes, or cybernetic processes, but does not utilize both at the same time, as per the requirements of the CPS domain.

Globally, research into CPSs has been driven in various directions, such as the description of a general architecture or a generic classification of the CPS designing principles for varying domains of the applications. Research has also included modeling of the CPS, where the dependability of CPS must be guaranteed, and finally the implementation of a CPS for critical infrastructure control and so on. In this section, we therefore enlist the future research challenges of the CPS domain which is still at an early stage.

Architectures and abstractions: In order to let any CPS communicate, control, compute and integrate between physical and cyber properties, we need to develop an innovative approach and define abstracts to formulate new architectures. Those architectures may allow the fusion and integration of a set of various heterogeneous systems with different properties that composed the CPS in an integrated, robust and efficient manner.

Network control and distributed computations: Here we refer to developing new frameworks, methods, algorithms and Software Development Kits (SDKs) related to event-driven and time-dependent calculations (computing). Moreover, we also lack software tools, various time-delay tolerance, malfunctions, runtime configuration and rapid decision support for distributed systems such as CPS. Therefore, we should take into account the satisfactory interaction of the physical environment with highly reliable and secure but varying components, which later on take decisions on runtime autonomously.

Validation and verification: Any CPS system expects to be composed of such hardware and software modules that should outperform their current stage and must achieve a high level of re-configurability, dependability and certification when it is required. Hereafter, new algorithms, models, tools and methods are expected in order to validate the software components. In addition, an entire system from its early design stage leads us toward the research directions to be addressed by the scientific and research community. Also, those research challenges in the CPS field were highlighted by the CPS Steering Group in their report [37] and are described in detail below:

1. The redesigning of abstract layers is required where those layers must include some physical concepts including energy and time. Those changes will relatively allow the fusion of computations/calculations together with the physical properties and physical system dynamics which are currently causing uncertainties in the implementation of CPS.
2. Semantic foundations need to be developed for heterogeneous models of CPS. Also, various modeling languages are also required in order to describe the numerous associated logics of physics.
3. Understanding and composing of a new heterogeneous system that tolerates the large amount of data. In addition, a networked system that fulfills a set of essential physical characteristics and has the ability to deliver the desired CPS functionalities in a more reliable fashion.

4. The development of a technology for achieving the predictability in partially compositional properties.
5. The modeling of a predictable and precise technology to be considered as a base foundation for integrating various systems in the future.
6. Proposing a new infrastructure for swift design and automation of CPS.
7. The design of a new series of flexible architectures, enabling CPS, ensuring the development of the nationwide and global effective systems.
8. From the existing but unreliable modules, developing the architectures and modeling tools for a reliable and resilient CPS that must be able to withstand malicious attack from either the physical or cyber world. Furthermore, the newly designed architectures may leverage an open system and other technologies with fluctuating design time while increasing the confidence interval of the CPS.

2.6   Conclusion

This chapter has provided a concept of CPSs together with their applications and current research progress in this area. The basics and fundamentals of CPS are then explained. In addition, the characteristics of CPS, such as real time, scalability and reliability, are introduced and which present a number of challenges in CPS design and implementation. There are numerous possible applications yet to be highlighted; however, this chapter also provides a technical overview of forthcoming applications such as smart grid, vehicular environments and agriculture. Along with that, we also highlight several research issues and architectures, including one explicit CPS architecture which was recently introduced in 2013, verifying the basic requirements for modeling and simulations of CPS environments. In short, 'dynamic communications' will be the future of networks and CPS is thus providing the necessary upcoming technologies.

Acknowledgments

The authors of this chapter would like to thank the editors and the editing staff of this book for giving them the opportunity to contribute to this forthcoming book.

References

[1]   R. Rajkumar, I. Lee, L. Sha and J. Stankovic. Cyber-physical systems: the next computing revolution. In *Proceedings of the 47th ACM Design Automation Conference*, pp. 731–736, 2010.
[2]   P. Tabuada, S. Y. Caliskan, M. Rungger and R. Majumdar. Towards robustness of cyber-physical systems. In *IEEE Transactions on Automatic Control*, 59(12): 3151–3156, December 2014.

[3]   M.Broy. Challenges in modeling cyber-physical systems. In *Proceedings of the 12th ACM International Conference on Information Processing in Sensor Networks*, pp. 5–6, 2013.

[4]   A. Mahmood, A. Ismail, Z. Zaman, H. Fakhar, Z. Najam, M. S. Hasan and S. H. Ahmed. A comparative study of wireless power transmission techniques. *Journal of Basic and Applied Scientific Research (JBASR)*, 4(1): 321–326, 2014.

[5]   L.-A. Tang, X. Yu, S. Kim, Q. Gu, J. Han, A. Leung and T. La Porta. Trustworthiness analysis of sensor data in cyber-physical systems. *Journal of Computer and System Sciences*, 79(3): 383–401, 2013.

[6]   S.-S. Lim, E.-J. Im, N. Dutt, K.-W. Lee, I. Shin, C.-G. Lee and I. Lee. A reliable, safe, and secure run-time platform for cyber physical systems. In *Proceedings of the 6th IEEE International Conference on Service-Oriented Computing and Applications (SOCA)*, pp. 268–274, 2013.

[7]   M. P. E. Heimdahl, L. Duan, A. Murugesan and S. Rayadurgam. Modeling and requirements on the physical side of cyber-physical systems. In *Proceedings of the 2nd IEEE International Workshop on the Twin Peaks of Requirements and Architecture (Twin Peaks)*, pp. 1–7, 2013.

[8]   J.-S. Choi, T. McCarthy, M. Yadav, M. Kim, C. Talcott and E. Gressier-Soudan. Application patterns for cyber-physical systems. In *Proceedings of the 1st IEEE International Conference on Cyber-Physical Systems, Networks, and Applications (CPSNA)*, pp. 52–59, 2013.

[9]   A. A. Cardenas, S. Amin and S. Sastry. Secure control: Towards survivable cyber-physical systems. In *Proceedings of the 28th International Conference on Distributed Computing Systems (ICDCS '08)*, pp. 495–500, 17–20 June 2008.

[10] L. Hu, N. Xie, Z. Kuang and K. Zhao. Review of cyber-physical system architecture. In *Proceedings of the 15th IEEE International Symposium on Object/Component/Service-Oriented Real-Time Distributed Computing Workshops (ISORCW)*, pp. 25–30, 2012.

[11] D. D. Hoang, H.-Y. Paik and C.-K. Kim. Service-oriented middleware architectures for cyber-physical systems. *International Journal of Computer Science and Network Security*, 12(1): 79–87, 2012.

[12] Y. Tan, S. Goddard and L. C. Perez. A prototype architecture for cyber-physical systems. *ACM Signed Review*, 5(1): 26, 2008.

[13] T. Sanislav and L. Miclea. Cyber-physical systems – concept, challenges and research areas. *Journal of Control Engineering and Applied Informatics*, 14 (2): 28–33, 2012.

[14] Y. Tan, S. Goddard and L. C. Perez. A prototype architecture for cyber-physical systems. *ACM SIGBED Review*, 5(1): 1–2, 2008.

[15] J. Wan, H. Yan, Q. Liu, K. Zhou, R. Lu and D. Li. Enabling cyber-physical systems with machine-to-machine technologies. *International Journal of Ad Hoc and Ubiquitous Computing*, 9(3/4): 1–9, 2012.

[16] S. Hassan Ahmed, G. Kim and D. Kim. Cyber physical system: Architecture, applications and research challenges. In *Proceedings of IEEE/IFIP Wireless Days (WD)*, Valencia, Spain, pp. 1–5, 13–15 November 2013.

[17] M.Stonebraker. SQL databases v. NoSQL databases. *Communications of the ACM*, 53(4): 10–11, 2010.

[18] A. Rezgui and M. Eltoweissy. Service-oriented sensor-actuator networks [ad hoc and sensor networks]. *IEEE Communications Magazine*, 45(12): 92–100, 2007.

[19] B. Hull, V. Bychkovsky, Y. Zhang, K. Chen and M. Goraczko. CarTel: A distributed mobile sensor computing system. In *Proceedings of the 4th ACM Conference on Embedded Networked Sensor Systems*, Boulder, pp. 125–138, 2006.

[20] M. Zhijun, Z. Chunjiang, W. Xiu, C. Liping and X. Xuzhang. Field multi-source information collection system based on GPS for precision agriculture. *Transaction of the CSAE*, 19(4): 13–18, 2003.

[21] I. Lee and O. Sokolsky. Health cyber physical systems. In *Proceedings of the 47th ACM/IEEE Design Automation Conference*, Anaheim, pp. 13–18, 2010.

[22] C. Kim, M. Sun, S. Mohan, H. Yun, L. Sha and T. F. Abdelzaher. A framework for the safe interoperability of health devices in the presence of network failures. In *Proceedings of the 1st ACM/IEEE International Conference on Cyber-Physical Systems*, Stockholm, pp. 149–158, 2010.

[23] M. E. Karim and V. V. Phoha. Cyber-physical systems security. In *Applied Cyber-Physical Systems*, Springer, New York, pp. 75–83, 2014.

[24] W. Ait-Cheik-Bihi, A. Nait-Sidi-Moh, M. Bakhouya, J. Gaber and M. Wack. TransportML platform for collaborative location-based services. *Service Oriented Computing and Applications*, 6(4): 363–378, 2012.

[25] A. Hahn, A. Ashok, S. Sridhar and M. Govindarasu. Cyber-physical security testbeds: Architecture, application, and evaluation for smart grid. *IEEE Transactions on Smart Grid*, 4(2): 847–855, 2013.

[26] S. Liu, S. Mashayekh, D. Kundur, T.Zourntos and K. Butler-Purry. A framework for modeling cyber-physical switching attacks in smart grid. *IEEE Transactions on Emerging Topics in Computing*, 1(2): 273–285, 2013.

[27] M. J. Stanovich, I. Leonard, K. Sanjeev, M. Steurer, T. P. Roth, S. Jackson and M. Bruce. Development of a smart-grid cyber-physical systems testbed. In *Proceedings of IEEE/PES Innovative Smart Grid Technologies (ISGT)*, pp. 1–6, Washington, DC, USA, 24–27 February 2013.

[28] A. AlMajali, E. Rice, A. Viswanathan, K. Tan and C. Neuman. A systems approach to analysing cyber-physical threats in the smart grid. In *Proceedings of IEEE International Conference on Smart Grid Communications (SmartGridComm)*, pp. 456 461, Vancouver, Canada, 21–24 October 2013.

[29] I. Stojmenovic. Machine-to-machine communications with in-network data aggregation, processing, and actuation for large-scale cyber-physical systems. *IEEE Internet of Things Journal*, 1(2): 122–128, 2014.

[30]   A. Anwar and A. Naser Mahmood. Cyber security of smart grid infra-structure. *arXiv preprint arXiv*:1401.3936, 2014.

[31]   C. Beasley, G. Kumar Venayagamoorthy and R. Brooks. Cyber security evaluation of synchrophasors in a power system. In *Proceedings of IEEE Power Systems Conference (PSC)*, Clemson University, pp. 1–5, 2014.

[32]   A. Nandanwar, M. Preetam Korukonda and L. Behera. A routing scheme for voltage stabilization in cyber physical energy systems. *Advances in Control and Optimization of Dynamical Systems*, 3(1): 812–818, 2014.

[33]   H. Jifeng. Cyber-physical systems journal, *China Computer Federation*, 6(1): 25–29, 2010.

[34]   R. West and G. Parmer. A software architecture for next-generation cyber-physical systems. In *Proceedings of the NSF Cyber-Physical Systems Workshop*, Austin, Texas, October 2006.

[35]   K. Wan, K. L. Man and D. Hughes. Specification, analyzing challenges and approaches for cyber-physical systems (CPS). *Engineering Letters*, 3, 2010.

[36]   Gartner.com. Gartner says worldwide device shipments to grow 1.5 per cent, to reach 2.5 billion units in 2015. See: http://www.gartner.com/newsroom/id/3088221 (accessed 20 August 2015).

[37]   Cyber-Physical Systems Virtual Organization. Cyber-Physical Systems – Executive Summary. See: http://cps-vo.org/node/204 (accessed 20 August 2015].

Chapter 3

Integrating wireless sensor networks and cyber-physical systems: challenges and opportunities

Teodora Sanislav[1], George Dan Mois[2], Silviu Folea[3] and Liviu Miclea[4]

Abstract

The integration of wireless sensor networks (WSNs) with cyber-physical systems (CPSs) has led to the emergence of a new framework which aims to respond to the need for 'smart everything' and to bring important benefits for the nation and indeed for society as a whole. The starting point in achieving this goal is represented by the thorough analysis of the research challenges posed by CPSs, whose inputs are provided by large numbers of wireless sensors grouped in sensor networks. The deployment of CPSs spanning multiple WSNs generates tremendous opportunities that justify research in this domain.

Following this approach, this chapter presents the main characteristics, application areas and design issues of WSNs, and the definitions, main characteristics, application domains and two types of architecture for CPSs. Scientific and technical barriers, and social and institutional problems are identified and enumerated. Further, the chapter gives an overview of the following research challenges posed by WSNs and CPSs and by their symbiosis: power consumption, communication, dependability, security, quality of services and Big Data analytics, among others. The chapter concludes with a discussion of the research, development, financial and market opportunities brought by the response to the synergy between these two domains.

[1]Technical University of Cluj-Napoca, Department of Automation, Cluj-Napoca, Romania, e-mail: teodora.sanislav@aut.utcluj.ro
[2]Technical University of Cluj-Napoca, Department of Automation, Cluj-Napoca, Romania, e-mail: george.mois@aut.utcluj.ro
[3]Technical University of Cluj-Napoca, Department of Automation, Cluj-Napoca, Romania, e-mail: silviu.folea@aut.utcluj.ro
[4]Technical University of Cluj-Napoca, Department of Automation, Cluj-Napoca, Romania, e-mail: liviu.miclea@aut.utcluj.ro

3.1   Introduction

The digital revolution began in the 1960s and 1970s with mainframe computing which involved large computers executing big data-processing applications. The next important step was taken in the 1980s and 1990s, when desktop computing and the internet appeared. This period had a huge influence on our lives, bringing computers to every desk to undertake business or personal activities. The 21st century began under the influence of an advanced computing concept, namely, 'ubiquitous computing', which assumes the use of numerous computing devices in any location and in any format. Ubiquitous computing deals with distributed, mobile, context-aware and location-aware computing, as well as with mobile networking, sensor networks, human–computer interaction and artificial intelligence. In the 2010s, a new generation of digital systems, called cyber-physical systems (CPSs), appeared as a response to our need to add new capabilities to physical systems, and to the technological and economic drivers' needs to decrease the cost of computation, networking and sensing. The rapid developments in embedded systems, sensors and wireless communication technologies had also led to the emergence of these systems. Further, social and economic bodies require more efficient use of national infrastructures, the development of technologies for improving energy efficiency and for reducing pollution, and accurate control within mission- and safety-critical systems. In this context, CPSs have to be intelligent distributed systems, which include cyber and physical components and act in real time, whose operations are monitored, coordinated, controlled and integrated by a computing and communication core, and which can be used in mission- and safety-critical applications. All of these features require a new understanding of computing as a physical act [1]. With the development of CPSs, information and communication technology (ICT) and sensing equipment, new demands and opportunities for the use of wireless sensor networks (WSNs) within these newly emerged systems are enabled. This will ultimately lead to CPSs composed of interconnected clusters of processing elements and large-scale wired and wireless networks gathering data from a variety of smart sensors and controlling different types of actuators [2]. These next-generation CPSs will raise new challenges for systems designers, in terms of constraints, requirements, implementation, deployment and maintenance, forcing manufacturers to reinvent their innovation and development processes and to consider a user-centric engineering approach [3].

3.1.1   Chapter overview

This chapter deals with the challenges posed by the CPSs, whose inputs are provided by large numbers of wireless sensors grouped in sensor networks. The large amount of future opportunities brought by the deployment of CPSs spanning over multiple WSNs are also discussed, along with the major forces and organizations that support the research in this domain. Following this approach, the structure of the chapter is as presented below.

Section 3.2 briefly presents the main characteristics, application areas and design issues of WSNs, while Section 3.3 covers some of the salient features of

CPSs. The definitions of CPSs along with the main characteristics and application domains are given here. Two types of architecture for these systems, namely, a basic one and one which adopts an information-centric view, are briefly described in Section 3.4. Section 3.5, 'Characteristics and challenges of WSN-CPS', outlines the issues encountered in this research area. It gives an overview of some of the main challenges posed by WSNs, CPSs and by their symbiosis. First, the scientific and technical barriers, and the social and institutional problems are enumerated. The matters concerned with power consumption, communication, dependability, security, quality of services, Big Data and a few other aspects, such as cost and analysis and modelling environments, are discussed in this section. Section 3.6 concentrates on the opportunities offered by the development of new WSN-CPS, while the conclusions are outlined in the final section.

3.2 Wireless sensor networks

WSNs are large networks of resource-constrained sensors, having processing and wireless communication capabilities, that implement different application objectives within a specific sensing field. Sensor networks significantly expand the internet, representing the cyber environment, into physical spaces [4]. These consist of spatially distributed autonomous devices, called nodes or motes, which use sensors to monitor physical or environmental conditions [5]. The nodes communicate either among each other or directly to an external base station, as can be seen in Figure 3.1, where a generic wireless sensor network is shown [6]. WSNs play a key role in pervasive computing, which envisions a world where computers are ubiquitous [7].

Typically, a WSN consists of a large number of small, low-cost nodes either deployed inside the phenomenon or close to it, and a number of base stations, or gateways, which forward the data gathered from the sensors to a monitoring centre or to a decision-making system using either high-quality wireless or wired links. The number of wireless sensor devices in a WSN has increased in recent years, reaching hundreds or thousands, and it is expected that this trend will increase even more in coming years [8]. The sensing nodes are represented by low-power multifunctional devices, having three basic functionalities, which consist of *sensing*, *data processing* and *communication* capabilities [9]. The components that

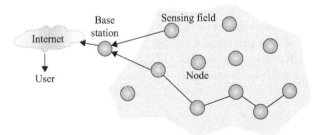

Figure 3.1   Wireless sensor network © IEEE 2004. Reprinted with permissions from [6]

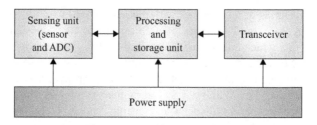

Figure 3.2   Sensor node components

generally make up a node in a WSN are represented by a sensor, a processor, a transceiver and a power supply (Figure 3.2). Depending on the application, these can be accompanied by other components, such as power generators, location-finding systems or mobilizers [10].

The sensing unit is in charge of collecting the data from the physical world, and typically consists of several micro-electro-mechanical-systems (MEMS) sensors, such as pressure, acceleration, humidity, temperature, proximity, radiation or seismic sensors, whose output is digitized with the help of an analog-to-digital converter (ADC), in order for it to be acquired by the processing module. The advances in MEMS technologies have made possible the development of these small, low-power sensors, which are best suited for integration into WSNs. The processing unit is generally made up of a microcontroller or a dedicated processor, associated with a small memory used for data storage. It is the central part of the sensor node, controlling the execution of tasks assigned to it and the communication with other nodes or with the base station. The transceiver, or the wireless module, is the component that provides the wireless interface for communication and which makes the connection to the network possible. With the majority of wireless sensors running on batteries, the fourth component of a node, the power unit, is of great importance. It manages and allocates the power resource, which can include generation or conversion, besides storage.

Acting as gateways between the WSN and the external world, base stations are usually well designed and more complex than the nodes [11]. Their tasks include the collection, processing and forwarding of field data to a remote application server over the internet. This is the reason why they are fitted with an in-field communication interface for receiving data from the sensor nodes, and with an out-of-field, long-range, communication channel for sending data to the server [12].

The main goal of sensor networks is represented by the delivery of sensed data from a large number of sensors to a sink, or base station. After data collection, further data processing actions can be performed at the base station or at a remote location, where the acquired data is forwarded. By combining sensors with data processing capabilities and with wireless connectivity, a wide range of new applications had been enabled. The following equation best describes the WSN concept [13]:

Sensing + CPU + Radio = Thousands of potential applications

The first field where WSNs had been used and where the potential is huge is represented by *environmental monitoring*, the primary purpose of deploying sensor networks being the observation of the physical world and the recording of physical quantities characterizing it [14]. Other domains include, but are not limited to, military, health, home and other commercial applications [10]. Each application area poses different technical issues and has specific requirements that have to be taken into consideration by researchers and developers. Some of the environmental conditions, that can be monitored with the help of WSNs, are related to crops and livestock, irrigation systems, forest fires and flooding, disasters and pollution levels. The *military field* includes target tracking, battlefield surveillance, intrusion detection and identification, reconnaissance of opposing forces and terrain, nuclear, biological and chemical attack detection, and others [10]. A few *health applications* consist in patient monitoring, the tracking of patients and medical staff in hospitals, the assistance of disabled patients or body sensor networks, where surgical implants of sensors can be used for monitoring patients' health. The use of WSNs in *home automation* or in *smart home* facilities can significantly improve the standard of living. *Other areas* include the monitoring and control of environmental conditions in office buildings, the monitoring of product quality, inventory management, vehicle tracking and detection, intelligent transportation systems, interactive environments and others. WSNs can also be used in industry for the local control of actuators or for controlling production processes. These are only a small fraction of applications examples and it is expected that the pace of deployment of wireless sensor networks will be accelerated even more in the following years, leading to a multi-billion market in the near future [15].

The *network topology* of a WSN is defined as a connectivity graph where the nodes are sensor nodes and the edges are communication links, each one representing a one-hop connection [4]. Typically, the sensor nodes within a WSN are organized in one of three network topologies, namely, the star topology, cluster tree topology and mesh network topology, each one of them having its own advantages and drawbacks in terms of communication reliability, latency and energy efficiency [5].

In general, WSNs are constrained regarding *computation, communication* and *energy resources*, and in order to provide proper solutions, specific care must be invested in their development [16]. Besides the selection of the best communication protocols, operating systems and hardware platforms, and the generation of high-quality code, the implementation of productive power management mechanisms and the development of efficient MAC (medium-access control) and routing protocols are aspects of great importance in the design phase of WSN applications. Power supplies play a key role in determining the lifetime of the sensor network and aggressive power-saving schemes have to be designed. The most common strategy for the saving of wireless sensor nodes' energy is sleep scheduling, but there are cases in which this alone is not enough and energy-harvesting components have to be added for extending the lifetime of the network [17, 18]. Some of these issues and challenges posed by WSNs will be presented and discussed in the following sections. These are added to those that arise in CPSs and consequently result in the emergence of new problems that have to be addressed.

3.3   Cyber-physical systems

CPSs integrate computing, communication and storage capabilities with monitoring and control of the entities in the physical world. Eventually, they will enable the development of a modern vision over the social services, which transcends time and space to new dimensions [19].

Several definitions of the CPSs can be found in the scientific literature, starting from the fact that these systems represent the intersection between computations and physical processes, and not their union [20]. The definitions emphasize a strong relationship between computational (cybernetic) and physical resources in order to exceed the current ICT systems in terms of autonomy, efficiency, functionality, adaptability, dependability and usability. For example, CPSs can be defined as follows: 'Cyber-physical systems refer to ICT systems (sensing, actuating, computing, communication, and others) embedded in physical objects, interconnected, including through the internet, and providing citizens and businesses with a wide range of innovative applications and services' [21].

CPSs are not merely desktop applications, distributed systems, control systems, traditional embedded and real-time systems, or wireless sensor networks. They have the following defining characteristics: (i) cybernetic capacities in all physical components, (ii) large and very large-scale networking, (iii) heterogeneity, (iv) dynamic reconfiguration or reorganization, (v) closed-loop control and high degrees of automation, (vi) operation must be dependable and certified in some cases [22–27]. Cybernetic and physical components are integrated for learning, cooperation and context adaptation, high performance, self-organization, self-assembly, while the process must be dependable and secure, and in certain cases certifiable [28]. CPSs present a set of advantages and bring many benefits: they are efficient and safe systems, they reduce the production and operational costs of systems, they allow individual entities to work together in order to form complex systems with new capabilities. Therefore, cyber-physical technology can be applied in a wide range of domains with specific features:

- Transportation (automotive, avionics, railways) – CPSs have to reduce the impact on the environment, to perform complex traffic control algorithms and to ensure a high degree of dependability for these services.
- Energy (electricity, gas, oil) – CPSs have to increase energy efficiency and high levels of security through new smart grid approaches.
- Civil infrastructure (water, dams, bridges) – CPSs have to fulfil a precise and reliable control and to provide warning and prevention mechanisms based on active monitoring, leading to application software methodologies to ensure the software quality.
- Environment monitoring – CPSs have to operate without human intervention for long periods of time with minimal energy consumption and to collect accurate and timely data from heterogeneous sensors.
- Defence – CPSs have to fulfil a precise control, high-security requirements, and last but not least high power computing demands.

- Smart buildings – CPSs have to reduce overall energy consumption to improve the security and safety of the buildings and also of their occupants.
- Health and biomedical – CPSs have to provide a new generation of analysis, synthesis and integration technologies, leading to the development of new interoperability algorithms and medical devices.
- Manufacturing and production – CPSs have to provide intelligent control, automatic assembly, environments where the robots work safely among humans, and to enhance global competitiveness.

3.4 WSN-CPS architecture

CPSs include embedded systems, sensor networks, actuators, coordination and management processes, services, to capture physical data and to act on the physical environment, all integrated under an intelligent decision system [27, 29]. In this context, WSNs can be used to collect physical information that is further exploited by CPSs [30]. With actuation being often integrated with the WSNs, these distributed monitoring and control systems become even more attractive for use as components in CPSs [31]. This is particularly reflected in Figure 3.3.

An architecture of a WSN-CPS is mainly composed of two layers: physical and cyber. The physical layer contains sensors and actuators to collect information from physical processes, respectively to control them. The collected information is sent to the cyber layer, as WSN-CPS inputs, where a real-time decision-making system executes the computations according to the received data, transforming them into high-level knowledge, and takes decisions which are transmitted then to the actuators, as WSN-CPS outputs [32]. To address issues regarding communication delays and the time differences between the physical and cyber layers, and the system heterogeneity, an abstraction layer, called middleware, can be included [33, 34]. Also, the middleware could ensure the dynamic reconfiguration of the WSN-CPS in an adaptive context in order for it to satisfy the system non-functional requirements, such as dependability and security. Figure 3.4 presents a basic WSN-CPS architecture model.

Starting from the architecture of the WSN-CPS mentioned above, more detailed architectures can be defined. In the following, an architecture based on a

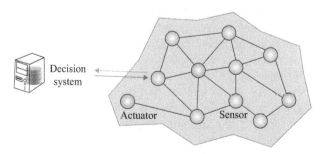

Figure 3.3   Cyber-physical system

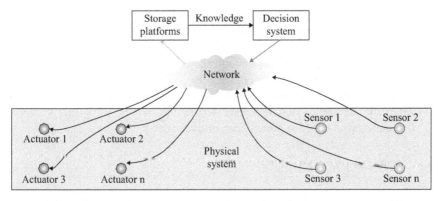

Figure 3.4   A basic WSN-CPS architecture model

new paradigm, which adopts an information-centric view instead of a device-centric one, is presented. The information-centric perspective is essential in applications where the extraction of knowledge from the gathered information is used to drive decision-making and control [35]. This approach can be used in a cloud-based WSN-CPS and leads to an architecture on three layers: (I) the physical layer, (II) the middle layer, consisting of data and knowledge, and (III) the top layer, represented by cyberspace services. The middle and top layers form an intelligent decision system. The physical layer, composed of large numbers of nodes with different kinds of sensors and actuators grouped into various WSNs, meets the following two functions: collects information and controls the physical processes. At this layer, the energy efficiency is important in terms of heterogeneity, data movement and communication. The middle layer of the CPS meets the functions: (i) stores the data acquired by the CPS physical layer; (ii) provides decisions for the CPS control and monitoring tasks, ensuring system dependability and security; (iii) performs complex analysis required by a third party (end-users). The middle layer has the following components: distributed databases containing the measurements provided by the sensor nodes; a knowledge base (ontology), used to provide decisions and negotiation rules to model the CPS behaviour in terms of cyber, physical and human aspects; a multi-agent society, with cooperation and negotiation capabilities supporting autonomous, context-adaptive, reactive and targeted problem solving to monitor and control the CPS and to provide various analysis to the end-users. The multi-agent approach is a perfect candidate for the CPS middle layer due to its capabilities to solve problems through the cooperation between heterogeneous and distributed components, to distribute tasks and to make decisions in the context of incomplete information [28]. The multi-agent society has several kinds of specialized agents: agents able to store the data, agents able to negotiate in case of possible conflicts, agents able to process and validate the data, agents able to test and diagnose the CPS and to take control decisions based on the rules stored into the knowledge database. At the middle layer, system complexity has to be addressed in terms of distributed computing, Big Data analytics, interoperability and dependability. The top layer of the CPS meets the functions: (i) provides CPS

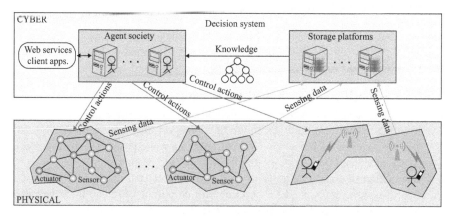

Figure 3.5   A CPS architecture including WSNs

monitoring, control and data analyses web services to the end-users (Software as a Service - SaaS); (ii) provides dedicated client applications to access web services. This proposed architecture (Figure 3.5) can be applied to monitor and control critical infrastructures, the environment, smart buildings and many others.

3.5   Characteristics and challenges of WSN-CPS

The present section intends to give an overview of some of the main challenges of WSN-CPS, noting that this is the author's point of view, starting from the main issues confronting the design of CPSs. Depending on application, only a part of the presented aspects are relevant. For example, for a CPS monitoring and controlling heavy machinery, power consumption does not raise any problems, because all the nodes can be powered from the power lines. One of the major problems here consists in the high levels of interference caused by electromagnetic fields and large obstructions. Another case in which power consumption is not of great importance is represented by power metering applications employing WSNs. Some of the industrial applications relying on CPSs and WSNs may be concerned not so much with flexibility and scalability but rather with high reliability and safety.

3.5.1   CPS challenges

The complexity of CPSs is beyond the state of current discoveries in terms of: abstraction (formalisms) and architectures able to allow the integration and the interoperability of various heterogeneous CPS's components; distributed computations and networked control that target time- and event-driven computing, time delays, failures, reconfiguration, distributed decision support to satisfy a high degree of reliability and security of the heterogeneous cooperating components that interact in a physical environment; verification and validation to guarantee the Quality of Services (QoS) of the software and hardware components and also of the entire CPS from its early design stages [36]. Many challenges have to be fulfilled at different layers of the CPS architecture to achieve the easy integration of the physical and cyber components. There are two categories of research challenges concerning CPSs: (i) scientific and technical, and (ii) social and institutional [27, 29, 34, 37]. A brief review of these challenges is presented below.

3.5.1.1 Scientific and technical challenges

- Analysis, design and verification methods and tools. Since the CPSs involve multiple disciplines, methods and tools to support the analysis, the design and the verification of these kind of systems, as well as to achieve automatic development process from modelling to code, are required.
- Comprehensive and integrated system and architectural models. To develop a hybrid system, such as a CPS, a number of comprehensive and formal models have to be developed in order to integrate the physical models with the digital ones, belonging to software and system engineering. These will take into account system behaviour, the real-time performance of the system, the semantic interoperability between its components, and the networking models. They will be domain-specific with reusable features.
- Networking. Regarding the sensors and the actuators of a CPS, future developments in the fields of virtualization, energy management, QoS management and locally/globally networked electronic devices with real-time management have to be resolved. These involve parallel and accurate data acquisition, sensor fusion, real-time physical data processing, data interpretation, prediction of faults, globally distributed and networked real-time control.
- Non-functional requirements. These requirements address important challenges in the fields of dependability, QoS and verification/validation, and represent important characteristics of the CPSs, as they operate in critical environments. Important from the dependability point of view are the maintainability, the safety, the integrity and the security attributes. These attributes should be considered starting from the design phase of the CPS. The maintenance process has to be a proactive and reactive one. Methods for guaranteeing the safety and the integrity of the acquired data have to be defined and implemented, such as system's threats identification methods, fault tolerance mechanisms and data protection levels definitions. In terms of security, new efficient cryptographic algorithms and protocols, and new privacy-preserving schemes are essential. The verification and validation of the CPSs involve the definition of new methods for test and analysis, as well as the definition of new metrics for certification/validation.
- Design and development standards. CPSs incorporate many technologies with application in various industries. This is why standardized abstractions and architectures of the CPSs, which permit modular design and implementation, are needed.

3.5.1.2 Social and institutional challenges

In addition to scientific and technical challenges, CPSs have to face social ones too. These imply inter- and trans-disciplinary research and development activities, interactive innovations by means of economic ecosystems and platforms, and by regional and international innovation systems [29]. Also, CPSs have to be designed for appropriate human–system cooperation beyond current man–machine interaction. The interface of the CPSs will answer the expectation of the users, will learn from the users' behaviour, will be usable, controllable and fault tolerant [29]. From the institutional challenges point of view, effective models of governance and CPS business models have to be implemented to provide directives, standards and protocols for systems that operate simultaneously in cyber and physical environments, to reduce the unforeseen liabilities, respectively to understand the benefits and to mitigate the business associated risks [37].

All the challenges presented so far represent aspects that become even more important in the context of CPSs including large numbers of sensor networks. Figure 3.6 presents challenges and opportunities of WSN-CPS, starting from the ARTEMIS matrix [38], in order to establish the main research directions and to ascertain the expected impacts of CPSs in various domains of application. The overcoming of several design issues posed by the symbiosis of WSNs and CPSs, including power consumption, communication, dependability, security, QoS, Big Data and others as described below, will lead to efficient solutions and to a wider acceptance of the new WSN-CPS paradigm.

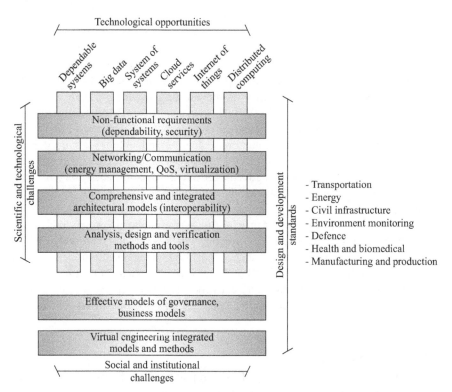

Figure 3.6 WSN-CPS challenges and opportunities

3.5.2   WSN-CPS challenges and characteristics

3.5.2.1   Power consumption

In the vast majority of cases, sensor nodes in WSNs are devices running on batteries, which are expected to function for multiple years without requiring replacement. This is the reason why energy efficiency and long battery lifetime are of crucial importance in this domain [39]. There is also a generalized trend of reducing the environmental impact of technology, and this also applies to WSN-CPS, where the achievement of green communication and data processing is targeted [40]. Although WSNs have already been demonstrated as a viable solution in a wide range of applications, if they are to become ubiquitous, power consumption still needs increased attention [18].

Another problem consists in the battery weight and dimension, an aspect that has to be considered in the sensor nodes' design phase. The reduction of their physical dimensions will lead to a shorter operation time, which has to be compensated by efficient power-saving strategies [41]. This also leads to the need for a thorough testing and evaluation of the power source used, the lowest volume and weight along with the maximum capacity being desired [42]. Depending on application, other characteristics, such as immunity to temperature variations (for example, in the case of environmental monitoring), or tolerance to high discharge pulses, without drastically reducing capacity, may be required. By observing the current consumption of a device in as many operation scenarios as possible and by studying the discharge curves of the battery, the lifetime of the system powered by it can be approximated. One of the approaches for extending a mote's lifetime consists in the addition of a secondary power source, for energy harvesting. This means that the node has the ability of replenishing its energy from an energy source, such as solar cells, vibration or fuel cells, acoustic noise or mobile suppliers (robots), solar energy harvesting being the current mature technique [43]. The use of a supercapacitor, also known as an ultracapacitor or a double-layer capacitor, is a common practice in the design of secondary power sources. These can store the harvested energy, offering several advantages over batteries, namely, higher power density, the elimination of special charging circuitry and long operational lifetimes, considered to be unrelated to the number of charge/discharge cycles [44].

In Yick *et al.* [43], the tasks in wireless sensor applications are divided into three different groups, namely (i) the system, which is, in fact, represented by the mote, (ii) the communication protocols, enabling communication between the application and the nodes, and (iii) the services, which enhance the entire solution and improve system performance and network efficiency. All these three components also affect energy consumption and the need for optimized solutions is imperative.

3.5.2.2   Communication

Typically, communication between CPSs' components is performed through wires. Besides high costs, their placement requires efforts for deployment and maintenance since these are systems covering vast areas and often operating in harsh environments. These merits, coupled with the continuous improvement of wireless sensing solutions, make the use of WSNs for data gathering and for the control of actuators in CPSs very attractive.

Communication is of great importance in CPS applications supported by WSNs, because, in addition to sensor networks' specific challenges, the problem of *interoperability between heterogeneous devices and technologies* arises [45]. The motes communicate over short distances and their position, in some cases is not pre-determined, leading to the need for developing efficient sensor network protocols and algorithms with self-organizing capabilities [10]. With the radio frequency (RF) transceiver being the biggest power consumer in wireless sensor or actuator devices, its operation time has to be kept to a minimum [46]. In general, wireless sensing applications are duty-cycled systems, where the sensor nodes are not required to work all the time, but by alternating activity periods with idle intervals (sleep). The duty-cycle is defined as the ratio between the active period and the full active/dormant period [47]. Another category of WSNs employs nodes that have to wake up from idle states and transmit the processed data only when certain conditions are met. While the use of the former type of WSN has been proven as an efficient technique for prolonging the lifetime of the application, it leads to a set of significant challenges, such as loss of synchronization between nodes, loss of clock synchronization, the proper management of different rates of data transmissions from different sensors (acceleration and temperature), loss of reliability due to message collisions and so on. Data is transmitted through a connection directly with the base station (single hop) or through multi-hop communication and therefore the actions performed by the nodes which cannot establish a network connection also need attention [48].

The basic data transport of the WSN is determined by the medium access control, or MAC, protocol, which also has a great impact on power consumption and determines the successful operation of the network [49, 50]. A large number of MAC protocols for WSNs have been developed, but most of them are purely designed to minimize energy consumption and only a small fraction consider aspects such as delay and reliability, to name but two of the requirements in CPSs [49]. Sensor-MAC (S-MAC), time division multiple access (TDMA), frequency division multiple access (FDMA), code division multiple access (CDMA) and IEEE 802.11 are some of the protocols used in WSNs. However, although a lot of research is taking place in this area, communication protocols are still trying to achieve reasonable throughput and low latency when the dimension of the network increases [43]. One of the most challenging domains to use WSNs consists in factory automation, where besides high reliability and availability levels, update frequencies between 100 ms and less than 1 ms are required [51]. The standards developed for these types of applications include the IEEE 802.15.4, ZigBee, WirelessHART and ISA-100.11a. IEEE 802.15.4 supports the star topology and the peer-to-peer topology and recognizes two types of devices, namely, a full-function device (FFD) and a reduced-function device (RFD) [52]. It offers low costs and low power at low data rates and represents the basis for the other three mentioned standards. WirelessHART is the first open and interoperable wireless communication standard which was designed specifically for real-world applications in process automation [51]. It can guarantee end to end delay and improve end to end reliability, characteristics that make it suitable for several mission-critical applications [49]. Until recently, Wi-Fi technology has not been considered for implementing wireless sensing solutions because of its unsatisfactory energy consumption. This has changed, since multiple companies have recently developed low-power Wi-Fi devices and new solutions can benefit from several significant advantages offered by this technology: easy integration with the existing infrastructure, built-in IP-network compatibility and the existence of familiar protocols and management tools [53]. Furthermore, it has been demonstrated that low-power Wi-Fi solutions can achieve similar performances to those provided by ZigBee (IEEE 802.15.4) in terms of years of battery lifetime [39, 54]. A lot of work still has to be done in this direction because the majority of Wi-Fi-based sensing solutions are implemented using infrastructure networks, where sensors communicate directly to the gateway, or the access point. CPSs can span from WSNs to the internet and the co-existence and collaboration between huge numbers of devices makes communication reliability even more important. This becomes a crucial problem where devices communication using Wi-Fi, Bluetooth and ZigBee, which work on the same 2.4 GHz ISM band, and viable channel management schemes have to be developed [30]. As a conclusion, the heterogeneity proposed by CPSs and the interoperability between hundreds or even thousands of devices seems to be the largest obstacle in developing solutions which include WSNs.

3.5.2.3 Dependability

CPSs represent more than traditional sensor networks and are made up of components provided with learning and adapting capabilities, self-organization and self-management functions, which have to accomplish their tasks with a high degree of dependability and security. Dependability represents the ability to deliver a service that can justifiably be trusted, and is an integrating concept that encompasses several attributes, namely availability, reliability, safety, confidentiality, integrity and maintainability [55]. Depending on application, it is possible that some of the aforementioned attributes have greater importance than the others.

As a feature of CPSs, dependability has to provide permanent system operation within evolving and adaptive contexts. In the case of WSN-CPS, the dependability attributes will have to be refined and classified and new methodologies for dependability assurance, specific to this application field, will have to be developed. These methodologies will constitute the base for new online dependability models and analysis, other than classical ones [56–58]. The metrics of the identified dependability (qualitative/quantitative indicators adequate for emphasizing a system's capacity of carrying out its associated task according to its specifications) are modelled for generating a probabilistic estimation of the WSN-CPS behaviour, influenced by the sensors and the actuators in the network. The modelling procedure consists of two phases. The first one is represented by the development of models for the dynamic adaptation of the system's dependability starting from elementary stochastic processes, models that will describe the behaviour of WSN-CPS components and the interactions between them. The second one implies model processing for obtaining the expressions and the values of quantifiable system attributes, which assure dependability. The development of dependability models is difficult due to the increased complexity, to the fundamental differences in the control of cyber and physical components, and to the interdependencies between them, especially in the case of CPS including several heterogeneous sensor networks. A possible modelling solution is represented by the definition of a dependability ontology, which interprets the raw streams, ensures the accessibility and communication between the various heterogeneous devices and terminal categories, and models the WSN-CPS's behaviour in terms of cyber and physical components and human aspects. The ontology describes the physical entities of the system (sensors and actuators), the analyses and processing, which occur in relation to these entities, and also the types of faults and failures, and emphasizes the possible causes leading to different system component failures [28]. The implemented models will be evaluated after the prior definition of dependability evaluation indicators and have to satisfy the following:

- Data and system components permanent availability – WSN-CPS have to rapidly detect all the fault categories (i.e. software, hardware, communication faults) and take the necessary measures to ensure system functionality (i.e. the system reconfiguration, proactive and reactive maintenance, fault-tolerance mechanisms).
- Data and system components reliability – WSN-CPS have to incorporate features that prevent failures from occurring or that limit failure consequences.
- Data and system components safety – WSN-CPS have to ensure the absence of disastrous consequences on users and on the environment [59].

Also, WSN-CPS have to be continuously monitored with the help of adaptive monitoring software tools, capable of detecting potential problems and of providing feedback to the implemented models, which will have to react accordingly.

Dependability, with all its components, is also an important aspect of WSNs within CPSs, since there are cases in which such sensor networks are used for enhancing exactly this characteristic for the latter. For example, in Shrestha *et al.* [60], a solution is presented, where a heterogeneous multi-hop network is used for real-time monitoring and alerting for rail stock with the objective of improving overall safety, security and reliability. Dependability assurance in CPS applications employing WSNs is a complicated task, because the 'best effort' communication within WSNs cannot guarantee reliability, latency and capacity [45]. However, reliability is a highly application-dependent property, and there are scenarios where it can be neglected as a trade-off between reliability and sensor network lifetime, such as monitoring solutions. At least two aspects of reliability have to be taken into consideration in the case of sensor networks, namely, communication reliability and the reliability of the wireless sensors themselves. Often being deployed in harsh and hostile environments, the possibility of permanent or intermittent faults is significantly high [61]. This is becoming a major challenge, since the technology is evolving and is miniaturized continuously, allowing for smaller chips with lower threshold voltages, which makes them even more vulnerable to extreme conditions. Furthermore, ageing effects, such as time-dependent dielectric breakdown or negative bias temperature instability, which are dependent on environmental conditions, are more susceptible to causing faults in such devices. To enhance the motes' reliability, local or remote test routines have to be developed, keeping in mind the fact that energy consumption has to remain as low as possible. In terms of communication, a high level of reliability assumes the successful delivery of packages. The difficulty for solving this problem increases as the size of the CPS increases and substantial research efforts are targeted in this direction. One of the fields that require a high level of reliability, accuracy and time criticality is factory automation, where trade-offs between stringent time deadlines, reliability and fidelity of data, and energy efficiency have to be carefully considered [62]. The research concerned with data transmission reliability mainly falls into two categories: packet-loss avoidance and packet-loss recovery [63]. The fact that some sensor nodes may fail or may die because of battery depletion, physical damage or heavy interference also has to be taken into account, and fault tolerance mechanisms at node and network levels have to be included in the design of the CPS.

3.5.2.4   Security

In the context of dependability, security represents the concurrent existence of availability, which is the readiness for correct service, integrity, which can be defined as the absence of improper system alteration, and confidentiality [55]. Two other important properties of sensor networks security, that have to be considered, are represented by authenticity and freshness. Since CPSs are often applied to mission-critical processes, securing these networked and complex systems against malicious attacks, such as passive interception of data, active injection of traffic, radio jamming or network overloading with garbage packets, becomes a challenging task. Furthermore, in the case of CPSs including sensor networks, communication itself makes them vulnerable to a wider range of threats and risks because of the open, shared nature of their medium [64]. The most common method for achieving security in wireless communication consists in message encryption, but implementing this is often difficult because of the motes' constraints in terms of storage, communication, computation and processing capabilities [43]. The nature of wireless communications, the increased size and density of the networks, in some cases the topology prior to deployment, which is unknown, and the high risk of physical attacks to unattended sensors complicate the issue even more [65]. The solutions for enhancing the security in WSNs imply the direct use of pre-distributed keys or the use of keying materials for the dynamic generation of pair-wise keys. However, the vast majority of the proposed schemes consider sensor nodes which are homogeneous, have the same capabilities and are tamper-proof [65]. This does not apply to CPSs spanning over multiple WSNs, where a high degree of system heterogeneity is assumed. Access control is another problem that has to be considered, especially in the cases where a large number of users from different groups, which sometimes have conflicting interests, are present. Therefore, as a starting point, the development of privacy-preserving access control mechanisms in the WSNs present in the CPS, satisfying requirements as user authentication, user privacy-preserving, node compromise tolerance, scalability, freshness, limits of access privileges, dynamic participation, amongst other things, is desirable [66]. From the security point of view, Wi-Fi-enabled sensors seem to have an advantage over those using other technologies, 802.11 providing several standard security schemes that provide data confidentiality, authentication and availability, such as WEP, WPA/TKIP-PSK and WPA2/AES-PSK. Obviously, there is a trade-off between the strength of the security mechanism used and the power consumption of the motes [39], showing that WPA2/AES-PSK is the best choice for this type of sensor.

Another component of dependability, which is also part of security and is of great importance in WSN-CPS design, is availability, or the readiness for correct service. The availability of services ensures that only authorized entities can access the data, services or other available resources when these are requested [62]. The most detrimental threat to a network's availability consists in the denial of service (DoS), caused by attackers through methods such as producing radio interference, disrupting network protocols or depleting sensor power sources [11]. Violation of availability can cause damage or affect safety in certain scenarios and should be prevented from occurring by efficient security mechanisms.

In conclusion, it can be stated that the design of secure protocols for WSNs in particular, and for CPSs in general, is a challenging task because these systems are still under development and the design of strong security protocols while maintaining low overheads is required [11].

3.5.2.5 Quality of services (QoS)

QoS aims to provide guarantees on the ability of a network/system to deliver predictable results. As mentioned in Section 3.5.1, QoS represents an important issue concerning the CPS communication layer (networking), as well as of the entire system.

CPSs are application-oriented systems. For that reason, WSNs have to provide QoS support in order to satisfy the service requirements of target applications [67]. QoS management in WSNs will play an important role in CPS design and implementation. Xia *et al.* [67] identified several research topics and challenges in supporting QoS in WSNs in order to meet the requirements and constraints of diverse applications and proposed a solution to network QoS management-feedback scheduling. The identified QoS challenges are the following: service-oriented architecture (SOA), QoS-aware communication protocols, resource self-management and QoS-aware power management. SOA supports the rapid, cost-effective composition of interoperable, scalable systems based on reusable services. These features make SOA appropriate for guaranteeing QoS in WSN components of CPSs. The communication protocols for WSNs need to be designed based on the heterogeneity of sensors and actuators within the CPSs. CPS features, such as the complexity and the dynamic nature due to the continuously changing and unpredictable environments, lead to the need for resource self-management [68]. This technique assumes the automatic adaptation of resource usage to obtain an optimized overall QoS. The feedback scheduling solution dynamically adjusts specific scheduling parameters of relevant traffic to maintain a desired QoS level and enhances CPS predictability, using control theory and technology [67]. This solution can be implemented in each node of a WSN. As mentioned earlier, the reliable delivery of gathered data with a minimum energy consumption is desired in sensor networks. According to the topologies and the QoS requirements, the transmission power management mechanisms for actuator nodes, respectively for sensors nodes within WSNs, may be different. A solution for QoS-aware power management can be the assignment of different transmission power levels to the same wireless node with respect to QoS requirements imposed by different types of traffic [67]. In the case of WSN nodes which include only sensors and no actuators, an energy-aware QoS provisioning mechanism is proposed [69]. This approach provides both the communication reliability and delay guarantees with minimum energy consumption, based on an addressed optimization problem and using a greedy algorithm to solve it.

Cloud computing technology, through its three types of services, Infrastructure as a Service (IaaS), Platform as a Service (PaaS) and Software as a Service (SaaS), can also represent a solution to minimize energy consumption and to maximize QoS [30]. The virtualization technology, specific to cloud computing, allows a dynamic provisioning of the resources by service isolation. Thus, IaaS isolates the physical layer from users' cloud infrastructures, PaaS provides 'virtual platforms' to facilitate application development, while SaaS provides software applications with manageable QoS [30].

The above two paragraphs present some QoS management challenges and possible solutions within a CPS, at the WSN level. However, QoS for the entire CPS also has to be addressed. The complex nature of a CPS creates difficulties in providing end to end QoS for such systems. Several issues – such as latency, sensor failures, energy efficiency, scalability, security, trust and data privacy – lead to the following QoS challenges: cross-platform sensor-actuator communication, energy efficiency and scalability at the same time, and high-level security and privacy data transmission and communication protocols [70]. A framework for CPSs needs to be developed to satisfy these QoS challenges. Dillon *et al.* [71] present a Web-of-Things (WoT) framework with five layers that provides end to end QoS at all CPS levels. The WoT framework assures the development of CPSs that connect cyber entities to the physical ones, and this is demonstrated successfully in two case studies (electricity smart grids and intelligent vehicle systems). An aspect-oriented QoS modelling method, based on UML and formal methods, is an alternative to the WoT framework, as can be seen in Liu and Zhang [72]. The method uses the UML extension of QoS and emphasizes the dependability characteristics of QoS in the form of QoS aspects. These two approaches demonstrate the need for the development of a unique framework for the entire CPS to support the following challenging QoS constraints: real-time requirements (low latency and bounded jitter), availability requirements (fault propagation, recovery across boundaries), security requirements (appropriate authentication and authorization) and physical requirements (limited weight, power consumption and memory footprint) [72].

3.5.2.6 Big Data

The newly emerging vision of the Internet of Things (IoT), which calls for the interconnection of an unmatched number of embedded computing devices, coupled with trends such as virtualization and cloud computing, generates a huge amount of data that has to be further processed to extract value. The increased rate of adoption of these systems leads to an unprecedented growth of interconnected and highly complex data produced by sensors, social media and mobile and location systems. This phenomenon is known today as the data deluge or Big Data [73]. An attributive definition of Big Data was given by International Data Corporation (IDC), as follows: 'Big data technologies describe a new generation of technologies and architectures, designed to economically extract value from very large volumes of a wide variety of data, by enabling high-velocity capture, discovery, and/or analysis' [74]. In the context of sensor-based data collection systems – WSN-CPS – information about the physical world is collected from a wide range of wired or wireless heterogeneous sensors, smart phones, tablets and sensor-based internet-enabled devices. There are great technical challenges posed by the collection, processing and analysis of this huge amount of data which require the emergence of new solutions and techniques embracing the latest advances in the information and communication technology field [75]. These challenges include data representation, redundancy reduction and compression, life-cycle management, and privacy and security for data collection and management, and the development of approximate and deep analytics technologies for achieving efficient interpretation, modelling, predicting and simulation methods. The development of large-scale parallel systems for managing the data also means we must confront several issues, with the most common consisting in energy management for reducing costs and carbon emissions, scalability for supporting continuously increasing data sets, and collaboration between professionals acting in a wide range of different fields [75].

The Big Data analytics within WSN-CPS is an iterative process that involves the following stages along all the architectural layers: data acquisition, information extraction and pre-processing, data integration, aggregation and representation, modelling and analysis, and interpretation [76]. Regarding the data acquisition stage, one challenge is represented by the definition of 'online' filters which do not discard useful information, given that the volume of raw data is too big. Another challenge consists in the loading of large data sets, especially when these are combined with online filtering and data reduction. Incremental ingestion techniques could be a potential solution for this issue. The acquired information is not always ready for analysis and, therefore, in the next stage of Big Data analytics, an information extraction process, that pulls out the necessary information from the original sources and expresses it in a suitable form for analysis, is required. This information extraction process also represents an important challenge, and its implementation depends on the application domain. A possible solution in this case could be the definition of declarative methods in order to specify the tasks of the information extraction process and the optimization of their execution when new data arrives. Another fact that has to be considered consists in data reliability, the acquired information being affected by sensor failures, magnetic interference, security attacks and other matters. This is addressed by studying the sources of errors and by developing efficient data pre-processing techniques. Data integration, aggregation and representation deal with the large-scale analysis which requires the collection of heterogeneous data from different sources. Therefore, for resolving these structural and semantic issues, the development of data transformation and integration tools is required. The modelling and analysis of Big Data differs from traditional statistical analysis, and for that reason methods for querying and mining the data are also demanded [76]. With WSN-CPS producing and consuming a continuously increasing amount of information, the problem of Big Data becomes a hot topic in this field.

3.5.2.7   Other challenges

Both WSNs and CPSs represent interdisciplinary research areas, where knowledge and expertise from different fields are combined. Some of these domains consist in signal processing, networking and protocols, databases and information management, distributed algorithms, embedded systems and architecture. The previous subsections briefly presented only a small group of characteristics and challenges encountered in the study and development of such systems. Probably one of the greatest issues that prevents the widespread adoption of WSN-based solutions is represented by *costs*, a topic addressed by Merrill [31]. Current research in this field reveals that a gap between academic papers and a feasible return of investment for WSNs consisting of large numbers of nodes still exists. Furthermore, application-specific software development is still costly and even with the existence of simulation tools, demonstration testbeds are required. In turn, these are also lagging behind research papers. The *development of environments for analysis and modelling*, which allows the selection of the best application configuration in terms of modules, software tools, network type, protocols and others, based on performance and cost criteria established by the user, needs attention also. The existing tools for design, verification and support are not suited for complex system design involving multiple disciplines, such as networking and embedded systems [27].

Characteristics such as *real-time performance* and *quality of service* can be neglected in systems used for monitoring purposes, but this changes in the case of industrial implementations. In addition to the features presented earlier, mission-critical applications require a *high level of safety*, considerable *standardization* efforts for easy system integration and development and *increased immunity* to large amounts of interference from harsh environments [51].

Some CPS applications might require knowledge of the *location information* for each sensor node in the network. This implies the determination of the current location of the nodes in the network within a given coordinate system [45]. There are several methods for localization within WSNs, including the use of a global positioning system (GPS) receiver, the use of beacon or anchor nodes or the use of neighbouring nodes for determining a node's location [43]. Each one of the methods mentioned has its own advantages and drawbacks. For example, a GPS receiver operates properly only outdoors and in the absence of large obstructions, and the use of other mechanisms such as inertial sensors are required to assist in indoor positioning. The problem of *fault diagnosis* in WSNs also needs to be addressed, because, in certain scenarios, the presence of undetected erroneous data can lead to severe consequences in terms of human life, environmental impact or economic losses [77]. Other aspects that have to be considered in WSNs, and consequently in CPSs, which are equally important to the ones discussed so far, consist in *coverage and deployment issues, fault tolerance, scalability, message delays* and *throughput*. Although a huge amount of research is carried out by both academia and industry, several challenges still have to be solved before WSNs will meet their promised potential, and the seamless integration of heterogeneous devices, protocols and design architectures with CPS solutions can become a reality [32].

3.6   Opportunities

The opportunities for deploying CPSs spanning large numbers of sensor networks are tremendous, the combination of advanced sensing, measurement and process control, including cyber-physical systems and wireless sensing technology having applicability across almost all industry domains. However, as the previous section outlined, there is still a lot of research work that needs to be devoted to this area.

It is expected that, in the near future, sensor networks will grow in size because of lower development and deployment costs, better protocols and higher processing power. Through the use of sensing networks, the new emerging CPS solutions will be able to close the loop via actuation. By addressing reliability, security and safety in the WSN design phase, it will be possible for these networks to provide timely inputs to CPSs, ensuring overall system reliability and predictability. In turn, sensors will become even more power efficient and accurate, and with the current developments in micromachines, they will be able to provide motes even with mobility. For example, the use of gas sensors in small battery-powered devices was problematic a short while ago, because of their large power consumption and dimensions. However, recently this has changed, since new technologies and designs are available on the market. Such a device is the low-power Cozir Ambient CO_2 NDIR sensor, which consumes 3.5 mW in continuous operation. Furthermore, power sources and energy harvesting schemes show significant improvements and further expand the application areas.

CPSs spanning multiple sensor networks support the idea of the IoT, a vision which calls for connectivity not only to consumer electronics and home appliances but also to small battery-powered devices which cannot be recharged [39]. Several sites presenting data from thousands or hundreds of thousands of sensors all over the world exist, one of them being www.xively.com, a 'Public Cloud for the Internet of Things'. The IoT paradigm brings the need for the convergence of existing networks, such as mobile ad-hoc networks (MANETs), mobile cellular networks (MCNs), wireless sensor networks (WSNs) and others, to provide human-to-human, human-to-object and object-to-object interactions between the physical world and the virtual world [32]. As conventional computing systems' resources increase almost exponentially as technology improves, this will be used to decrease the cost of the nodes in sensor networks, enabling the development of small high-performance *disposable devices* [78]. The use of the IEEE 802.11 standards in wireless sensing applications is now possible with the development of power-efficient Wi-Fi components, promising multiple years of battery lifetime while maintaining compatibility with standard Wi-Fi [40]. This type of connectivity leads to reduced infrastructure costs while improving the total cost of ownership, fast sensor deployment on pre-existing 802.11 networks, the use of network management tools, a vast knowledge base and protocol familiarity – advantages which increase the attractiveness of this technology [53]. Another technology that could not be considered until recently in the design of WSN-CPS solutions, because of its high power consumption, is represented by FPGAs (field-programmable gate arrays). However, with the development of *low-power FPGA chips*, wireless sensing applications are now able to benefit from their undeniable advantages, such as flexibility and natural parallelism.

The overcoming of the system integration and interoperability challenges posed by such complex and innovative heterogeneous systems, which integrate computing and communication capabilities with sensing and actuation in the physical world, will generate unique opportunities for economic growth and will help ensure population health, safety and security while *improving quality of life* [37]. Although several concepts such as autonomous cars, robotic surgery, intelligent buildings, smart electric grid, smart manufacturing and implanted medical devices already exist, the field of CPS is only at the beginning and new revolutionary and innovative solutions in many industries are expected to emerge. These applications include smart manufacturing and production, transportation and mobility, energy, civil infrastructure, healthcare, social networking and gaming, defence and emergency response [37].

CPSs spanning multiple sensor networks will become critical to the improvement of business sectors' competitiveness and success, and of government services. The development of new WSN-CPS offers *R&D opportunities* to academia and business environments in order to response to the technological, institutional and social challenges enforced by this category of systems. These opportunities target frequent and strategic themes that could have a significant impact if they are addressed, and cover the spectrum of WSNs in CPSs and CPS challenges: robust, effective design and construction of systems and infrastructure; improved performance and quality assurance; effective, reliable system integration and interoperability; and dynamic, multi-disciplinary education and training [37, 79]. *Research and innovation programmes*, funded by public/government bodies around the world, will come to meet these R&D opportunities. The United States has placed CPSs at the top of its list of priorities for investment in research. Several reports of the President's Council of Advisors on Science and Technology (PCAST) demonstrate a huge interest in the United States to lead this field and to enhance its competitiveness in global markets [80–82]. The main federal research agencies are involved in this process. The European Union ARTEMIS and ECSEL programmes, which specialize in embedded systems, electronic components, systems and CPSs development, are investing billions of euros to make Europe a global leader in this field by the year 2020 [83]. Also, the Horizon 2020 EU programme supports research and innovation activities in this field through the 'Smart Cyber-Physical Systems' topic and helps ARTEMIS and ECSEL to achieve their goals [84]. These three programmes encourage the establishment of partnerships between academic communities, industry, decision-making bodies and stakeholders. The great potential of CPSs motivates countries such as Japan, China and India to invest in this sector. The intense competition between these countries, to achieve CPS research and innovation activities and to obtain significant results, will provide new discoveries and cross-cutting technologies and will address the CPS and WSN market requirements that span many sectors of the economy, such as the development of smart vehicles, smart electric power grid, smart gas and oil distribution grid, smart water and waste-water grid, smart factories, wireless body area networks, wearable sensors, resilient communications networks, emergency response systems. The CPS market will encompass other markets, including the WSN market, since CPSs integrate sensor and actuator networks. A recent WSN market analysis shows its rapid growth, which is expected to reach $1.8 billion by 2024 [85]. This more than satisfactory market forecast attracts many industry actors interested in CPS and WSN businesses. The forecast will be reached and exceeded only if the new cost-effective technologies are reliable, standardized and certified, and if they bring significant advantages over current knowledge to eliminate existing and future barriers. For the moment, these requirements are not fully met, but researchers and specialists are very optimistic.

The R&D, financial and market opportunities create a framework for mastering the WSNs in CPS challenges and the CPS challenges themselves, and for satisfying the need for 'smart everything' that will bring important benefits for the nation and indeed for society as a whole.

3.7 Conclusion

CPSs enable interactions between the cyber and the physical worlds in order to improve safety, convenience and comfort in our everyday lives [32]. WSNs play a significant role in CPSs, as they can act as a human-physical world interface with the cyber world through sensing and actuation [86]. This chapter has provided a guided tour of WSN-CPS and emphasized the following aspects: WSNs – definition, application areas, architecture and constraints; CPSs – definition, their main characteristics, benefits, application domains, architectures and research challenges; characteristics and challenges of WSN-CPS – power consumption, communication, dependability, security, quality of services, Big Data analytics and a few other aspects (location, coverage and deployment issues, fault tolerance, scalability, message delays and throughput); and finally, research, development, funding and market opportunities.

The following conclusions can be summarized from the survey presented in this chapter:

- There are many challenges to be addressed, especially in CPSs spanning multiple WSNs, where, first of all, a framework for analysis, design, modelling and testing is required. These challenges have to respond to the issues posed by CPS networking and to non-functional requirements. Networking involves new techniques for minimizing power consumption and for increasing information accuracy in WSN-CPS, and new communication protocols that ensure the interoperability between heterogeneous devices. The non-functional requirements target new dependability models to satisfy the permanent avail ability, reliability and safety of large amounts of data and system components (including wireless sensors), new communication reliability techniques and security protocols, and new QoS management. Other aspects to be considered are Big Data analytics, location awareness, time, scalability and fault tolerance.
- The identified challenges create exciting future R&D opportunities, the vast majority of them funded by national or multinational programmes to address the market needs, given that the WSN and CPS markets are very favourable.

This chapter offers an overview of the current research state in WSN-CPS, identifying a great number of challenges and future opportunities brought by the response to the synergy between these two vast domains.

Acknowledgments

This chapter was supported by the PARTING project, ctr. no. POSDRU/159/1.5/S/137516, co-funded from the European Social Fund through the Human Resources Sectorial Operational Program 2007–2013.

References

[1]   W. Wolf. Cyber-physical systems. *Computer*, 42(3): 88–89, 2009.
[2]   S. Gao, H. Luo, D. Chen, S. Li, P. Gallinari, Z. Ma and J. Guo. A cross-domain recommendation model for cyber-physical systems. *IEEE Transactions on Emerging Topics in Computing*, 1(2): 384–393, 2013.
[3]   M. Broy and A. Schmidt. Challenges in engineering cyber-physical systems. *Computer*, 47(2): 70–72, 2014.
[4]   F. Zhao and L. Guibas. *Wireless Sensor Networks: An Information Processing Approach*. Morgan Kaufmann, San Francisco, CA, USA, 2004.
[5]   National Instruments, *What is a wireless sensor network?* Online, 2012. Available from: http://www.ni.com/white-paper/7142/en/ (Accessed 20 May 2014.)
[6]   J. Al-Karaki J. and A. Kamal. Routing techniques in wireless sensor networks: A survey. *IEEE Wireless Communications*, 1(6): 6–28, 2004.

[7]   M. Weiser. The computer for the 21st century. *Scientific American Special Issue on Communications, Computers, and Networks*, 265(3): 94–104, 1991.

[8]   R. Yan, H. Sun and Y. Qian. Energy-aware sensor node design with its application in wireless sensor networks. *IEEE Transactions on Instrumentation and Measurement*, 62(5): 1183–1191, 2013.

[9]   D. Puccinelli and M. Haenggi. Wireless sensor networks: Applications and challenges of ubiquitous sensing. *IEEE Circuits and Systems Magazine*, (3): 19–31, 2005.

[10] I. F. Akyildiz, W. Su, Y. Sankarasubramaniam and E. Cayirci. Wireless sensor networks: a survey. *Computer Networks*, 38(4): 393–422, 2002.

[11] Y. Zhou, Y. Fang and Y. Zhang. Securing wireless sensor networks: A survey. *IEEE Communications Surveys Tutorials*, 10(3): 6–28, 2008.

[12] M. Lazarescu. Design of a WSN platform for long-term environmental monitoring for IoT applications. *IEEE Journal on Emerging and Selected Topics in Circuits and Systems*, 3(1): 45–54, 2013.

[13] J. L. Hill. *System Architecture for Wireless Sensor Networks*. PhD dissertation, University of California, Berkeley, 2003.

[14] D. Larios, J. Barbancho, G. Rodriguez, J. Sevillano, F. Molina and C. Leon. Energy efficient wireless sensor network communications based on computational intelligent data fusion for environmental monitoring. *IET Communications*, 6(14): 2189–2197, 2012.

[15] Research and Markets. Wireless sensor networks (WSN) 2014–2024: Forecasts, technologies, players Online, 2014. Available from: http://www.research andmarkets.com/reports/2782659/wireless\s\do5(s)ensor\s\do5(n)etworks\s \do5(w)sn\s\do5(2)0142024 (Accessed 20 May 2014.)

[16] A. Fraboulet, G. Chelius and E. Fleury. Worldsens: Development and prototyping tools for application specific wireless sensors networks. In *Proceedings of the 6th International Symposium on Information Processing in Sensor Networks (IPSN 2007)*, pp. 176–185, Cambridge, MA, USA, 2007.

[17] J. Hao, B. Zhang and H. Mouftah. Routing protocols for duty cycled wireless sensor networks: A survey. *IEEE Communications Magazine*, 50(12): 116–112, 2012.

[18] Y. Li, H. Yu, B. Su and Y. Shang. Hybrid micropower source for wireless sensor network. *IEEE Sensors Journal*, 8(6): 678–681, 2008.

[19] CPS Steering Group. *Cyber-physical systems – executive summary*. Online, 2008. Available from: http://varma.ece.cmu.edu/summit (Accessed 20 May 2014.)

[20] E. A. Lee and S. A. Seshia. Introduction to embedded systems – a cyber-physical systems approach. UC Berkeley, 2011.

[21] M. Lemke. Research on embedded and cyber-physical systems – towards *HORIZON 2020. Online 2013*. See http://www.indin2013.org/n/wp-content/uploads/2013/05/INDIN-2013-Bochum-LEMKE.pdf (Accessed 21 May 2014.)

[22] B. H. Krogh. *Cyber Physical Systems: The Need for New Models and Design Paradigms*. Carnegie Mellon University, Technical Report, 2008.

[23]   B. X. Huang. *Cyber Physical Systems: A Survey*. Technical Report. Online presentation, access via www.archive.org. See http://www.ux.uis.no:80/atc08/smarthome/CPS_A_Survey.pdf (Accessed 20 September 2010.)

[24]   Z. Liu, D.-S. Yang, D. Wen, W.-M. Zhang and W. Mao. Cyber-physical-social systems for command and control. *IEEE Intelligent Systems*, 26(4): 92–96, 2011.

[25]   D. Chen and G. Chang. A survey on security issues of M2M communications in cyber-physical systems. *KSII Transactions on Internet & Information Systems*, 6(1): 24–45, 2012.

[26]   S. Barnum, S. Sastry and J. Stankovic. Roundtable: Reliability of embedded and cyber-physical systems. *IEEE Security Privacy*, 8(5): 27–32, 2010.

[27]   J. Wan, M. Chen, F. Xia, D. Li and K. Zhou. From machine-to-machine communications towards cyber-physical systems. *Computer Science and Information Systems*, 10(3): 1105–1128, 2013.

[28]   T. Sanislav and L. Miclea. An agent-oriented approach for cyber-physical system with dependability features. In *Proceedings of the 2012 IEEE International Conference on Automation Quality and Testing Robotics (AQTR)*, pp. 356–361, Cluj-Napoca, Romania, May 2012.

[29]   M. Broy, M. V. Cengarle, and E. Geisberger. Cyber-physical systems: Imminent challenges. In R. Calinescu and D. Garlan (eds.), *Large-Scale Complex IT Systems. Development, Operation and Management*. Lecture Notes in Computer Science Series, pp. 1–28. Springer, Berlin, Heidelberg, 2012.

[30]   F.-J. Wu, Y.-F. Kao and Y.-C. Tseng. From wireless sensor networks towards cyber physical systems. *Pervasive and Mobile Computing*, 7(4): 397–413, 2011.

[31]   W. Merrill. Where is the return on investment in wireless sensor networks? *IEEE Wireless Communications*, 17(1): 4–6, 2010.

[32]   C.-Y. Lin, S. Zeadally, T.-S. Chen and C.-Y. Chang. Enabling cyber physical systems with wireless sensor networking technologies. *International Journal of Distributed Sensor Networks*, 2012: 1–21, 2012.

[33]   A. Dabholkar and A. Gokhale. An approach to middleware specialization for cyber physical systems. In *Proceedings of the 9th IEEE International Conference on Distributed Computing Systems, ICDCS Workshops '09*, pp. 73–79, Montreal, Canada, 2009.

[34]   L. Zhang, J. He and W. Yu. *Challenges and Solutions of Cyber-Physical Systems*. Online, 2012. Available from: http://onlinepresent.org/proceedings/vol4\s\do5(2)012/8.pdf (Accessed 20 May 2014.)

[35]   G. Denker, N. Dutt, S. Mehrotra, M.-O. Stehr, C. Talcott and N. Venkatasubramanian. Resilient dependable cyber-physical systems: A middleware perspective. *Journal of Internet Services and Applications*, 1(1): 41–49, 2010.

[36]   R. Baheti and H. Gill. Cyber-physical systems. In *The Impact of Control Technology*. Online, 2011. Available from: www.ieeecss.org (Accessed 20 May 2014.)

[37] Steering Committee for Foundations in Innovation for Cyber-Physical Systems. *Foundations for Innovation: Strategic R&D Opportunities for 21st Century Cyber-physical Systems. Workshop Report*, 2013.

[38] IDTechEx. *Embedded/Cyber-Physical Systems ARTEMIS Major Challenges: 2014–2020*. ARTEMIS Industry Association, Technical Report, Online, 2013. See https://artemis-ia.eu/publication/download/910-sra-addendum (Accessed 15 April 2014.)

[39] S. Tozlu, M. Senel, W. Mao and A. Keshavarzian. Wi-Fi enabled sensors for internet of things: A practical approach. *IEEE Communications Magazine*, 50(6): 134–143, 2012.

[40] G. Palem and S. Tozlu. On energy consumption of Wi-Fi access points. In *Proceedings of the 2012 IEEE Consumer Communications and Networking Conference (CCNC)*, pp. 434–438, Las Vegas, NV, USA, 2012.

[41] G. Welch. A survey of power management techniques in mobile computing operating systems. *Operating Systems Review*, 29(4): 47–56, 1995.

[42] S. Folea, G. Mois, L. Miclea and D. Ursutiu. Battery lifetime testing using LabVIEW. In *Proceedings of the 9th International Conference on Remote Engineering and Virtual Instrumentation (REV)*, 1–6, Bilbao, Spain, 2012.

[43] J. Yick, B. Mukherjee and D. Ghosal. Wireless sensor network survey. *Computer Networks*, 52(12): 2292–2330, 2008. Online, 2008. Available from: http://www.sciencedirect.com/science/article/pii/S1389128608001254 (Accessed 20 May 2014.)

[44] A. Weddell, G. V. Merrett, T. Kazmierski and B. Al-Hashimi. Accurate supercapacitor modeling for energy harvesting wireless sensor nodes. *IEEE Transactions on Circuits and Systems II: Express Briefs*, 58(12): 911–915, 2011.

[45] P. Misra, L. Mottola, S. Raza, S. Duquennoy, N. Tsiftes, J. Höglund and T. Voigt. Supporting cyber-physical systems with wireless sensor networks: An outlook of software and services. *Journal of the Indian Institute of Science*, 93(3): 441–462, 2013.

[46] X. Shi and G. Stromberg. SyncWUF: An ultra low-power MAC protocol for wireless sensor networks. *IEEE Transactions on Mobile Computing*, 6(1): 115–125, 2007.

[47] F. Wang and J. Liu. Networked wireless sensor data collection: Issues, challenges, and approaches. *IEEE Communications Surveys Tutorials*, 13(4): 673–687, 2011.

[48] S. Maheswararajah, S. Halgamuge, K. Dassanayake and D. Chapman. Management of orphaned-nodes in wireless sensor networks for smart irrigation systems. *IEEE Transactions on Signal Processing*, 59(10): 4909–4922, 2011.

[49] P. Suriyachai., U. Roedig and A. Scott. A survey of MAC protocols for mission-critical applications in wireless sensor networks. *IEEE Communications Surveys Tutorials*, 14(2): 240–264, 2012.

[50] W. Ye, J. Heidemann and D. Estrin. An energy-efficient MAC protocol for wireless sensor networks. In *Proceedings of the IEEE INFOCOM 2002, 21st*

Annual Joint Conference of the IEEE Computer and Communications Societies, vol. 3, pp. 1567–1576, 2002.

[51] J. Åkerberg, M. Gidlund, T. Lennvall, K. Landerns and M. Bjökman. Design challenges and objectives in industrial wireless sensor networks. In G. P. Hancke (ed.), *Industrial Wireless Sensor Networks*. CRC Press, Boca Raton, 2013.

[52] P. Huang, L. Xiao, S. Soltani, M. Mutka and N. Xi. The evolution of MAC protocols in wireless sensor networks: A survey. *IEEE Communications Surveys Tutorials*, 15(1): 101–120, 2013.

[53] S. Tozlu. Feasibility of Wi-Fi enabled sensors for Internet of Things. In *Proceedings of the 7th International Wireless Communications and Mobile Computing Conference (IWCMC)*, pp. 291–296, Istanbul, Turkey, 2011.

[54] T. Sanislav, G. Mois, S. Folea, L. Miclea, G. Gambardella and P. Prinetto. A cloud-based cyber-physical system for environmental monitoring. In *Proceedings of the 3rd Mediterranean Conference on Embedded Computing*, pp. 15–19, Budva, Montenegro, June 2014.

[55] A. Avižienis, J.-C. Laprie and B. Randell. Dependability and its threats: a taxonomy. In R. Jacquart (ed.), *Building the Information Society*. Springer US, pp. 91–120, Toulouse, France, 2004.

[56] S. Bernardi, J. Merseguer and D. Petriu.*A UML Profile for Dependability Analysis and Modeling of Software Systems*. University of Zaragoza, Spain, Technical Report, rR-08-05, 2008.

[57] S. Bernardi, J. Merseguer and D. C. Petriu. Dependability modeling and assessment in UML-based software development. *The Scientific World Journal*, 2012: 1–11, 2012.

[58] A. Bertolino, A. Calabrò, F. D. Giandomenico and N. Nostro. An approach to adaptive dependability assessment in dynamic and evolving connected systems. *International Journal of Adaptive, Resilient and Autonomic Systems*, 4(1), 1–25, 2013.

[59] T. Sanislav and L. Miclea. An ontology-driven dependable water treatment plant CPS. *Journal of Computer Science and Control Systems*, 6(1): 99–104, 2013.

[60] P. Shrestha, M. Hempel, H. Sharif and H.-H. Chen. Modeling latency and reliability of hybrid technology networking. *IEEE Sensors Journal*, 13(10): 3616–3624, 2013.

[61] A. Merentitis, N. Kranitis, A. Paschalis and D. Gizopoulos. Low energy online self-test of embedded processors in dependable WSN nodes. *IEEE Transactions on Dependable and Secure Computing*, 9(1): 86–100, 2012.

[62] K. Islam, W. Shen and X. Wang. Wireless sensor network reliability and security in factory automation: A survey. *IEEE Transactions on Systems, Man, and Cybernetics, Part C: Applications and Reviews*, 42(6): 1243–1256, 2012.

[63] Y. Liu, Y. Zhu and L. Ni. A reliability-oriented transmission service in wireless sensor networks. In *Proceedings of the IEEE Internatonal*

Conference on Mobile Adhoc and Sensor Systems (MASS 2007), pp. 1–8, Pisa, Italy, 2007.

[64] M. Li, I. Koutsopoulos and R. Poovendran. Optimal jamming attack strategies and network defense policies in wireless sensor networks. *IEEE Transactions on Mobile Computing*, 9(8): 1119–1133, 2010.

[65] K. Lu , Y. Qian, M. Guizani and H.-H. Chen. A framework for a distributed key management scheme in heterogeneous wireless sensor networks. *IEEE Transactions on Wireless Communications*, 7(2): 639–647, 2008.

[66] D. He, J. Bu, S. Zhu, S. Chan and C. Chen. Distributed access control with privacy support in wireless sensor networks. *IEEE Transactions on Wireless Communications*, 10(10): 3472–3481, 2011.

[67] F. Xia, L. Ma, J. Dong and Y. Sun. Network QoS management in cyber-physical systems. In *Proceedings of the International Conference on Embedded Software and Systems Symposia (ICESS Symposia '08)*, pp. 302–307, Sichuan, China, 2008.

[68] A. Ganek and T. A. Corbi. The dawning of the autonomic computing era. *IBM Systems Journal*, 42(1): 5–18, 2003.

[69] M. M. Alam, M. A. Razzaque, M. Mamun-Or-Rashid and C. S. Hong. Energy-aware QoS provisioning for wireless sensor networks: Analysis and protocol. *Journal of Communications and Networks*, 11(4): 390–405, 2009.

[70] T. Dillon, V. Potdar, J. Singh J and A. Talevski. Cyber-physical systems: Providing quality of service (QoS) in a heterogeneous systems-of-systems environment. In *Proceedings of the 5th IEEE International Conference on Digital Ecosystems and Technologies Conference (DEST)*, pp. 330–335, Daejeon, South Korea, 2011.

[71] T. S. Dillon, H. Zhuge, C. Wu, J. Singh and E. Chang. Web-of-things framework for cyber-physical systems. *Concurrency and Computation: Practice and Experience*, 23(9): 905–923, 2011.

[72] J. Liu and L. Zhang. QoS modeling for cyber-physical systems using aspect-oriented approach. In *Proceedings of the 2nd International Conference on Networking and Distributed Computing (ICNDC)*, pp. 154–158, Beijing, China, 2011.

[73] W. Tan, M. Blake, I. Saleh I and S. Dustdar. Social-network-sourced big data analytics. *IEEE Internet Computing*, 17(5): 62–69, 2013.

[74] R. L. Villars, C. W. Olofson and M. Eastwood. *Big Data: What It Is and Why You Should Care*. White Paper, June 2011.

[75] H. Hu, Y. Wen, T.-S. Chua and X. Li. Toward scalable systems for big data analytics: A technology tutorial. *IEEE Access*, 2: 652–687, 2014.

[76] H. V. Jagadish , J. Gehrke, A. Labrinidis, Y. Papakonstantinou, J. M. Patel, R. Ramakrishnan and C. Shahabi. Big data and its technical challenges. *Communications of the ACM*, 57(7): 86–94, 2014.

[77] A. Mahapatro and P. Khilar. Fault diagnosis in wireless sensor networks: A survey. *IEEE Communications Surveys Tutorials*, 15(4): 2000–2026, 2013.

[78] M. Healy, T. Newe and E. Lewis. Wireless sensor node hardware: A review. In *Proceedings of IEEE Sensors 2008*, pp. 621–624, Lecce, Italy, 2008.

[79] Executive Roundtable on Cyber-Physical Systems. *Foundations for Inno-vation: Strategic Vision and Business Drivers for 21st Century Cyber-physical Systems*. Executive Roundtable on Cyber-Physical Systems, Workshop Report, Report of the Executive Roundtable on Cyber-Physical Systems, online 2013. See http://www.nist.gov/el/upload/Exec-Roundtable-SumReport-Final-1-30-13.pdf (Accessed 18 May 2014.)

[80] PCAST 2012 (USA). *Report to the President on Capturing Domestic Com-petitive Advantage in Advanced Manufacturing*. Executive Office of the President, President's Council of Advisors on Science and Technology (PCAST). Technical Report, 2012. Available from: http://www.whitehouse.gov/sites/default/files/microsites/ostp/pcast\s\do5(a)mp\s\do5(s)teering\s\do5(c)ommittee\s\do5(r)eport\s\do5(f)inal\s\do5(j)uly\s\do5(2)7\s\do5(2)012.pdf (Accessed 20 May 2014.)

[81] PCAST 2011 (USA). *Foundations for Innovation: Strategic Vision and Busi-ness Drivers for 21st Century Cyber-physical Systems*. Executive Office of the President. Technical Report 2011. Available from: http://www.whitehouse.gov/sites/default/files/microsites/ostp/pcast-advancedmanufacturing-june2011.pdf (Accessed 20 May 2014.)

[82] PCAST 2010 (USA). *Designing a Digital Future: Federally Funded Research and Development in Networking and Information Technology*. Executive Office of the President. Technical Report, 2010. Available from: http://www.whitehouse.gov/sites/default/files/microsites/ostp/pcast-nitrd-report-2010.pdf (Accessed 20 May 2014.)

[83] EU 2012. *The ARTEMIS Embedded Computing*. Online, 2012. Available from: http://www.artemis-ju.eu/ (Accessed 20 May 2014.)

[84] *Horizon 2013*. Online, 2013. Available from: http://ec.europa.eu/research/participants/portal/desktop/en/opportunities/h2020/topics/78-ict-01-2014.html (Accessed 20 May 2014.)

[85] IDTechEx. *Wireless Sensor Networks (WSN) 2014–2024: Forecasts, Tech-nologies, Players*. IDTechEx, Technical Report, March 2014.

[86] S. K. Das. Cyber-physical and networked sensor systems: Challenges and opportunities. In *Proceedings of the 1st International Conference on Wire-less Technologies for Humanitarian Relief*, p. 9, 2011 Amritapuri, India. Online, 2011. See http://doi.acm.org/10.1145/2185216.2185226 (Accessed 20 May 2014.)

Chapter 4

Enabling cyber-physical systems architectural design with wireless sensor network technologies

Nafaâ Jabeur[1]

Abstract

Several emergent technologies, including mobile ad-hoc networks (MANETs), machine to machine (M2M), wireless sensor networks (WSNs), and cyber-physical systems (CPSs), are actually sharing fundamental features and complementing each other towards allowing spatially distributed devices to interoperate regardless of their software and hardware components. In this chapter, we make a rough comparison between some of these technologies with the aim of clarifying the position of CPS with respect to the others. We particularly outline the challenges related to CPS design while focusing on the main components of these systems by depicting some high-level architectures found in the literature. Since a CPS is basically better understood within its context of use, we review several examples of architectures proposed in a variety of domains, including healthcare, smart space, emergency and real time, control, environment, and intelligent transportation. To meet the requirements of these domains, the role of WSN technologies and their contributions to CPS are highlighted. Finally, we propose a new architecture where WSN technologies, multi-agent system paradigm, and natural ecosystems are combined to enable CPS design and operations.

4.1 Introduction

Cyber-physical systems (CPSs) are being used in several application domains, including healthcare, environment monitoring, real-time and emergency systems, and transportation. These systems have recently emerged as a promising tool where the operations of the physical and engineered systems are monitored, controlled, coordinated, and integrated by means of a computing and communication core [1]. The integration of commonly heterogeneous hardware and software facilities within the same CPS infrastructure is currently fuelling intense research and development works, thereby causing opinions to diverge when it comes to defining

[1]Computer Science Department, German University of Technology in Oman (GUtech), P.O. Box 1816, Athaibah, PC 130, Sultanate of Oman, e-mail: nafaa.jabeur@gutech.edu.om

a CPS. Opinions are also divided regarding the exact links of CPS with other technologies, such as mobile ad-hoc networks (MANETs), machine-to-machine (M2M), and wireless sensor networks (WSNs).

Due to the absence of a common definition of CPS, the literature review includes numerous architectures which are generally specific to the proposed CPS applications. Several initiatives have proposed generic architectures. However, the lack of detail in these architectures does not allow readers to understand the exact components and mechanisms of a CPS.

In this chapter, we highlight, in Section 4.2, the main features of MANET, M2M, WSN, and CPS technologies and propose a rough comparison between them. The idea is to clarify the position of CPS with respect to the other technologies. In Section 4.3, we outline the challenges related to CPSs' design. In Section 4.4, we present the main components of a CPS by presenting some high-level architectures found in the literature. Since a CPS could be better understood within its context of use, we also present and explain several examples of architectures proposed in a variety of domains such as healthcare, smart space, emergency and real time, control, environment, and intelligent transportation. In Section 4.5, we focus on the role of WSN technologies and their contribution to CPS. In Section 4.6, we propose a new architecture where WSN technologies, namely, multi-agent system paradigm and natural ecosystems are combined to enable CPS representation and operations. Section 4.7 rounds off with conclusions.

4.2   Distinguishing WSN, MANET, M2M, and CPS

The past two decades have witnessed considerable advances in the fast-growing wireless communication and network research areas. Considerable progress has been, indeed, recorded in the fields of, M2M, and WSN technologies. MANET is a self-configuring infrastructure consisting in mobile devices having the ability to freely and independently move in any direction. Devices, commonly playing the role of routers to forward traffic unrelated to their individual use, must be well equipped with capabilities to communicate with other devices using heterogeneous protocols. MANET has particularly benefited from the significant progress in wireless communication and embedded computing technologies. By enabling better interactions between a network and its surrounding environment with the use of micro-sensing MEMS (micro-electro-mechanical systems) technologies, WSNs have emerged as a special form of MANETs. For instance, WSNs commonly comprise spatially distributed sensing devices to monitor physical or environmental conditions. Because of their limited energy and processing capability, sensors commonly cooperate to achieve goals which are beyond their own individual capabilities. The introduction of mobility to WSNs has then created considerable opportunities towards tracking, collecting, and reporting relevant data on events and objects of interest.

M2M is a generic label describing any technology where networked devices are able to exchange information and perform actions without manual human assistance. Some references are restricting these devices to be of the same type and/ or capability. The main components of an M2M system include RFID (radio frequency identification), sensors, and ZigBee, WiMAX, Wi-Fi, or cellular communication links. They also include a computing software component hosted on a remote device to interpret data and make decisions.

Similar to WSNs, M2M systems have distinctive characteristics, including the support of a huge amount of nodes, seamless domain interoperability, autonomous operation, and self-organization [2]. Furthermore, although M2M approaches and technologies are being used in practical automated applications (e.g. smart homes and smart grids), their support for intelligent information processing, for example by means of data fusion, distributed real-time control, and artificial neural networks, remains neglected [2]. Because of the heterogeneity of devices and their mobility patterns, as well as the huge amount of data being collected, communication remains a challenging issue for M2M technologies.

More recently, the CPS concept has emerged as a promising direction to enrich object-to-object, human-to-object, and human-to-human interactions in the virtual world as well as in the physical world [3]. CPS has a very complicated configuration characterized by a close combination and coordination between the computational and physical elements of the system [4]. Generally, the actuator and the wireless sensor networks are seen as the originator of CPS. Generally, CPSs use sensors and actuators to enforce smartness at several levels, including physical components' interactions, distributed real-time control, cross-domain optimization, and data routing and safety.

In the current literature, opinions are still divided concerning the exact definitions and limits of some concepts (e.g. M2M and CPS) and how they are linked together. However, in addition to security and privacy, four components seem to be common to MANET, WSN, M2M, and CPS: sensing, information processing, heterogeneous access, and applications and services [2]. Nevertheless, the difference lies in the share of the design between the four components. Although MANET, WSN, M2M, and CPS have some similarities in many networking aspects, they have some major differences. Indeed, MANET is generally for extending the coverage of infrastructure networks or supporting ad-hoc communications [5]. A WSN is specifically designed for conveying sensor-related data. Because of the absence of a homogeneous connected device platform, M2M systems are generally built to be device- or task-specific. In addition, these systems are still focusing on enabling machine-type communication between devices using a variety of communication protocols. A CPS normally involves multiple sensor networks and the internet, multiple dimensions of sensing data, and aims at building intelligence across these domains [2]. Table 4.1 outlines the differences between MANET, WSN, M2M, and CPS at the functional, logical, and dynamic levels.

4.3 Cyber-physical system design challenges

Recent advantages in research on the CPS domain has mainly concentrated on several concerns, including data transmission and management, network security, energy management, model-based design, system resource management, distributed

Table 4.1 Comparing MANET, WSN M2M, and CPS

Criterion	MANET	WSN	M2M	CPS
Functional level				
Goals	MANET aims to support ad-hoc communications or extend the coverage of infrastructure networks	WSN is basically designed to deliver in situ data about events and objects of interest	M2M systems are generally built to be task- or device-specific. They are still focusing on enabling machine-type communication between devices using a variety of communication protocols	CPS normally involves multiple sensor networks and the internet, multiple dimensions of sensing data and targets the building of intelligence across these domains
Logical level				
Composition	Mobile communicating devices	Sensors and sensor nodes which could be static or mobile. Sensors have limited processing and communication capabilities	Sensors, RFID, and any networked device	Physical side (including sensors, RFID, networked devices, and application-related devices) and cyber side (including processing, storage, decision making, and monitoring capabilities)
Topology	Random network formation and highly changing topology	Random, predefined and/or field-specific network formation. Highly dynamic topology. Clustering techniques are commonly used	Random, predefined and/or field-specific network formation	Random, predefined and/or field-specific network formation. Highly dynamic topology. Clustering techniques are commonly used
Dynamic level				
Communication	Supports arbitrary communication patterns, including unicast, multicast, and broadcast	Supports collective communication	Supports communication between heterogeneous devices and protocols	Supports inter- and intra-WSN communications as well as cross-domain communications

Power management	Emphasizes energy saving. Nodes go to sleep by opportunity	Emphasizes energy saving. Nodes have prolonged sleep/wake-up cycles	Energy is not generally a major issue	Activation of sensors is mission-oriented
Mobility	Node mobility is arbitrary and unrestricted	Limited controllable and uncontrollable mobility	Mobility is field-specific	Controllable and uncontrollable mobility that engages sensors and physical devices
Network coverage	Needs to fulfil precise connectivity requirements	Needs to fulfil connectivity and sensing coverage requirements	Needs to fulfil connectivity requirements to allow devices to communicate	Needs to meet several levels of connectivity and coverage requirements
Knowledge mining	Not an issue except to facilitate networking issues	Focuses on acquiring and handling sensing data	Not an issue except to facilitate machine-type communications	Focuses on ways to extract new knowledge from multiple sensing domains and appropriately use intelligence
Quality of service	Quality of data transmission is essential	Quality of data sensing is essential	Machine communication is essential	Focuses on higher-level quality of services, such as availability, security, confidentiality

real-time control technique, and platforms and systems [6]. As a whole, although researchers have made important progress on these concerns, CPSs remain in the nascent stage [2]. A variety of concerns need, indeed, to be solved from different perspectives of system design to facilitate the integration of the cyber and the physical worlds. These research and development issues can be summarized from various viewpoints as follows [7, 8]:

- **Networking issues**. Since CPS includes a variety of devices, potentially using heterogeneous communication technologies and protocols (e.g. Wi-Fi, Bluetooth, and ZigBee), interworking problems must be solved in order to allow efficient communication within the physical environment as well as bridging the physical and cyber environments appropriately.
- **Design and verification tools**. Existing tools supporting the simulation and co-design of CPSs are not suitable for this field. Achieving an automatic CPS development process from modelling to coding is still necessary.
- **Real-time capabilities**. In some scenarios, many factors, including hardware platform and control methods, affect the response time of the CPS. Additional research and development efforts are needed to meet the time constraints of real-time and critical applications.
- **Cross-domain optimization**. Due to their multidisciplinary nature, CPS applications involve the information fusion of multiple domains and hierarchical architectures. To ensure system performance, optimization techniques are crucial.
- **QoS and cloud computing**. Minimizing energy consumption and meeting required QoS are expected to remain challenging concerns for future CPSs. Emergent technologies and techniques, including cloud computing techniques and ubiquitous connectivity and virtualization, can significantly help in this issue.
- **Location-based services**. As CPS intends to remotely monitor physical facilities, efficient techniques should be used to track the location of these facilities in indoor and outdoor contexts. In the indoor context, accuracy of positioning technologies is still questionable.
- **Monitoring services**. CPS-based monitoring services may outspread to cross-WSN, cross-M2M, and cooperative models. The way these services are generated and provided to a wide variety of devices should allow service composition, customization, extension, etc.
- **Security and privacy**. Thanks to their ability to engage and control distributed devices remotely and to track events in real time, CPSs are facing privacy challenges as well as security challenges to protect private data collected by and from individuals as well as private data owned by individual devices (since devices could be part of independent networks and/or belong to independent/competing parties).
- **Standards development**. CPS applications greatly depend on technologies provided by several industries. Therefore, standardization challenges are undoubtedly greater than in any traditional standard development.
- **Design support tools**. Some existing tools are of good help for networking simulations (e.g. NS2, OMNET++, and OPNet) and embedded system design (e.g. MATLAB and TrueTime). However, these tools should be extended to support the specific requirements of CPSs.

4.4 Cyber-physical systems architecture

4.4.1 Introduction

Although there are various definitions of CPSs, depending on their application domains, there is no agreement on what exactly a CPS is. CPS is, indeed, intersecting with WSN, MANET, M2M, and the Internet of Things on several aspects. It is also going beyond the simple application of advanced information technology by constructing new systems integrating information processes and physical processes with the help of a variety of communication technologies [9]. For this reason, many architectures have been proposed in the literature. We can here report the high-level architecture of Figure 4.1, where a decision-making system located in the cyber world is getting inputs from spatially distributed sensors and actuators and returns decisions based on current situations.

Figure 4.2 depicts another high-level architecture where the CPS bridges a physical world (including infrastructure, hardware, humans, and processes) and a cyberspace (including networks, software, computing, and storage facilities). In the CPS, sensors are collecting and reporting data about the physical environment to a controller, whereas actuators are enforcing the decisions of the controller to monitor this environment. This vision is very restrictive as it limits the CPS to a simple bridge with limited awareness on the cyber and physical environments.

Ahmed *et al.* [10] have also proposed a generic architecture (Figure 4.3) that includes five main modules: sensing module (data collection from the physical world); data management module (data processing and storage), next-generation

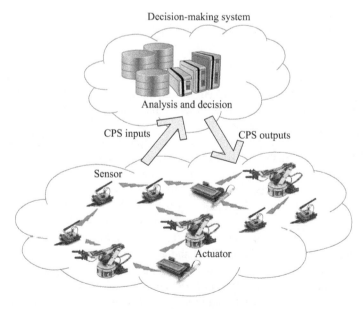

Figure 4.1   First example of a high-level CPS architecture [9]

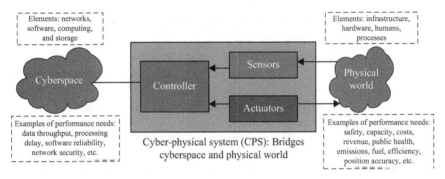

Figure 4.2   Second example of a high-level CPS architecture © IEEE 2013. Reprinted with permissions from [11]

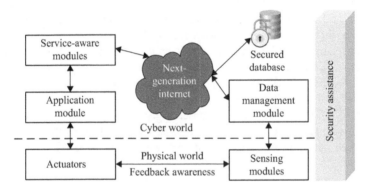

Figure 4.3   Third example of a high-level CPS architecture © IEEE 2013. Reprinted with permissions from [10]

internet (communication); service-aware module (data analysis and decision-making); application module (service deployment); and sensors and actuators (devices receiving and executing commands from the application module). All these modules include security mechanisms in order to maintain security requirements as per the application and users' needs.

Another high-level architecture of CPS is depicted in Figure 4.4, where one or more WSNs collect data about physical entities and events of interest and then report these data to a storage platform. A computing and/or intelligent decision module extracts data as per the current requirements, carries out the necessary processing, and then returns the required services to the users. This module consists of multiple integrated static and mobile sensors and actuators. It is also responsible for federating the exchanges between the different parties of the whole system as well as monitoring the data acquisition and analysis process.

We summarize in Figure 4.5 what we think are the main components of a high-level CPS architecture. In this architecture, an interfacing module is used to allow communication between the CPS and the users. The cyber environment includes a storage space where relevant data – for example, about the physical environment,

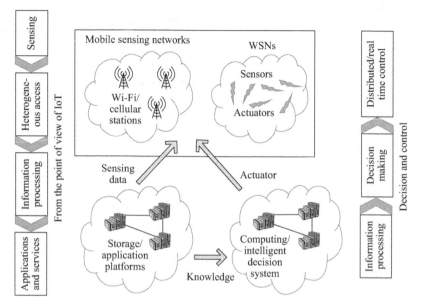

Figure 4.4   Fourth example of a high-level CPS architecture [2]

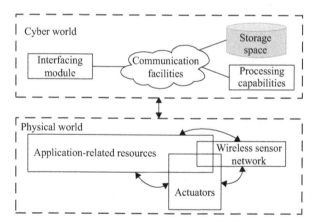

Figure 4.5   Proposed high-level CPS architecture

the geographic space, the current and recently delivered services – are stored. Available data and data collected from the physical environment are used by a processing and decision-making module to assess current situations and deliver appropriate decisions at the right time. This module may include mechanisms to enforce privacy and safety. The flow of information and decisions between the cyber and the physical environment is facilitated thanks to the communication facilities. In the physical environment, there exist application-related resources (e.g. specific assets and equipment for the envisioned application), actuators, and sensors. The communication facilities are used by the interfacing module to communicate the requirements of the users to the processing capabilities, which in turn use the same facilities to send control instructions to the physical world. In the physical world, the application-related resources are monitored by means of sensor devices and actuators. Sensors collect in situ data, carry out preliminary processing (e.g. filtering and aggregation), and report results to the cyber world. The actuators allow the cyber environment to remotely control and monitor the physical facilities. We reflect in our vision (see Figure 4.5) the fact that sensors and actuators can be tools for the CPS to control and monitor the physical facilities as well as part of these facilities (application-related resources).

The above-mentioned architectures do not inform us much on several issues, including which data are collected, by means of which sensors, with which frequency, and according to which format and granularity. Neither do they inform us on which processing is performed on data, which control is implemented, and which components are controlling which other components. Additional details on CPS architecture are thus needed to clarify these issues, among others. To this end, it is important to consider the context of use of the CPS.

Roughly, the CPS applications can be classified into smart space, healthcare, emergency real-time system, environmental monitoring and control, as well as smart transportation [9, 12]. We are not aiming in this chapter to survey the different CPS architectures proposed in each of the application domains. Rather we are aiming to highlight the requirements of each application field through selected architectures.

4.4.2   Healthcare applications

In the healthcare domain, a CPS can be seen as the combination of active user input (such as smart feedback system), digital records of patient data, and passive user input (such as biosensors and/or smart devices) [8]. The healthcare CPS particularly benefits from the capability of a variety of devices for the acquisition of timely data on different hardware facilities. Because these devices can interact autonomously, they can particularly contribute towards an efficient decision-making process.

Devices could be of a variety of sizes. For example, small devices could be implanted in a patient's body to collect parameters of interest. Other sensors may be placed in the surrounding environments of patients (such as walls) to track medical resources (equipment and personal) and collect contextual information for an improved decision-making process. To control and benefit from the spatially distributed devices, appropriate computation modules are necessary to process data and recommend appropriate decisions to patients and medical staff.

The current use of CPS in the healthcare domain is still in the early stages. The number of CPS architectures proposed is also limited [3]. For the sake of illustration, a service-oriented architecture and a secured CPS architecture utilizing a WSN–cloud integrated framework can be found in Hu *et al.* [13] and Wang *et al.* [14], respectively. A more consistent architecture compiled from existing CPS healthcare initiatives is depicted in Figure 4.6 [8]. In this architecture, data collected by a variety of sensors on patients and infrastructure are sent to a cloud

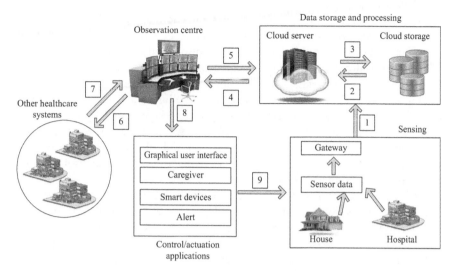

Figure 4.6   An example of a CPS architecture for healthcare applications [8]

storage via a gateway (step 1). A dedicated cloud server will then extract and process this data in real time as per specific current requirements (step 2). The cloud server sends the processed data back to the cloud storage (step 3) and conveys the results of data processing while generating appropriate alarms to an observation centre (step 4). Clinicians and specialists in the observation centre access (step 5), process patient data, and approach, when necessary, other healthcare systems for consultation (steps 6 and 7). Decisions are then sent to an actuation component (step 8) to make appropriate interventions to patients (step 9).

4.4.3   Smart space applications

The rapid expansion of pervasive computing and telecommunication systems is allowing CPSs to efficiently query, control, and monitor machinery, equipment, and personnel states. Smarter living spaces are then being created where the synergy between hardware and software facilities and human beings is boosting a more accurate data collection and analysis of several events of interest. However, one major handicap preventing the achievement of this goal results from the different communication protocols being executed by electronic devices, such as Bluetooth, ZigBee, and radio frequency.

Many companies, such as Google, Samsung, and Medusa, have created proprietary technologies for home appliance control systems where a selection of daily-life electronic products are integrated. Medusa has created a home appliance control platform and used a peer-to-peer architecture to control networked multimedia devices [15]. The Samsung Company has developed an intelligent home control system called Home Vita [16]. This system monitors and controls various home electronic devices by integrating internet and home networks and using remote connection technologies. Google has released the Android@Home concept [17], which extends the Android from a mobile or tablet device to household appliances to build smart living spaces. To control appliances, users can send signals through Android devices to a signal converter box. Once the signal is received, the box sends a corresponding signal to control the appliances.

Furthermore, Chun *et al.* [18] developed intelligent devices which are convenient for smart space applications. They also designed and implemented an Intelligent Control Box that converts different wireless signals coming from a variety of devices. This box could be seen as a multiple control platform which integrates facilities, including the systems of lighting, access control, air conditioning, alarm, and video surveillance.

Within the research community that addressed CPS-based applications for smart living spaces, some initiatives have addressed the challenging issue of energy conservation. In this regard, Han and Lim [19] proposed a smart home energy management system consisting mainly of three components: sensing infrastructure, context-aware component, and service management module. The sensing data (e.g. temperature, noise level, and light intensity) acquired by sensor nodes are gathered by the sensing infrastructure and then communicated to the context-aware module for analysis. The results of the analysis are then communicated to the service management component which is responsible for taking decisions for the control of appliances in the physical world. In order to make daily appliances more intelligent and energy efficient, Byun and Park [20] developed a self-adapting system consisting of two main modules: a self-adapting intelligent gateway (SIG) and a self-adapting intelligent sensor (SIS). The SIG is in charge of several tasks, such as service decisions, power/environmental information collection and analysis, appliance and sensor node management, and provision of energy management services. These tasks are carried out based on contextual and energy information collected by the SIS.

Chun *et al.* [21] proposed an agent-based self-adapting architecture for smart space applications that aims to ensure the reliability and predictability requirement in CPS designs. The architecture proposes a self-adaptive robot equipped with a variety of sensors (including electronic compass, motor, and web camera) for detecting events (e.g. equipment crash, temperature exceeding acceptable limits). Data collected will then be used by a decision-making system responsible for executing self-adaptation processes to control the robot's behaviour. Because of the heterogeneity of communication standards used by different CPS devices, Bai and Huang [18] proposed to design and implement an intelligent control box (ICB) to convert different wireless signals (Figure 4.7). The ICB can, for example, be used to convert ZigBee signals coming from temperature sensors to Wi-Fi signals recognized by a decision-making system. After analyzing the data, the decision system sends an instruction via Wi-Fi communication protocol to the ICB, which in turn converts the signal, for example, to infrared signals recognized by air-conditioning equipment. The ICB could be extended to support and convert other signals depending on the devices used as well as the application requirements.

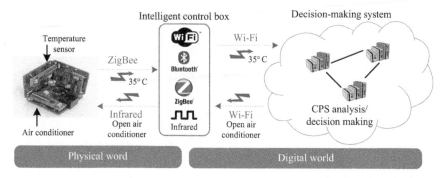

Figure 4.7   An example of a CPS architecture for smart space applications [18]

4.4.4   Emergency and real-time applications

In addition to helping people avoid natural disasters (such as tsunami, volcanic eruption, or mudslide), emergency and real-time systems provide potential escape solutions for people [9]. When an unanticipated event happens in a public environment, decision-makers need to collect information and plan and carry out convenient actions concerning, for example, environment monitoring, event warnings, mitigation operations, and safety plans. Knowledge that could be revealed from such actions could help us understand how perception and actuation services can be exploited to assist in the procedures of emergency handling [22]. With an approach that appropriately utilizes CPS, a solution for emergency control could provide timely response to events and contribute by minimizing damage.

For emergency response operations, pervasive sensing facilities should be installed adequately in the physical environment to get a global view of the affected areas. Timely, efficient, and secure communication mediums are necessary to establish real-time data exchange between sensors and control and computing facilities. Control and computing facilities should be able to identify the priorities of events and set up responding plans, such as trajectory navigation and task allocation. Decisions should be sent to networked static and mobile devices to perform the decided tasks. New data should be collected and communicated to control and computation facilities to evaluate the results of actions and improve performance.

Figure 4.8 [22] gives an example of a CPS-based system for environment perception and emergency handling. The system consists of several layers that establish a hierarchical structure of the CPS. The wireless communication networks layer contains sensors capable of acquiring data on many heterogeneous entities and events from the physical world and according to specific perceptions defined by the environment perception layer. The integration of sensing data, which covers both cyber and physical aspects, assists the CPS in creating an overall environment perception as well as constructing a global map of the current situation. The heterogeneous networks convergence layer is used to enable services composition as well as to enable providing subscribers with services in both network coverage and QoS dimensions. The architecture envisions a group of autonomous mobile robots

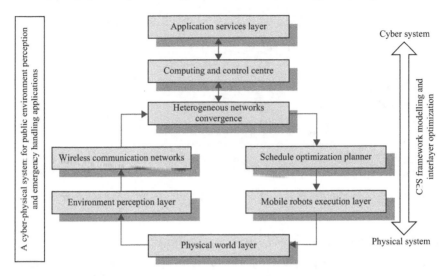

*Figure 4.8   An example of a CPS architecture for environment perception and
emergency handling © IEEE 2011. Reprinted with permissions from [22]*

that are equipped with interaction and perceptions capabilities. The coordination
and relocation of robots in a CPS typically involve significant challenges in highly
dynamic conditions. Efficient and accurate reactions of CPS workloads are then
necessary to fulfil the requirements of emergency handling [22].

4.4.5   Control applications

In order to smoothly integrate computation, networking, and physical processes
within CPS solutions, serious problems related to the heterogeneity of software
components and hardware facilities must be solved. In control applications, current
control, communications, and software theory may not be efficient within the
multi-domain context of CPSs [23]. Additional research and development efforts
are therefore needed in order to bring isolated subsystems of a production processing line into a unified large-scale system capable of adequately managing problems in industry process control. Conventional control theory does not include the
necessary mechanisms to analyse interconnected systems of heterogeneous components in a large-scale context. It is therefore important to revise communication
and control theory for the purpose of guaranteeing real-time and reliable control.
To address these problems, Wang *et al.* [24] proposed a CPS architecture for the
control network (Figure 4.9) where several interconnected CPS components (e.g.
cameras, actuators, sensors, and control units) are considered. The solution includes
units with extended processing capabilities and capable of communicating with
different actuators and sensors to monitor control tasks. Other units do communicate with actuators or sensors but afford computing power to activate on demand.
The use of wired network or wireless communications is decided based on current
environmental conditions. The key idea of the proposed architecture is the use of

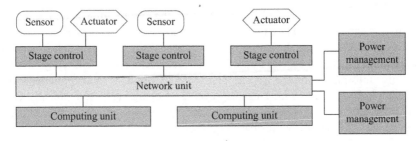

Figure 4.9   An example of a CPS architecture for control applications [24]

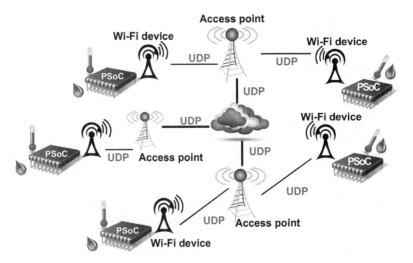

Figure 4.10   An example of a CPS architecture for environmental applications
© IEEE 2014. Reprinted with permissions from [25]

heterogeneous networking units, which abstract the reliability and timing issues of the various components of the CPS. Additional units are linked to the network unit via local on-board network adapters [24].

4.4.6   Environmental applications

In the environmental monitoring field, sensors could be deployed in outdoor environments to monitor numerous parameters, such as air quality and soil moisture. When specific events of interest are detected and reported by sensors, a decision-making system can be used to communicate instructions to actuators to carry out required tasks. In this context, where knowledge is extracted from the gathered information and used to drive decision making and/or control, an information-centric perspective could be followed for the creation of a CPS-based solution. This perspective was adopted by Sanislav *et al.* [25] who proposed a CPS architecture (Figure 4.10) consisting of three layers. The bottom layer (physical layer) contains several spatially distributed sensors communicating through UDP messages. The middle layer (data knowledge layer) contains the information acquired by the bottom layer and is stored in a measurement database. The database is then used by a multi-agent system for decision-making, data analysis, and negotiation process. The top layer includes services based on a cloud architecture. Basically, the services give information on the CPS monitoring activities as well as on-demand data analysis results. By combining agents capable of operating in dynamically changing environments with cloud computing, the system can automatically adjust its physical and software resources based on user needs.

CPS-based solutions including pervasive networks of actuators and sensors can allow for easy access to large scale areas of the environment at exceptional spatial and temporal resolutions. Such fine-grained real-time data can considerably transform scientific approaches. Furthermore, CPS platforms could be seen as a crucial solution against environmental catastrophes since they are able to collect and report data about a wide range of events, from harsh and remote areas, and then act appropriately while taking into consideration current contextual developments.

4.4.7   Intelligent transportation applications

The road transportation system is a large-scale human-made engineering system. It constructs a complex social scale of the physical world. It includes a large amount of critical infrastructure expanded over wide natural geographical environments as well as all kinds of man-made environments, such as large bridges across the sea or rivers, and long and large tunnels. It also includes a massive variety of vehicles, people, and goods in the complex road environment [26]. Because of critical safety issues, involving human beings as well as physical equipment (including cars and road assets), efficient remotely controlled mechanisms are being investigated within the context of intelligent transportation solutions. These mechanisms go beyond simple advanced electronic devices and information systems, especially since the scale and the intensive usage of the traffic system is expanding rapidly.

Researchers and developers are giving increasing attention to CPS-based solutions in order to improve road transportation safety, as CPSs possess capabilities to manage heterogeneous information, physical assets, and social features [27]. In a CPS-based smart transportation application, sensor nodes could be embedded in vehicles to improve traffic safety and efficiency. For example, an onboard accelerometer can be used to infer information on the existence of fissures on the road. When a fissure is detected, its location (collected by a GPS receiver) is sent to neighbouring vehicles, thus enhancing traffic efficiency and safety [9].

The traffic control system is one of the most complex systems within CPS research. Some studies have attempted to apply CPS to the analysis and design of intelligent transportation systems (ITSs), traffic control, and guidance systems. In this regard, Jianjun *et al.* [26] recommended that effective CPS-based traffic control should be built on the basis of traffic control system analysis and the design of the characteristics of information systems. They also recommended that the CPS should be composed of three levels (see Figure 4.11): the application of CPS theories for the integration of information processing into transportation processes; traffic detection and control of information on the implementation of technical solutions; and support

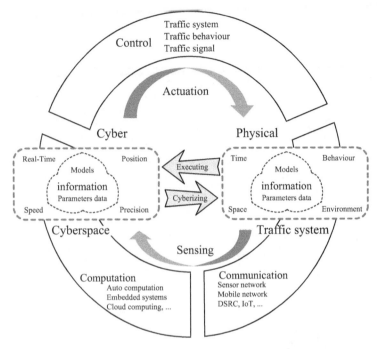

Figure 4.11   An example of a CPS architecture for intelligent transportation applications [26]

of modern computing, communication, and control technology. Effective mapping should be constructed for the road traffic system in cyberspace to establish traffic control system models and parameter sets. Control is realized for the road traffic system based on the results from the computation of the model-based and the data-driven information space. The data collected by sensors in real-time are used for setting traffic control devices and implementing control of the traffic system and behaviour. The computing, communication, and control technologies serving traffic control should be sufficiently applied and developed, and information should be shared between road users and other devices collaboratively and efficiently [26].

Considering the characteristics of transportation systems, Yongfu *et al.* [23] proposed a service-oriented architecture for CPS-based intelligent transportation consisting in perception, communication, computation, control, and service layers. The architecture, that we present in Figure 4.12, includes a large number of inter-connected physical perception devices such as infrared detectors, microwave detectors, ultrasonic detectors, radio frequency identification (RFID) tags, video sensors, ETC, VMS, and LED as well as interconnected computing devices such as servers and computers.

4.4.8   Summary

Based on the above-mentioned architectures as well as an extended literature review, we can summarize the specific needs of individual application domains in Table 4.2.

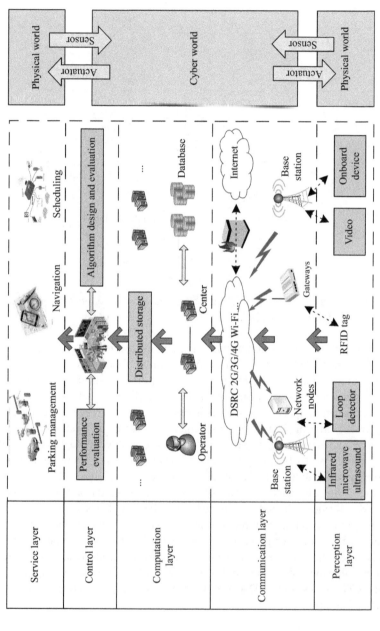

Figure 4.12 A service-oriented architecture for a CPS-based intelligent transportation solution. Adapted from [23]

Table 4.2 Summary of CPS requirements for different application domains

Application	Requirements of physical environment	Requirements of cyber environment
Healthcare applications	Several technologies should be used to maintain an accurate localization of patients and medical resources (personnel and equipment) in indoor and outdoor contexts. Efficient mechanisms should be used to enable efficient communication between potentially hetero-geneous devices. This commu-nication should be priority based to allow a better use of shared resource and a better response to patients' cases	Data pushed from the physical world to the cyber world must be processed in a way which pro-vides patients and the concerned medical persons with timely assistance. This assistance should be personalized as per the current requirements and con-textual information
Smart space applications	An efficient query, control, and monitoring of machinery, equip-ment, and personnel states requires a more accurate data collection and analysis of several events of interest. To this end, the physical facilities should be endowed with capabilities to run a variety of communication pro-tocols, and exchange data with optimal delays, energy, and pro-cessing overheads	In the cyber environment, data should be analysed and pro-cessed in timely fashion and appropriate actions should be sent to the right physical facil-ities depending on current events and requirements. Decided actions must ensure reliable and secure operations of physical devices while promoting colla-boration and efficient informa-tion/resource sharing
Emergency and real-time applications	Physical devices must be able to communicate smoothly regard-less of their communication protocols. They must also be able to take some decisions to react on the fly to some events. Data routing and processing activities must be reliable, fault-tolerant, and optimized	Analysis and processing activities in the cyber side must be accu-rate and carried out in real time. The decisions taken must ensure reliable and secure operations of physical devices while promot-ing an efficient information/ resource sharing
Control applications	In control applications, several isolated subsystems must be allowed to communicate securely and reliably while respecting their mutual qualities of services, regardless of their heterogeneity	Data processing and decided actions must take into con-sideration the heterogeneity of physical facilities and their locations. Decisions should pro-mote security, reliability, and fault-tolerance of operations
Environment applications	Specific events of interest must be detected and analysed. The spa-tially distributed devices must adapt their operations accord-ingly. Local adaptation requires	Decisions of the cyber side must be based on the analysis of the col-lected data as well as on the awareness of current locations of physical devices and their

(Continues)

Table 4.2   (Continued)

Application	Requirements of physical environment	Requirements of cyber environment
	embedded intelligence and autonomy mechanisms. Collective adaptation requires reliable mechanisms to share available information/resource	contextual information. They should promote collaboration between physical facilities while preserving their energy
Intelligent transportation	Devices must be able to collect and share data as well as receive decisions at the right time. Because spatial events may happen in sporadic ways and objects (vehicles) may be moving very fast, embedded mechanisms should allow local decisions to be taken	Data analysis and decisions of the cyber side should be taken and communicated within the time restrictions of the physical side. Decisions should be prioritized and intelligently taken to predict and avoid potential problems

4.5   The role of WSN technologies in CPSs

In CPSs, sensors are being seen as integral and invaluable components to enable an efficient and optimal control of the physical world. Sensors are also being deployed as interfaces through which in situ data are collected about/from the physical environment and then transferred to the cyber environment, as well as interfaces through which new instructions/parameters are injected from the cyber environment to the physical environment. An integration of both technologies is therefore very important towards implementing an efficient control and monitoring of the physical world. Within this integration, sensor technologies must not just be seen as data collecting devices. They are also, indeed, intelligent, autonomous, and self-adapting facilities that would make decisions to improve the CPS response time. In addition, sensors could be considered as mini-CPSs that would act and react whenever and wherever the cyber side of the CPS cannot provide the physical side with appropriate, timely decisions and data. This is particularly happening in remote and harsh environments.

Enabling WSN and CPS integration is actually not straightforward. Several challenges must be addressed, including the integration of appliances with different communication protocols, the mobility of sensor nodes, and the on-time delivery of sensor data to the cyber system. Specific requirements and constraints of WSN must also be taken into consideration. In this chapter, we are interested in this integration at the level of the architectural design. To this end, let us list the contributions of WSN to any CPS:

- Collect in situ, timely data for the envisioned applications from the right geographic area and about the right events of interest.
- Report relevant contextual data to neighbouring physical devices (application-related resources) at the appropriate time.
- Route data between physical facilities.

- Support physical facilities by carrying out specific processing on behalf of, or for, the benefits of these facilities when and where needed.
- Watch over the current conditions/capabilities of physical devices.
- Control the operations of physical devices in specific applications.
- Aggregate, filter, analyse, and process collected data.
- Forward relevant decisions and information from the cyber world to the physical devices.
- Embed intelligence and autonomy mechanisms within the physical environment.
- Allow for multi-level control of the physical environment.
- Contribute in the process of securing data processing and routing within the physical side.
- Contribute in saving the energy of specific appliances of the physical environment.

In short, the WSN permits to close the loop between the physical and the cyber worlds. It is, indeed, bridging both environments and facilitating an efficient and timely communication between both of them. Ultimately, the WSN extends our DAST principle, which is to acquire the right data, from the right area, with the right sensors, at the right time [28]. While this definition of DAST remains the same from the cyber-side point of view, the DAST principle from the physical point of view is now to deliver the right decision to the right area, with the right sensors, at the right time.

In Section 4.4, we outlined several CPS architectural details within the context of a variety of application domains. Since we believe that, in general, the role of WSN was not adequately esteemed and reflected in existing architectures, we propose, in Table 4.3, what we think are important contributions of WSN technologies in several CPS application domains as well as the challenges that could be encountered to meet these expected contributions.

4.6 Towards a new CPS architecture

CPSs are very complex systems where several software and hardware facilities coexist and collaborate, despite usual heterogeneity surrounding their goals and communication protocols. In a previous work [29], we proposed to enable CPSs with WSN technologies, multi-agent system paradigm, and natural ecosystems. For instance, because of their limited battery lifetimes, context-awareness, and processing capabilities, sensor actions would not meet the requirements of CPSs while monitoring, controlling, and coordinating the operations of physical and engineered systems. To overcome these constraints, we considered the ecosystem metaphor for WSNs and CPSs with the aim of taking advantage of the efficient adaptation behaviour and strong communication mechanisms used by living organisms. Use of the multi-agent system technology is motivated by its proven flexibility, autonomy, and intelligence to solve complex problems within highly dynamic, constrained, and uncertain environments [30]. This technology is particularly used wherever and whenever the WSN fails and needs autonomy, flexibility, intelligence, and cooperation mechanisms to adequately meet the expectations of the CPS. We believe that agents can also be efficiently used to strengthen the link between the cyber and

Table 4.3   Contributions of WSN technologies in several CPS application domains

Application	WSN contributions	Challenges
Healthcare applications	Collect and report real-time data about patients and medical resources equipment. Communicate information, decisions, and personalized alerts to concerned persons at the right time	Because patients and medical resources commonly switch between indoor and outdoor locations, identifying and using appropriate communication approaches could be energy consuming. Mobility may also disconnect sensors from the rest of the network
Smart space applications	Control of current operational context of equipment and devices. Timely report appropriate information on devices and events of interest. Contribute in smart monitoring of appliances. Enable data, information, and decisions routing	Intelligent and self-adaptation mechanisms should be embedded to sensors in order to reduce communications and determine the right time to collect and report data. This is difficult to meet because of current limited sensors' capabilities
Emergency and real-time applications	Collect and report timely data on events of interest. Carry out on-site processing and analysis of data. Share relevant data and decisions with appropriate actors (other sensors, physical appliance, human operator, etc.) in real time. Enforce security and fault-tolerance mechanisms	Intelligent and self-adaptation mechanisms should be embedded to sensors as well as mechanisms to optimize energy consumption. These facilities generally go beyond the capability of individual sensors
Control applications	Bridge several heterogeneous devices and isolated subsystems and facilitate their secure and reliable communication. Enable efficient communication between facilities in the physical side and the cyber side	Devices in control applications could be heterogeneous and belonging to different systems. Connecting these devices requires extended ability to use different communication protocols
Environment applications	Collect data on a variety of events of interest. Adapt sampling rates, processing, and communication pathways depending on current contextual conditions. Enable collaborative and distributed decision-making process	Intelligent and self-adaptation mechanisms should be embedded to sensors in order to deal with dynamic and unpredictable spatio-temporal events
Intelligent transportation	Collect, analyse, share, and report information in timely fashion. Enable to deal with highly dynamic configuration of events and moving objects (vehicles)	Extended capabilities are needed for an efficient, secure, and intense communication, especially when events require immediate actions

physical worlds. We thus propose a new CPS architecture (Figure 4.13) where agents are deployed on the cyber and physical sides of the CPS. Some of our agents (Service Subnet Agent, Area of Interest Agent, Cluster Head Agent, Ranger Agent, Support Agent, etc.) are used to implement an embedded multi-level control over the WSN and the application resources (atom, micro, meso, and macro control levels). These agents are hosted on physical sensors and can be supported by some mobile agents in collecting specific data (such as the availability and state of application resources, sensors' connectivity, routing paths). At any moment, these mobile agents can move to the cyber system wherein extended communication and processing facilities and data are available. They can also be used to make soft copies of some of the capabilities of a given sensor and share or implement them on other sensors.

In addition to the communication infrastructure, the cyber side of our architecture includes a data module, a service module, and a major control module. The data module contains a spatial database for the identification of areas of interest, the location of sensors, and the application resources and a database containing several bio-inspired algorithms that could be used as needed by the Agent Controller. It also contains a record of the recent services it has provided and the ongoing services to reduce user queries' response times. The service module contains the Agent Input/Output Interface which is responsible for collecting and responding to the user requests. After checking the ongoing and recent services provided, this agent sends a request to the Agent Controller in the major control module. This agent is supported by two special agents: RCA (Resource Chasing Agent) and ECA (Event Chasing Agent). The ECA is used to track and detect events of interest within the spatial areas where the WSN and the application resources are located. This agent is important especially since sensors as well as the application resources may be heavily affected by certain events (e.g. heavy rain) or are tracking events of interest (e.g. level of water in specific areas). The RCA is used to track current available resources (e.g. energy levels of sensors, redundant sensors). Additional details about the architecture below can be found in Jabeur *et al.* [29].

4.7   Conclusion

In this chapter, we highlighted the main components of a CPS architecture. After surveying these components from some existing high-level architectures, we presented and explained selected architectures proposed within the context of the current main application domains of CPS. We also put the focus on the contributions of the WSN technologies in CPS and presented how these contributions could be better used within the context of several application fields. Because of the current limited capabilities of sensor nodes as well as the complexity of CPSs, we proposed to combine the use of WSN, multi-agent system paradigm, and natural ecosystems in order to enable the design of CPS and a better support of its operations. We believe that our architecture could be a good starting point for students, researchers, and developers towards creating a scalable CPS infrastructure that is able to manage internal processing and interaction with other CPS infrastructures in an intelligent, secure, and optimized way.

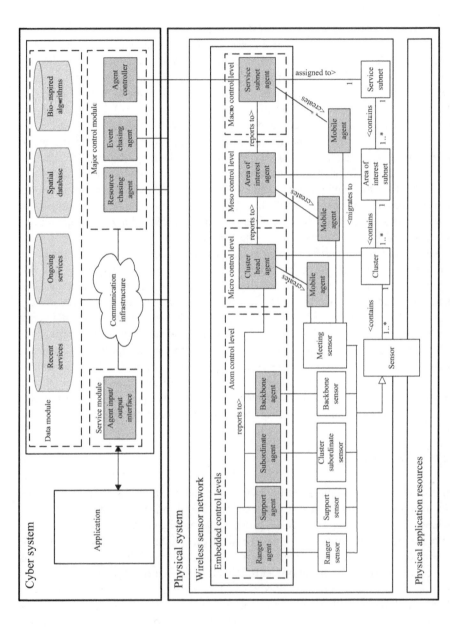

Figure 4.13   Enabling CPSs with WSN technology, multi-agent system paradigm, and natural ecosystems (modified from Jabeur et al. [29])

References

[1]   R. Rajkumar, I. Lee, L. Sha and J. Stankovic. Cyber-physical systems: The next computing revolution. In *Proceedings of the 47th Design Automation Conference*, pp. 731–736, ACM, New York, USA, 2010.

[2]   J. Wan, M. Chen, F. Xia, D. Li, and K. Zhou. From machine-to-machine communications towards cyber-physical systems. *Computer Science and Information Systems*, 10(3): 1105–1128, 2013.

[3]   F. Wu, Y. Kao and Y. Tseng. From wireless sensor networks towards cyber physical systems. *Pervasive and Mobile Computing*, 7(4): 397–413, 2011.

[4]   F. Cuomo, E. Cipollone and A. Abbagnale. Performance analysis of IEEE 802.15.4 wireless sensor networks: an insight into the topology formation process. *Computer Networks*, 53(18): 3057–3075, 2009.

[5]   M. Chen, V. Leung, X. Huang, I. Balasingham, I. and M. Li. Recent advances in sensor integration. *International Journal of Sensor Networks*, 9(1): 1–2, 2011.

[6]   J. Wan, H. Yan H. Suo and F. Li. Advances in cyber-physical systems research. *KSII Transactions on Internet and Information Systems*, 5(11): 1891–1908, 2011.

[7]   T. Qiu, L. Feng, F. Xia, G. Wu, and Y. Zhou. A packet buffer evaluation method exploiting queuing theory for wireless sensor networks. *Computer Science and Information Systems,* 8(4): 1027–1049, 2011 .

[8]   S. A. Haque, S. M. Aziz and M. Rahman. Review of cyber-physical systems in healthcare. *International Journal of Distributed Sensor Networks*, vol. 2014, article ID 217415, 20 pp., 2014. doi:10.1155/2014/217415.

[9]   C. Lin, S. Zeadally, T. Chen and C. Chang. Enabling cyber physical systems with wireless sensor networking technologies. *International Journal of Distributed Sensor Networks*, 21pp, 2012.

[10]   S. H. Ahmed, G. Kim and D. Kim. Cyber physical system: architecture, applications and research challenges. In *Proceedings of IFIP Wireless Days* 2013, pp. 1–5, 13–15 November 2013. doi: 10.1109/WD.2013.6686528.

[11]   K. Sampigethaya and R. Poovendran. Aviation cyber–physical systems: foundations for future aircraft and air transport. *Proceedings of the IEEE*, 101(8): 1823–1855, 2013.

[12]   E. M. Shakshuki, H. Malik and T. R. Sheltami. WSN in cyber physical systems: enhanced energy management routing approach using software agents. *Future Generation Computer Systems*, 31: 93–104, 2014.

[13]   L. Hu, N. Xie, Z. Kuang and K. Zhao. Review of cyber-physical system architecture. In *Proceedings of the 15th IEEE International Symposium on Object/Component/Service-Oriented Real-Time Distributed ComputingWorkshops (ISORCW)*, pp. 25–30, April 2012.

[14]   J. Wang, H. Abid, S. Lee, L. Shu and F. Xia. A secured health care application architecture for cyber-physical systems. *Control Engineering and Applied Informatics*, 13(3): 101–108, 2011.

[15] S. Wray, T. Glauert and A. Hopper. Networked multimedia: The Medusa environment. *IEEE Multimedia*, 1(4): 54–63, 1994.

[16] Y. Son, S. Ko, J. Jang, H. Lee, J. Jeon and J. Kim. Halfpush/half-polling. In *Proceedings of the 16th Conference on Pattern Languages of Programs (ACM PLoP '09)*, ACM, New York, USA, August 2009. DOI 10.1145/1943226.1943244. See http://doi.acm.org/10.1145/1943226.1943244.

[17] M. Karch. *Android Tablets Made Simple: For Motorola Xoom, Samsung Galaxy Tab, Asus, Toshiba, and Other Tablets*. Apress, New York, NY, USA, 2011.

[18] I. Chun, J. Park, H. Lee, W Kim, S. Park and E. Lee. An agent-based self-adaptation architecture for implementing smart devices in smart space. *Telecommunication Systems*, 52(4): 2335–2346, 2013.

[19] D. M. Han and J. H. Lim. Design and implementation of smart home energy management systems based on ZigBee. *IEEE Transactions on Consumer Electronics*, 56(3): 1417–1425, 2010.

[20] J. Byun and S. Park. Development of a self-adapting intelligent system for building energy saving and context-aware smart services. *IEEE Transactions on Consumer Electronics*, 57(1): 90–98, 2011.

[21] I. Chun, J. Park, H. Lee, W. Kim, S. Park and E. Lee. An agent-based self-adaptation architecture for implementing smart devices in smart space. *Telecommunication Systems*, 2011.

[22] W. Meng, Q. Liu, W. Xu and Z. Zhou. A cyber-physical system for public environment perception and emergency handling. In *Proceedings of the IEEE 13th International Conference on High Performance Computing and Communications (HPCC)*, pp. 734–738, 2–4 September 2011.

[23] Y. Li, D. Sun, W. Liu and X. Zhang. A service-oriented architecture for the transportation cyber-physical systems. In *Proceedings of the 31st Chinese Control Conference (CCC)*, pp. 7674–7678, 25–27 July 2012.

[24] Y. Wang, M. C. Vuran and S. Goddard. Cyber-physical systems in industrial process control. *SIGBED Review*, 5: 1–2, 2008.

[25] T. Sanislav, G. Mois, S. Folea, L. Miclea, G. Gambardella, and P. Prinetto. A cloud-based cyber-physical system for environmental monitoring. In *Proceedings of the 3rd Mediterranean Conference on Embedded Computing (MECO)*, pp. 6–9, 15–19 June 2014.

[26] S. Jianjun, W. Xu, G. Jizhen and C. Yangzhou. The analysis of traffic control cyber-physical systems. *Procedia – Social and Behavioral Sciences*, 96: 2487–2496, 6 November 2013. ISSN 1877-0428, http://dx.doi.org/10.1016/j.sbspro.2013.08.278.

[27] E. Lee. *Cyber Physical Systems: Design Challenges*. University of California, Berkeley, Technical Report No. UCB/EECS-2008-8, 2008.

[28] N. Jabeur, P. Graniero, J. McCarthy and X. Xing. A knowledge-oriented meta-framework for integrating sensor network infrastructures. *International Journal of Computers & Geosciences*, 35(4): 809–819, 2009.

[29] S. Bandyopadhyay and E. J. Coyle. An energy efficient hierarchical clustering algorithm for wireless sensor networks. In *Proceedings of INFOCOM 2003, IEEE Societies*, vol. 3, pp. 1713–1723, 30 March–3 April 2003.

[30] N. Jabeur, N. Sahli and S. Zeadally. Enabling cyber-physical systems with wireless sensor networking technologies, multi-agent paradigm, and natural ecosystems. *International Journal of Mobile Information Systems*. See http://www.hindawi.com/journals/misy/aa/908315/cta/.

Chapter 5

Cyber security in cyber-physical systems: on false data injection attacks in the smart grid

Jie Lin[1], Wei Yu[2], Xinyu Yang[3] and Guobin Xu[4]

Abstract

The cyber-physical system (CPS) is a system that can efficiently integrate both cyber and physical components by leveraging modern sensing, computing, and networking technologies. As an important role of CPS, sensors can obtain measurements of physical components and transmit the measurement data to the control center through communication networks. Nonetheless, sensor networks can pose serious security challenges to CPSs as well. In this chapter, we first review CPSs and wireless sensor networks in CPSs. We then take the smart grid as an example of a CPS and investigate the impacts of one category of attacks, namely false data injection attacks, against the operation of the smart grid. Through a combination of theoretical analysis and performance evaluation, we find that false data injection attacks can disrupt the electricity market operations, posing an increased electricity price to consumers. We also discuss the impacts of false data injection attacks on other key functions in the smart grid, including energy distribution, state estimation, etc.

5.1 Introduction

Embedded systems have been widely used to control physical components for decades [1, 2]. With the advance of sensing, computing, and communication network technologies, next-generation embedded systems will be designed based on the principle of CPSs, which emphasizes the efficient interaction between cyber components and physical components. Generally speaking, a CPS is a complex

[1]Xi'an Jiaotong University, Xi'an, Shaanxi, China, e-mail: jielin@mail.mail.xjtu.edu.cn
[2]Department of Computer and Information Sciences, Towson University, MD 21252, USA, e-mail: wyu@towson.edu
[3]Xi'an Jiaotong University, Xi'an, Shaanxi, China, e-mail: yxyphd@mail.xjtu.edu.cn
[4]Department of Computer Sciences and Information Technologies, Frostburg State University, MD 21532, USA

system that integrates advanced computing and communication techniques into the effective monitoring and control of physical systems [3, 4]. CPSs have been extended to numerous areas [1], including smart grid, smart transportation, smart building, smart health, etc.

The components in a CPS, including sensors, actuators, distributed control centers, and wired and wireless communication networks, should be effectively integrated to enable effective monitoring and control of physical systems [5, 6]. Figure 5.1 [5] shows a generic deployment structure for a CPS. As we can see, to measure the states of physical components, sensors are deployed to measure and send measured reports to the control center through wired and wireless networks [7]. It is worth noting that when considering cost-efficient deployment, wireless sensor networks have been commonly used. Based on the measurements sent from sensors, the control center obtains the states of physical components and then sends control commands to actuators through wired or wireless networks. The control commands will ensure desirable operations of physical components. As we can see, sensors and sensor networks play a very important role in the operations of a CPS [8]. If sensors send bad or misleading measurements of physical devices to the control center through wireless networks, the control center can then send control commands with misleading information to actors. In this way, the operations of physical components could be disrupted dramatically.

Nonetheless, sensors and wireless sensor networks in CPSs may be deployed in open network environments and the wireless communications could increase attack interfaces to an adversary. The adversary can disrupt the operations of CPSs through compromising sensors and wireless communications [9–11]. Based on compromised sensors, the adversary can launch various attacks against CPSs. These attacks

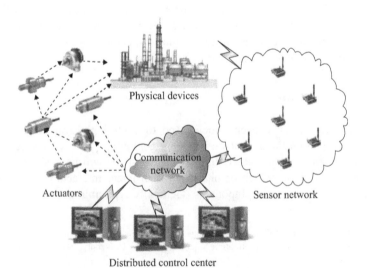

Figure 5.1   The architecture of CPSs © IEEE 2008. Reprinted with permissions from [5]

are different from physical failures and cyber threats over the internet. In the context of CPSs, cyber threats can be launched from cyber components by attacking the data and the interaction between cyber components and physical components, posing the disrupted impact on the operations of physical components [12].

In this chapter, we focus on one class of cyber threats against the CPS, which is denoted as false data injection attacks [5, 13]. In this attack, the adversary can forge and manipulate the measurements of physical components through the compromised sensors and wireless communications, posing the disrupted impact on physical components. As an example, Liu *et al.* [13] showed that false data injection attacks can bypass the detection of existing bad data detection schemes and maliciously manipulate the estimated states of the power grid, which is a typical energy-based CPS. Therefore, the security risks of cyber threats need to be seriously studied before massively deploying CPSs for different applications.

In this chapter, we first review CPSs and highlight the importance of the wireless sensor networks in CPSs and the difference between wireless sensor networks in CPSs and traditional wireless sensor networks. We then investigate various cyber threats in CPSs. To show the impact of cyber threats on CPSs, we take the smart grid as an example to systematically study the impact of false data injection attacks on the smart grid. To be specific, we present a number of false data injection attacks, in which the adversary can forge and manipulate the quantities of energy supplies and energy requests. We then model and analyze the impact of those attacks on electricity market operations. We also conduct a simulation study to validate our theoretical findings. Our evaluation results consistently confirm our theoretical findings and show that false data injection attacks can effectively increase the electricity price to consumers. For example, the electricity price for energy-demand consumers increases when the adversary manipulates the quantity of energy supplies and energy requests through false data injection attacks. We also discuss the impact of false data injection attacks on other functions of the smart grid, including energy distribution processes, state estimation, etc.

The rest of the chapter is organized as follows. In Section 5.2, we review CPSs and the wireless sensor networks in CPSs. In Section 5.3, we explore present attacks on CPSs from the data integrity aspect. In Section 5.4, we present a case study of false data injection attacks on the smart grid. The general guideline for defending against the false data injection attacks is discussed in Section 5.5. Finally, we give the final remarks in Section 5.6.

5.2 CPSs and wireless sensor networks

With the advances of sensing, communication, networking, and computing technologies, CPSs have been introduced [14]. Through the integration of both cyber components and physical components, CPSs can make the monitoring and control of physical systems more efficient [1].

In CPSs, 'cyber' means to use computing, communication, and networking techniques to connect and control physical components, 'physical' represents physical components, and 'system' reflects complexity and diversity [1]. This means that a

CPS may consist of several distributed subsystems, which will be heterogeneous and independent. These subsystems may need to interact with each other. Therefore, the key for designing an effective CPS is the effective integration and interaction between electrical engineering and computer science [4]. Nonetheless, because of different theories and technologies in different disciplines, the development of the science and engineering of CPSs remains a challenging and ongoing research issue.

CPSs have been extended to a number of fields. For example, smart health, as a health-based CPS, collects and monitors health-related information of patients through body sensors and networks. The smart grid, as an energy-based CPS, integrates distributed renewable energy resources and enables two-way communication between consumers and providers in both information and electrical power flows [15]. In addition, other CPSs, such as smart transport, smart building, and smart agriculture, are being developed.

CPSs have changed the interaction among the human world, cyber world, and physical world, just as the operating system (OS) changed the interaction between humans and computers. Instead of collecting measurements through inputs from humans, the control center in a CPS can collect measurements by sensors through communication networks. In this way, the control center can obtain accurate measurements of physical components remotely. In this regard, CPSs can possess great intelligence and can be considered as a human body, where sensor networks can be considered as sensory organs, actuators can be considered as muscles and bones, communication networks can be considered as meridians, and the control center can be considered as a brain. The collaboration of these components can improve the operation efficiency of the physical components.

As stated above, sensor networks, considered as the sensory organs of a CPS, play a very critical role in the operations of a CPS. If no measurement is collected by sensors and transmitted to the control center, the control decision cannot be made. From this aspect, sensor networks in CPSs can bridge the cyber components and the physical components. In CPSs, sensor networks support sensing, communication, and computation with a low overhead [8]. Different from the traditional wireless sensor networks used to transmit sensing data, sensor networks in the CPS should include multidimensional sensory data across multiple sensing networks and/or the internet with the goal of supporting intelligence [3]. Therefore, sensor networking technologies in the CPS focus not only on function design and energy consumption, but also on the interaction among heterogeneous physical components along with the security issues raised by these interactions. Specifically, cyber threats, which use sensor nodes and wireless sensor networks in CPSs as entry points to disrupt the operations of CPSs, need to be seriously addressed in CPS research and development.

5.3   Cyber threats in CPS

To obtain the measurements of physical devices accurately in real time, sensor nodes in CPSs may be deployed in an open and hostile environment and communicate with other components through wireless communications. In addition, sensor nodes lack tamper-resistant hardware and can be compromised by the adversary [9]. The adversary may capture sensor nodes physically and compromise sensor nodes by launching code injection attacks [11] and node replication attacks [10] through wireless interfaces. As a result, the compromised sensor nodes can be controlled by the adversary and further attacks can be launched to disrupt the operations of CPSs.

As we can see, cyber threats in CPSs raise security challenges caused by integrating computation, communication, and networking into the control of physical systems. To increase the attack impact, one objective of cyber threats is to disrupt the operations of CPSs by attacking the data and network infrastructures [5]. Figure 5.2 [5] shows threats that appear in CPSs. Here, Attack A is a physical attack that aims to disrupt the physical processes and components, and Attacks B, C, and D can be considered as cyber threats aiming to disrupt system operations. Specifically, Attacks C and D aim to disrupt the availability and timing of communication networks so that measurements and feedback control comments cannot be delivered to the control center and actuators, respectively, in time. Examples of these types of attacks include denial-of-service (DoS) attacks, jamming attacks, and others. Because Attacks C and D are launched against the communication networks in CPSs, network security mechanisms designed for communication networks should be deployed to mitigate Attacks C and D.

Attack B is one type of cyber threat specific to CPSs and aims to disrupt the integrity of measurement data. We generally denote this type of attack as the data

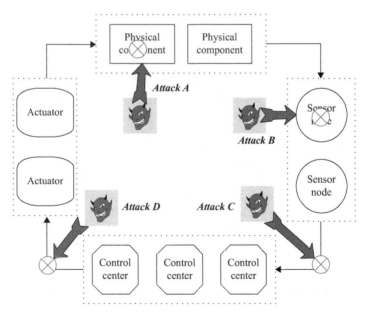

Figure 5.2   The space of attacks © IEEE 2008. Reprinted with permissions from [5]

integrity attack. To launch Attack B, the adversary should first compromise a number of sensor nodes in a CPS. Then, based on the compromised nodes, the adversary can forge or modify measurements of physical components sensed by compromised sensor nodes and send false measurements to the control center. In this way, the state of a CPS can be mistakenly estimated, leading to the wrong control information and putting the operations of a CPS at risk. Because the key of Attack B is to inject false measurements, we denote such attacks as false data injection attacks [13]. Because the injected false data only manipulates the measurements of physical components and has no obvious characteristics, standard network security mechanisms cannot defend against such attacks effectively.

In the past few years, false data injection attacks have been investigated with a number of CPS applications such as the smart grid [13]. Generally speaking, the smart grid, as the next-generation energy-based CPS, is a critical infrastructure and its security can have a significant impact on our daily life. The adversary can launch attacks to disrupt the operations of the smart grid, including energy generation, transmission, distribution, and consumption. A number of research efforts have been devoted to study false data injection attacks in the smart grid [12, 13, 16–24]. In the following, we use the smart grid as an example to investigate the impact of false data injection attacks. Note that such an attack can be generally applied to security studies on other CPSs.

5.4   Case study: false data injection attacks in smart grid

In this section, we will first review the smart grid. We then systematically investigate the impact of false data injection attacks on electricity market operations, including modeling, analysis, and simulation results. Finally, we discuss the impacts of false data injection attacks on other smart grid components.

5.4.1   An overview of the smart grid

Generally speaking, the smart grid is one typical example of a CPS [15] which integrates the physical energy delivery system with the cyber computing and communication networks. The development of the next-generation electrical energy grid can improve energy efficiency by integrating renewable energy resources [25]. The smart grid supplies electrical power from generators through the energy delivery grid to large geographical areas. Figure 5.3 illustrates a domain model for the smart grid [26]. In the smart grid, a number of users are connected to the grid through the communication and energy delivery links. Each user in the smart grid can transform distributed renewable energy resources such as wind energy and solar energy into energy and store the energy locally. The energy-rich users, denoted as energy-supply nodes, can provide energy to users who need energy, denoted as energy-demand nodes.

In the smart grid, all electric appliances of users can obtain energy from the energy-storage equipment. Smart meters deployed at the user-side are developed to measure the energy consumption caused by electric appliances, the energy generation, and the energy storage. To connect smart meters to the control center, the

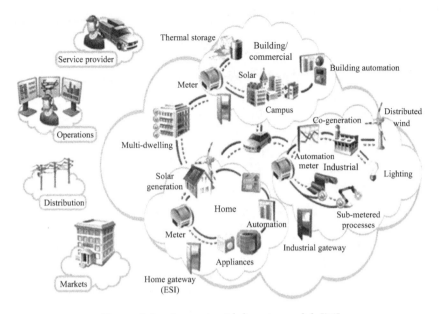

Figure 5.3   A smart grid domain model [26]

advanced metering infrastructure (AMI) has been developed [27]. One example of AMI is shown in Figure 5.4. As we can see from this figure, the measurements of consumers can be measured by smart meters and smart meters can connect to the operation center through multi-hop wireless communications. All smart meters in the smart grid can interact with each other through wireless communication. These smart meters can establish a wireless sensor network, where smart meters can be considered as sensor nodes with great computational capabilities. Cyber threats against the operations of a CPS can also be launched through smart meters (sensors) to disrupt the operations of the smart grid. Note that in the case study from this chapter, the smart grid is a typical CPS and a wireless sensor network connects smart meters in the smart grid. Therefore, the analysis of attack impacts of cyber threats on the operations of the smart grid can help us to have a better understanding of the consequences of cyber threats on CPSs.

Smart meters play a very critical role in the operation of the smart grid and are vulnerable to false data injection attacks. With compromised smart meters, the adversary can forge quantities of demanded energy and supplied energy of the customer and launch false data injection attacks against the smart grid. In the following, we define the two false data injection attacks, which were originally presented in Lin *et al.* [28].

- **Energy-request deceiving attack:** When the adversary compromises the demand-node u, he or she can forge a large quantity of demanded energy, say D_u^*, and send the energy-request messages to the control center in the smart grid through the AMI communication network.

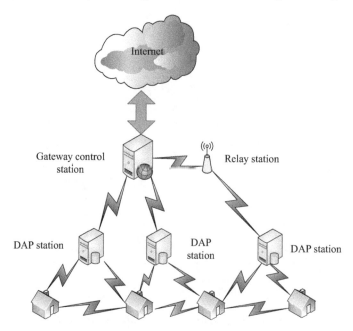

Figure 5.4   An example of AMI

- **Energy-supply deceiving attack:** When the adversary compromises the supply-node v, he or she can forge a false quantity of energy that the node can truly provide, say P_v^*, and send the energy-response messages to control centers in the smart grid through the AMI communication network.

In this chapter, if we say a node u is compromised, we mean that the smart meter deployed in customer u is compromised. With many compromised smart meters, the adversary can forge the quantity of demanded energy and supplied-energy of the customer and launch false data injection attacks against the operations of the smart grid. It is worth noting that to launch a false data injection attack, the adversary may either capture smart meters physically or compromise smart meters remotely by launching code injection attacks and node replication attacks [10] through wireless network interfaces.

5.4.2   Impact of false data injection attacks on electricity market operations

In the following, we present the attack impact of the aforementioned attacks on the electricity market operation, which is a very important application in the smart grid. Note that in the deregulated electricity market, the nodal price for the energy is determined by the regional transmission organizations (RTOs) [23]. Electricity markets consist of the forward market and the spot market. The market operation can compute the ex-post locational marginal price (ex-post LMP). Then, the results can be used as the settlement price [29, 30]. The nodal price is related to the load on each bus in the smart grid. Therefore, false data injection attacks can affect the nodal settlement price because of the imbalance of energy supply and demand.

The LMP is a mechanism for using the market-based price to manage delivery congestion. The LMP for energy at each node is determined by adding 1 MW of fixed load on the designated node and determining the least change of total system cost while satisfying all delivery and other operational constraints [31]. Based on this definition, the LMP on node u in the energy distribution process is determined by:

$$Pri_u = Min(Cost^{+1}) - Min(Cost) + Pri_G \tag{5.1}$$

$$Objective. \ Min\left\{ Cost_n^{+1} = \frac{1}{2}\sum_{L_{ij}\in L}\left(|E_{ij}|Cost_{ij}\right)\right\} \tag{5.2}$$

S.t.

$$\begin{cases} \forall v \in N_P, \ \sum_{i \in N_v} E_{vi} \le P_v \\[2mm] \forall u \in N_D, \ \sum_{j \in N_u} E_{uj} = -D_u - 1 \\[2mm] \forall L_{ij} \in L, \quad E_{ij} = -E_{ji} \end{cases} \tag{5.3}$$

Here, Pri_u is the LMP of node u, $Cost$ is the normal delivery cost, $Cost^{+1}$ is the least delivery cost for transmitting the extra 1 MW on node u, and Pri_G is the generation cost for the extra 1 MW. Without loss of generality, we assume that the energy generation costs of all nodes are the same in our model. Then, the cost of generating the extra 1 MW, say Pri_G, will be constant no matter which nodes generate extra energy. Therefore, in our model, the nodal price on each node will be changed only with the entire delivery cost. That is, the charge to the user is related to the change of the delivery cost. When a false data injection attack is launched, the energy delivery cost will be high and the least delivery cost of transmitting the extra 1 MW for node u will be manipulated, which are denoted as $Cost^*$ and $Cost^{+1*}$, respectively. Therefore, by launching a false data injection attack, the LMP change on node u^* is:

$$\Delta Pri_{u^*} = P_{u^*}^* - P_{u^*} \tag{5.4}$$

$$= Min(Cost^{+1*}) - Min(Cost^*)$$
$$- \left[Min(Cost^{+1}) - Min(Cost)\right] \tag{5.5}$$

Obviously, if $\Delta Pri_{u^*} > 0$, the LMP on node u^* will increase due to false data injection attacks. Note that u^* represents that node u was compromised by the adversary. If $\Delta Pri_{u^*} < 0$, the LMP on node u^* will decrease because of the attack. If $\Delta Pri_{u^*} = 0$, the LMP on node u^* will not be affected by the attack. In the following, we analyze the LMP changes at all demand-nodes, which are affected by the energy-request deceiving attack and the energy-supply deceiving attack.

5.4.2.1   Impact of energy-request deceiving attacks on LMP

As we stated, when the adversary compromises demand-nodes and launches an energy-request deceiving attack, the claimed quantity of demanded energy of the compromised demand-node u^* will increase from D_{u^*} to $D^*_{u^*}$ and the energy delivery cost will increase by $Cost^*$ as well. In this case, the formalization of the false delivery cost for transmitting an extra 1 MW to the compromised node u^* can be derived from Equations (5.6) and (5.7), which are listed as follows:

$$Objective.\ Min \left\{ Cost^{+1*} - \frac{1}{2} \sum_{L_{ij} \in L} \left(|F_{ij}| Cost_{ij} \right) \right\} \tag{5.6}$$

S.t.

$$\begin{cases} \forall v \in N_P, \ \sum_{i \in N_v} E_{vi} \leq P_v \\[2mm] \forall u \in N_D, \ \sum_{j \in N_u} E_{uj} = -D_u \\[2mm] \forall u \in N_{D^*}, \ \sum_{j \in N_{u^*}} E_{u^*j} = -D_{u^*} - 1 \\[2mm] \forall L_{ij} \in L, \ E_{ij} = -E_{ji} \end{cases} \tag{5.7}$$

Based on the definition of LMP, the exchanged LMP on the compromised node u^* can be listed below when the adversary launches an energy-request deceiving attack.

$$Objective.\ Pri_{u*} = Min \left(Cost^{+1*} \right) - Min \left(Cost^* \right) + Pri_G$$

$$= Min \left\{ \frac{1}{2} \sum_{L_{ij} \in L} \left(|E_{ij}| Cost_{ij} \right) \right\} + Pri_G \tag{5.8}$$

S.t.

$$\begin{cases} \forall v \in N_P, \ \sum_{i \in N_v} E_{vi} \leq P_v - P'^*_v \\[2mm] \forall u \in N_D, \ \sum_{j \in N_u} E_{uj} = 0 \\[2mm] \forall u^* \in N_{D^*}, \ \sum_{j \in N_{u^*}} E_{u^*j} = -1 \\[2mm] \forall L_{ij} \in L, \ E_{ij} = -E_{ji} \end{cases} \tag{5.9}$$

As the energy generation cost is assumed as a constant in our model, the false LMP on node u^* is related only to the delivery cost of transmitting the extra 1 MW energy to the node u^*. In the following, we will analyze the cost change of transmitting the extra 1 MW energy for the node u^* in the normal case and in the attack case, respectively. By doing this, we can understand the impact of false data injection attacks on the LMP on each demand-node.

Because of the balance of energy supply and demand, we have:

$$
\sum_{v\in N_P}\left(\sum_{u\in N_D}P'^*_{vu}+\sum_{u^*\in N_{D^*}}P'^*_{vu^*}\right)
$$

$$
=\sum_{v\in N_P}P'^*_v=\sum_{u\in N_D}D_u+\sum_{u^*\in N_{D^*}}D^*_{u^*}
$$

$$
>\sum_{u\in N_D}D_u+\sum_{u^*\in N_{D^*}}D_{u^*}=\sum_{v\in N_P}P'_v \tag{5.10}
$$

$$
=\sum_{v\in N_P}\left(\sum_{u\in N_D}P'_{vu}+\sum_{u^*\in N_{D^*}}P'_{vu^*}\right)
$$

where P'^*_{vu} and $P'^*_{vu^*}$ are the transmitted energy from the supply-node v to the normal demand-node u and the compromised demand-node u^* in the attack case, respectively, and P'_{vu} and P'_{vu^*} are the transmitted energy from the supply-node v to the normal demand-node u and the compromised demand-node u^* in the normal case, respectively. Therefore, for each supply-node v, we have:

$$
P_v\geq\sum_{u\in N_D}P'^*_{vu}+\sum_{u^*\in N_{D^*}}P'^*_{vu^*}\geq\sum_{u\in N_D}P'_{vu}+\sum_{u^*\in N_{D^*}}P'_{vu^*} \tag{5.11}
$$

Based on Theorem 2 in Lin *et al.* [28], each supply-node uses the lowest cost path to transmit energy to demand-nodes. For a demand-node u, we can obtain an order of all supply-nodes, say $\{v_1,\ldots,v_y\}$, in which y is the number of supply-nodes in the smart grid. In this order of supply-nodes, the energy delivery cost between the supply-nodes and the demand-node u increases. Recall that the basic idea of the energy distribution process is to find an optimal set of P'_{vu}, which leads to the minimum cost of energy delivery. Therefore, in the normal case, the demand-node u receives as much energy as possible from the former supply-nodes in $\{v_1,\ldots,v_y\}$ to minimize the total energy delivery cost. Then, we can find a supply-node v_i and have:

$$
\begin{cases}
D_u=\sum_{k=1}^{i}P'_{v_k u} \\[2mm]
\forall j\in[0,i-1],\ P_{v_j}=\sum_{u\in N_D}P'_{v_j u}+\sum_{u^*\in N_D}P'_{v_j u} \\[2mm]
P_{v_i}>\sum_{u\in N_D}P'_{v_i u}+\sum_{u^*\in N_D}P'_{v_i u} \\[2mm]
\forall g\in[i+1,y],\ P'_{v_g u}=0
\end{cases} \tag{5.12}
$$

Similarly, in the attack case, the compromised node u^* will receive as much energy as possible from the former supply-nodes in $\{v_1,\ldots,v_y\}$, even if the adversary claims the demanded energy of compromised demand-node u^* from D_{u^*} to $D^*_{u^*}$. Hence, we can find a supply-node v_{i^*} when the adversary compromises the demand-nodes and launches an energy-request deceiving attack. The formalization is listed as follows:

$$
\begin{cases}
D^*_{u^*} = \displaystyle\sum_{k=1}^{i^*} P'^*_{v_k u^*} \\[2mm]
\forall j \in [0, i^* - 1], \ P_{v_j} = \displaystyle\sum_{u \in N_D} P'^*_{v_j u} + \sum_{u^* \in N_{D^*}} P'^*_{v_j u^*} \\[2mm]
P_{v_{i^*}} > \displaystyle\sum_{u \in N_D} P'^*_{v_{i^*} u} + \sum_{u^* \in N_{D^*}} P'^*_{v_i u^*} \\[2mm]
\forall g \in [i^* + 1, y], \ P'^*_{v_g u} = 0
\end{cases}
\tag{5.13}
$$

Based on the above analysis, although the demanded energy of u^* increases from D_{u^*} to $D^*_{u^*}$, the energy supplied by the supply-nodes in $\{v_1, \ldots, v_i\}$ in the attack case will be equal to the energy supplied by these supply-nodes in the normal case. Therefore, the increased demanded energy forged by the adversary should be supplied by supply-nodes in $\{v_1, \ldots, v_{i^*}\}$ and can be denoted as follows:

$$
\begin{cases}
\forall j \in [0, i - 1], \ P'^*_{v_j u} = P'_{v_j u} \\[2mm]
D^*_{u^*} - D_{u^*} = \displaystyle\sum_{k=1}^{i^*} P'^*_{v_k u^*} \\[2mm]
\forall g \in [i^* + 1, y], \ P'^*_{v_g u} = 0
\end{cases}
\tag{5.14}
$$

According to the above analysis, in the attack case, the extra 1 MW of demanded energy of the demand-node u^* will be supplied by the supply node v^*_i and in the normal case, the extra 1 MW of demanded energy of the demand-node u^* will be supplied by the supply-node v_i. Also, the supply-node v^*_i and the v_i should be different supply-nodes. Recall that in our analysis, the LMP on each demand-node is related only to the delivery cost of transmitting the extra 1 MW energy to the demand-nodes. Hence, the LMP changes on node u^* should be the difference in cost between transmitting the extra 1 MW to node u^* from node v^*_i and transmitting the same amount of energy from node v_i to node u^*. In the supply-nodes order $\{v_1, \ldots, v_y\}$, the energy delivery cost between the supply-nodes and demand-node u increases and thus, the delivery cost from node v^*_i to node u^* is larger than that from node v_i to node u^* if i^* is larger than i. In this way, the LMP change on node u^* will be larger than zero (i.e. $\Delta Pri_{u^*} > 0$).

Because all demand-nodes are the same in the smart grid, the adversary compromises a number of demand-nodes and launches an energy-request deceiving attack to increase the demanded energy of compromised demand-nodes. This makes

$$
D^*_{u^*} = \sum_{k=1}^{i^*} P'^*_{v_k u^*} > \sum_{k=1}^{i} P'^*_{v_k u^*}
\tag{5.15}
$$

In this way, i^* will be larger than i and the LMP change on node u^* will be larger than zero. That is, the LMP on node u^* will increase due to false data injection attacks. In addition, the higher the demanded energy from the compromised demand-nodes, the larger the LMP on compromised demand-nodes.

With the above analysis, we can summarize the quantified analytical results shown in Proposition 1.

Proposition 1: With an energy-request deceiving attack, as the adversary manipulates the quantity of energy requests through the compromised demand-nodes, the LMP on demand-nodes will increase as the quantity of energy requests is increased by the adversary.

5.4.2.2 Impact of energy-supply deceiving attacks on LMP

When the adversary compromises the supply-nodes and launches an energy-supply deceiving attack, the claimed quantity of energy that the compromised supply-node u^* can provide will be manipulated from P_{v^*} to $P_{v^*}^*$. Different from an energy-request deceiving attack, in an energy supply deceiving attack, the adversary compromises the supply-nodes and manipulates the amount of energy which the compromised supply-nodes can actually provide to affect the LMP on the normal demand-nodes.

The compromised $P_{v^*}^*$ may be either larger or smaller than the normal P_{v^*}. As shown in Lin *et al.* [28], when the adversary claims more energy than the compromised supply-nodes can provide (i.e. $P_{v^*}^* > P_{v^*}$), the cost of energy delivery may not increase, making the LMP on the demand-nodes suffer no effect. Therefore, we analyze the impact of an energy-supply deceiving attack on the LMP on the demand-nodes in only one case: claiming less energy than the supply-node can provide.

When the adversary claims less energy than the compromised supply-nodes can provide, the energy delivery cost will increase to $Cost^*$ based on Proposition 2. In this scenario, the formalization of the compromised delivery cost of transmitting an extra 1 MW energy to demand-node u is listed as follows:

$$\text{\textit{Objective. Min}} \left\{ Cost^{+1*} = \frac{1}{2} \sum_{L_{ij} \in L} \left(|E_{ij}| Cost_{ij} \right) \right\} \tag{5.16}$$

S.t.

$$\begin{cases} \forall v \in N_P, \ \sum_{i \in N_v} E_{vi} \leq P_v \\ \forall u \in N_D, \ \sum_{j \in N_u} E_{uj} = -D_u - 1 \\ \forall v* \in N_{p^*}, \ \sum_{i \in N_{v^*}} E_{v^*i} \leq P_{v^*}^* \\ \forall L_{ij} \in L, \ E_{ij} = -E_{ji} \end{cases} \tag{5.17}$$

Based on Theorem 2 proved in Lin *et al.* [28], for each normal demand-node u, we can obtain the order of all supply-nodes $\{v_1, \ldots, v_y\}$ in which the energy delivery cost between the supply-nodes and the demand-node u is incremental. In the normal case, the energy distribution process makes the demand-node u receive as much energy as possible from the former supply-nodes in $\{v_1, \ldots, v_y\}$ to minimize the total energy delivery cost, which is similar to Equation (5.12). Similar to the analysis in the previous subsection, an order of supply-nodes exists in the smart grid, say $\{v_1, \ldots, v_i\}$, in which v_i is the certain supply-node that satisfies $D_u = \sum_{k=1}^{i} P'_{v_k u}$. When the adversary compromises the supply-nodes in $\{v_1, \ldots, v_i\}$, and claims less energy than these supply-nodes can provide to the demand-nodes, these supply-nodes will not be able to supply enough energy to the demand-nodes u (i.e. $D_u = \sum_{k=1}^{i} P'^{*}_{v_k u}$). In addition, there is a supply-node v_{i^*} so that

$$D_u = \sum_{k=1}^{i^*} P'^{*}_{v_k u} \tag{5.18}$$

Hence, i^* is larger than i and the extra 1 MW demanded energy of demand-node u will be provided by supply-node v_{i^*}. In a supply-node order of $\{v_1, \ldots, v_y\}$, the energy delivery cost between the supply-nodes and demand-node u is incremental. Then, the LMP change on node u will be larger than zero. Because all demand-nodes are the same in the smart grid, the LMP on the demand-nodes will be affected by claiming less energy than the supply-node can provide.

With the above analysis, we summarize the quantified analytical results as Proposition 2.

Proposition 2: With an energy-supply deceiving attack, as the adversary manipulates the quantity of energy supply through the compromised supply-nodes, the LMP on the demand-nodes will increase when the adversary claims less energy than the compromised supply-nodes can provide.

5.4.3 *Performance evaluation*

To demonstrate the impact of energy-deceiving attacks on the energy distribution process, we conducted performance evaluations based on a simplified version of the US power grid [28]. In our simplified model, the major cities of individual states are considered as nodes in the topology. The backbone of the interstate energy delivery is based on the connection between these nodes. We use the 2009 US Energy Information Administration State Electricity Profiles [32] as the data set to perform simulations. To measure the impact of false data injection attacks on the electricity market operation, we consider the LMP on demand-nodes, which are affected by the energy-request deceiving attack and the energy supply deceiving attack, respectively. It is worth noting that in our evaluation, '+(−)50, +(−)100, +(−)150 demand (supply)' means that the demanded (supplied) energy of each compromised node is increased (decreased) by 50, 100, and 150 units, respectively, through the investigated false data injection attack. All simulations in this chapter were completed using Matlab. It is worth noting that to measure the impact of false data injection attacks, we consider the increased delivery cost, the user outage rates, the supplied energy loss, and LMP change at demand-nodes. Due to limited space, we show only the result on LMP change at demand-nodes below, whereas other evaluation results can be found in Lin *et al.* [28].

To investigate the impact of false data injection attacks on the LMP on the demand-nodes, we show the relationship between LMP changes and the increased energy demand of the compromised demand-nodes in Figure 5.5, where $*$ represents the nodes compromised by the adversary. When the number of forged energy requests increases, the LMP of most demand-nodes increases. As we can see, as the forged energy-request increases by 100 and then 150 units, the LMP on demand-nodes u_{13}^* increases by \$300 per million megawatts and then \$700 per million megawatts, respectively. We can also observe that the LMP on demand-nodes u_7-u_{19} is manipulated rapidly by the attack and that the LMP on the other nodes has made no change. The reason is that when the number of forged energy requests increases, the demanded energy of demand-nodes u_7-u_{19} is constrained by Equation (5.15) and the demanded energy of the other nodes cannot meet the condition. Hence, the evaluation results match our analytical results well.

We show the LMP changes on the demand-nodes, which are affected by the energy-supply deceiving attack in Figure 5.6. Here, we assume that the demand-nodes are legitimate and 30% of the supply-nodes are compromised by the

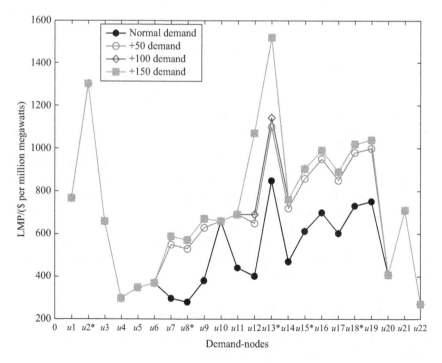

Figure 5.5   LMP versus demand-nodes

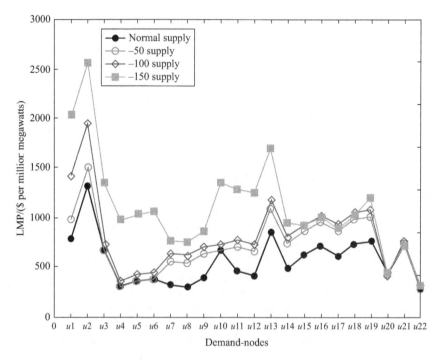

Figure 5.6   LMP versus demand-nodes

adversary. When the forged energy supply declines, the LMP on most demand-nodes decreases. As we can see, when the attack reduces the supply by 100 and then 150 energy units from the compromised supply-nodes than what they can actually provide, the LMP on all the demand-nodes will increase. The more the supplied energy declines, the more the LMP on the demand-nodes increases. Hence, the LMP on the demand-nodes increases when the adversary claims less energy than what the compromised supply-nodes can provide.

Figure 5.7 shows the LMP changes on the demand-nodes affected by both the energy-request deceiving attack and the energy-supply deceiving attacks, in which 50% of demand-nodes and 30% of supply-nodes are compromised by the adversary. As we can see, the LMP on the demand nodes increases rapidly when the smart grid suffers from the energy-deceiving attack. From Figures 5.5, 5.6, and 5.7, we can observe that the LMP on the demand-nodes increases more significantly after the smart grid is attacked by either the energy-request deceiving attack or the energy-supply deceiving attack.

5.4.4   Impact of false data injection attacks on other applications

In addition to impact on the LMP, false data injection attacks can have significant impacts on other functions in the smart grid.

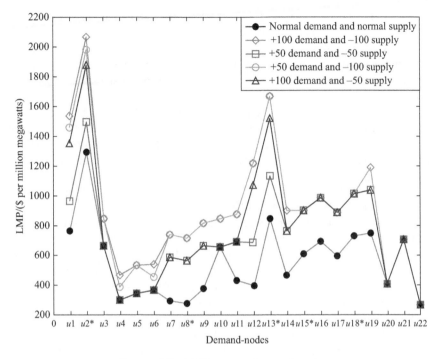

Figure 5.7   LMP versus demand-nodes

5.4.4.1   Impact on energy delivery

Energy-request deceiving attacks and energy-supply deceiving attacks can also disrupt the energy delivery process in the smart grid. In the smart grid, the two-way communication between customers and utility providers can be enabled. The energy generation dispatch and energy delivery process will be determined by the measurements collected from smart meters. Hence, the energy delivery process can also be disrupted if the measurements are manipulated by compromised smart meters from launching false data injection attacks.

In our previous work [28], we modeled and analyzed the attack impact of energy-request deceiving attacks and energy-supply deceiving attacks on the energy delivery process. Based on our modeling, analytical results, and simulation study, we found that the forged data injected by the attacks can pose an imbalanced demand response and incur increased costs for power delivery and disrupt the stability of energy delivery processes in the smart grid. To be specific, we found that the supplied power loss, the power delivery cost, and the number of outages users experience can increase when the adversary manipulates the quantity of power requests. The power delivery cost and the number of outage users may increase as well when the adversary manipulates the quantity of power supply. The number of outage users may increase when the adversary manipulates the state of power delivery links. Due to limited space, this detailed analysis can be found in

Lin *et al.* [28]. The summary of the analytical results is shown in Propositions 3 and 4, which can be found in Lin *et al.* [28].

Proposition 3: With energy-supply deceiving attacks, as the adversary manipulates the quantity of energy supply through the compromised supply-nodes, the supplied energy loss may not be affected, the number of outage users in the smart grid may increase, and the energy delivery cost may decrease (or increase) when the adversary claims more (or less) energy than what the supply-nodes can provide [28].

Proposition 4: With energy-request deceiving attacks, as the adversary manipulates the quantity of energy requests through the compromised demand-nodes, the supplied energy loss, energy delivery cost, and the number of outage users in the grid will increase [28].

5.4.4.2   Impact on state estimation and others

False data injection attacks can disrupt the state estimation of the smart grid as well. In the past few years, a number of research efforts have been conducted on this topic [13, 16–21]. For example, Liu *et al.* [13] introduced false data injection attacks against the state estimation of the power grid and demonstrated that the adversary could manipulate the state estimation of the power grid without being detected by existing bad data detection schemes.

Based on the findings in Liu *et al.* [13], the approach of finding an optimal attack strategy was proposed [33], in which a set of smart meters are selected to be compromised so that the maximum damage can be caused to grid operations. In addition, two indices are proposed in Sandberg [16] to quantify the efforts required to implement false data injection attacks. Based on these two indices, the least effort required to achieve false data injection attacks without being detected by existing bad data detection schemes was studied. All the above research efforts have shown that false data injection attacks can cause serious impacts on the state estimation in the smart grid.

The study of false data injection attacks has been extended to other areas as well [12, 22–24]. One example is malicious data attacks against the real-time electricity market [24]. In particular, the damage of malicious data attacks to the real-time electricity market is studied [24], in which an optimization of a quasi-concave objective function was found to support the understanding that the adversary is capable of obtaining the optimal attacking strategy to obtain financial benefits.

5.5   Discussion

We now discuss the general guideline for defending against false data injection attacks from the following aspects: attack prevention and detection.

5.5.1   Prevention

We need to enhance the network configuration to improve the resilience of the smart grid to cyber threats. One way is to fully protect some of the critical sensors and make them difficult to attack. From our analysis, we can see that false data injection attacks will become more complex and less efficient when a smaller number of sensors are compromised. Nevertheless, protecting all of the sensors is impossible to realize in real-world practice due to deployment costs. Hence, we shall study the subsequent problem: given a limited number of sensors to be protected due to cost constraints, how can we find the set of sensors to protect to make false data injection attacks difficult to deploy? To this end, we shall further investigate the effectiveness of this protection mechanism against attacks when critical and redundant measurements are provided.

5.5.2   Detection

To accurately detect attacks, we shall study effective and low-cost detection schemes. We shall develop schemes to defend against remote malicious code propagation and injection attacks, which are related to compromised sensors. To attack smart meters, as they are key pieces of equipment in the smart grid, the adversary can launch code to maliciously propagate traffic over the network and to inject malicious code into other devices in the smart grid. For this purpose, we shall consider the correlation and integration of both software behavior and network traffic to improve detection accuracy. For example, memory authentication protocols (MAPs) can be employed to detect compromised meters by detecting whether memory codes have been changed [34]. One general approach is to compute the checksum of the memory codes and then decide whether the obtained checksum is equal to the expected value defined by the verifier. Two types of MAP can be developed: bidirectional MAPs and one-way MAPs. The former has a high resilience only against local attacks whereas the latter has a high resilience against both local attacks and remote attacks. In addition, the modeling and forecasting of smart meter measurements and filtering mechanisms can be used to defend against injected false data [35, 36].

5.6   Conclusion

In this chapter, we first reviewed CPSs and highlighted the importance of wireless sensor networks in CPSs. To investigate the cyber threats in a CPS, we took the smart grid as an example and investigated the impacts of false data injection attacks on electricity market operations in the smart grid. We classified attacks and formally analyzed the impact of several attacks on electricity market operations with the consideration of the LMP on the electricity-demand nodes. Through extensive simulations, our data show that our investigated attacks can significantly disrupt the effectiveness of electricity market operations and lead to a high cost to consumers. As ongoing research, we are investigating the performance impacts of our investigated attacks on other CPSs and are designing effective mechanisms to defend against those attacks.

Acknowledgments

The work was supported in part by the Natural Science Foundation of China (NSFC) under grants 61373115 and 61402356. This work was also supported in part by the US National Science Foundation (NSF) under grants CNS 1117175 and 1350145. Any opinions, findings, and conclusions or recommendations expressed in this material are those of the authors and do not necessarily reflect the views of the funding agencies.

References

[1]   R. Helps and S. Pack. Cyber-physical system concepts for IT students. In *Proceedings of the 14th Annual ACM Conference on Information Technology Education (SIGITE '13)*, pp. 7–12, ACM, New York, USA, 2013.

[2]   R. Poovendran. Cyber-physical systems: Close encounters between two parallel worlds. *In Proceedings of IEEE*, 98(8): 1363–1366, 2010.

[3]   F. Wu, Y. Kao and Y. Tseng. From wireless sensor networks towards cyber physical systems. *Pervasive and Mobile Computing*, 7(4): 397–413, August 2011.

[4]   P. Derler, E. Lee, M. Torngren and S.Tripakis. Cyber-physical system design contracts. In *Proceedings of the ACM/IEEE 4th International Conference on Cyber-Physical Systems (ICCPS '13)*, pp. 109–118, ACM, New York, USA, 2013.

[5]   A. A. Cardenas, S. Amin and S. Sastry. Secure control: Towards survivable cyber-physical systems. In *Proceedings of the 1st International Workshop on Cyber-Physical Systems (WCPS)*, pp. 495–500, 2008.

[6]   M. Pajic, A. Chernoguzov and R. Mangharam. Robust architectures for embedded wireless network control and actuations. *Transactions on Embedded Computing Systems (TECS)*, 11(4): 24 p., December 2012.

[7]   A. Albur and A. G. Exposito. *Power System State Estimation: Theory and Implementation*. CRC Press, Boca Raton, March 2004.

[8]   P. Misra, L. Mottola, S. Raza, S. Duquennoy, N. Tsiftes, J. Hoglund and T. Voigt. Supporting cyber-physical systems with wireless sensor networks: An outlook of software and services. *Journal of the Indian Institute of Science*, 93(3): 441–462, 2014.

[9]   H. Chan and A. Perrig. Security and privacy in sensor networks. *Computer*, 36(10): 103–105, 2003.

[10] Y. Younan, P. Philippaerts, F. Piessens, W. Joosen, S. Lachmund and T. Walter. Filter-resistant code injection on ARM. In *Proceedings of the 16th ACM Conference on Computer and Communications Security (CCS '09)*, pp. 11–20, ACM, New York, USA, 2009.

[11] A. Francillon and C. Castelluccia. Code injection attacks on Harvard architecture devices. In *Proceedings of the 15th ACM Conference on Computer and Communications Security (CCS)*, pp. 15–26, 2008.

[12] Y. Mo and B. Sinopoli. False data injection attacks in control systems. In *Preprints of the 1st Workshop on Secure Control Systems*, CPSWEEK, April 2010.

[13]   Y. Liu, M. K. Reiter and P. Ning. False data injection attacks against state estimation in electric power grids. In *Proceedings of the 16th ACM Conference on Computer and Communications Security (CCS)*, pp. 21–32, ACM, New York, USA, 2009.

[14]   S. Ahmed, G. Kim and D. Kim. Cyber physical system: Architecture, applications and research challenges. In *Proceedings of 2013 IFIP Wireless Days (WD)*, pp. 1–5, Valencia, CA, USA, November 2013.

[15]   NSF workshop on new research directions for future cyber-physical energy systems, http://www.ece.cmu.edu/nsf-cps/, Baltimore, MD, Technical Report, 2009.

[16]   H. Sandberg, A. Teixeira and K. H. Johansson. On security indices for state estimators in power networks. In *Preprints of the First Workshop on Secure Control Systems*, CPSWEEK, April 2010.

[17]   G. Dan and H. Sandberg. Stealth attacks and protection schemes for state estimators in power systems. In *Proceedings of the 1st IEEE International Conference on Smart Grid Communications (SmartGridComm)*, pp. 214–219, IEEE, Gaithersburg, MD, USA, October 2010.

[18]   R. B. Bobba, K. M. Rogers, Q. Wang, H. Khurana, K. Nahrstedt and T. J. Overbye. Detecting false data injection attacks on dc state estimation. In *Preprints of the First Workshop on Secure Control Systems*, CPSWEEK, April 2010.

[19]   H. Yi, H. Li, K. Campbell and H. Zhu. Defending false data injection attack on smart grid network using adaptive CUSUM test. In *2011 45th Annual Conference on Information Sciences and Systems (CISS)*, pp. 1–6, IEEE, Baltimore, MS, USA, 23–25 March 2011.

[20]   H. Li, L. Lai and S. Djouadi. Combating false reports for secure networked control in smart grid via trustiness evaluation. In *2011 IEEE International Conference on Communications (ICC)*, pp. 1–5, IEEE, Kyoto, Japan, June 2011.

[21]   R. Akella and B. M. McMillin. Information flow analysis of energy management in a smart grid. *Lecture Notes in Computer Science*, 6351: 263–276, 2010.

[22]   A. Teixeira, G. Dan, H. Sandberg and K. H. Johansson. A cyber security study of a SCADA energy management system: Stealthy deception attacks on the state estimator. In *Proceedings of the 18th IFAC World Congress*, 2011.

[23]   L. Xie, Y. L. Mo and B. Sinopoli. False data injection attacks in electricity markets. In *Proceedings of the 1st IEEE International Conference on Smart Grid Communications (SmartGridComm)*, pp. 226–231, IEEE, Gaithersburg, MD, USA, October 2010.

[24]   L. Jia, R. J. Thomas and L. Tong (eds.). Malicious data attack on real-time electricity market. In *Proceedings of the 2011 IEEE International Conference on Acoustics, Speech and Signal Processing (ICASSP)*, pp. 5952–5933, IEEE, Prague, Czech Republic, May 2011.

[25]   See: http://www.nist.gov/smartgrid/nistandsmartgrid.cfm.

[26] RM 11-2000. Smart Grid Interoperability Standards, July 2011.

[27] M. Line, I. Tondel and M. Jaatun. Cyber security challenges in smart grids. In *Proceedings of the 2011 2nd IEEE PES International Conference Innovative Smart Grid Technologies (ISGT Europe)*, pp. 1–8, IEEE, Manchester, UK, December 2011.

[28] J. Lin, W. Yu, X. Yang, G. Xu and W. Zhao. On false data injection attacks against distributed energy routing in smart grid. In *Proceedings of the 2012 IEEE/ACM 3rd International Conference on Cyber-Physical Systems (ICCPS)*, pp. 183–192, ACM/IEEE, Beijing, China, April 2012.

[29] F. C. Schweppe, M. C. Caramanis, R. D. Tabors and R. E. Robn. *Spot Pricing of Electricity*. Kluwer, Norwell, MA, 1998.

[30] F. E. R. Commission. Remedying undue discrimination through open access transmission service and standard electricity market design. In *Notice of Proposed Rulemaking, IV FERC Stats and Regs*. 32,563, Docket No. RM01-12-000.

[31] California ISO. Locational marginal pricing (LMP): basics of nodal price calculation. CAISO Market Operation, http://www.caiso.com/docs/2004/02/13/200402131607358643.pdf, 2004.

[32] See: http://www.oe.energy.gov/smartgrid.htm.

[33] Q. Yang, J. Yang, W. Yu, D. An, N. Zhang and W. Zhao. On false data-injection attacks against power system state estimation: Modeling and countermeasures. *IEEE Transactions on Parallel and Distributed Systems*, 25(3): 717–729, 2014.

[34] K. Song, D. Seo, H. Park, H. Lee and A. Perrig. OMAP: One-way memory attestation protocol for smart meters. In *Proceedings of the 9th IEEE International Symposium on Parallel and Distributed Processing with Applications Workshops (ISPAW)*, pp. 111–118, May 2011.

[35] W. Yu, D. An, D. Griffith, Q. Yang and G. Xu. On statistical modeling and forecasting of energy usage in smart grid. *ACM International Journal of Applied Computing Review (ACR)*, 15(1): 1231–1234, March 2015.

[36] X. Yang, J. Lin, W. Yu, P. Moulema, X. Fu and W. Zhao. A novel en-route filtering scheme against false data injection attacks in cyber-physical networked systems. *IEEE Transactions on Computers (TC)*, 64(1): 4–18, 2015.

Chapter 6

Data management in cyber-physical systems with wireless sensor networks

Hedi Haddad[1] and Nabil Sahli[2]

Abstract

A WSN-CPS is a cyber-physical system (CPS) where the communication between the physical and cyber components is guaranteed through wireless sensor networks (WSNs). In this chapter we aim to give a comprehensive overview of the data management problem in WSN-CPSs. For the sake of clarity, the overview is organized into three main parts. Given that data management in WSNs has been the subject of substantial research works, and given the overlap between WSNs and WSN-CPSs, in the first part we take a comparative approach in order to identify what makes data management in WSN-CPSs different from data management in WSNs. Based on the state of the art, we identified three main constraints that need to be dealt with by data management solutions in the context of WSN-CPSs: (1) mobility of sensing and actuating devices, (2) high-level knowledge extraction from heterogeneous low-level data, and (3) real-time reaction to events of interest. In the second part we discuss how these three constraints are shaping current solutions proposed in the literature to address WSN-CPSs' data management activities, such as mobile data collection, data processing, storage, and querying. In the third part, we discuss the opportunities and challenges of using cloud computing to support data management in WSN-CPSs.

6.1 Introduction

Defining the concept of cyber-physical systems (CPSs) is still the subject of debate in the literature, mainly because of the multidisciplinary character of the concept, the heterogeneity of the technologies that it involves and the lack of international standards. Without entering into these debates, a CPS can be thought of as a system

[1]Computer Science Department, Dhofar University, Salalah, Sultanate of Oman, e-mail: hhaddad@du.edu.om
[2]Computer Science Department, German University of Technology (GUtech), Athaibah, Sultanate of Oman, e-mail: Nabil.sahli@gutech.edu.om

that integrates the cyber and physical worlds in order to automatically and intelligently react to real-world situations [1]. A WSN-CPS is a CPS where the communication between the physical and cyber components is guaranteed through wireless sensor networks (WSNs). WSN-CPSs are thus complex systems where the physical world is integrated with sensing, computing, actuation, and communication components (the physical world here refers to a natural or an engineered system that needs to be monitored and/or controlled, e.g. a machine or a geographic area in an environmental monitoring system). Figure 6.1 illustrates the complex interactions between those components [2]. The sensing component is composed of a set of sensor nodes which are commonly deployed in a geographic environment and interconnected through wireless communication means. The main role of sensor nodes is to take measurements about the state of the physical world and to transmit them to the cyber system, represented by the computation component. The cyber system analyses the collected measurements, decides about how to react, and assigns the actuation component with relevant actions to change the status of the physical world. Similarly to the sensing component, the actuation component is composed of a set of actuators which are geographically distributed and interconnected through wireless communication networks. The communication component corresponds to the wireless network that ensures that communication between the sensing, computation, and actuation components is efficient and secure. The WSN closes the loop that integrates the cyber and physical worlds [2].

Depending on the application scenario, the scale of a CPS-WSN ranges from small to large, but in general both sensors and actuators are equipped with some

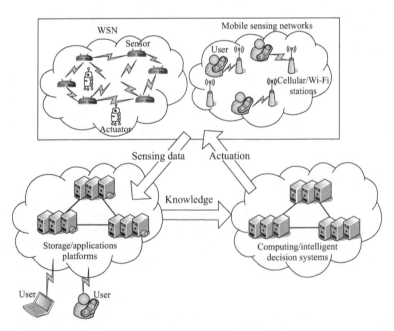

Figure 6.1   A typical CPS architecture [2]

data processing and wireless communication capabilities. Even though the computation functionalities can be distributed, the cyber system is typically an embedded computer that takes the sensor measurements in input, and generates an output according to the target scenario. The cyber system usually acts as a controller for the physical system.

In this chapter, we look to WSN-CPSs from a data management perspective. The rest of the chapter is organized as follows. Section 6.2 highlights the differences between WSNs and WSN-CPSs within the context of data management. Section 6.3 briefly describes the different common activities of data management in WSN-CPSs, with a particular focus on data collection in a mobile context. Finally, and before concluding the chapter, Section 6.4 identifies the main opportunities and challenges that WSN-CPSs have to deal with.

6.2 Data management: WSN vs. WSN-CPS

Even though the difference between WSNs and CPSs has not been always clear, their primary functionalities are quite different. While the main goal of WSNs is to passively sense and monitor the state of the physical world, i.e. without changing the physical environment, WSN-CPSs' main functionality is to control the physical world. CPSs can then be thought of as WSN systems augmented with intelligent actuation (control) functionality. Data management in WSN-CPSs shares similar characteristics with data management in WSNs. Both WSNs and CPSs deal with high-volume streaming and stored data, and demand low latency processing. Although the energy and processing constraints in a CPS are generally more relaxed compared to a WSN, time is even more important [2]. Furthermore, privacy is a more critical issue in CPSs, compared to WSNs.

6.2.1 Data management in WSNs

Although data management in WSNs has been the subject of substantial research works (detailed reviews are given in [3–8]), there is no clear definition in the literature of what exactly data management is. For example, in [3], the authors mention that sensorial data management typically includes four tasks: data acquisition, data cleaning, query processing, and data compression. In [4], data management in mobile sensor networks 'refers to a collection of centralized and distributed algorithms, architectures and systems to handle (store, process and analyze) the immense amount of spatio-temporal data that is cooperatively generated by collections of sensing devices that move in space over time'. In [9], data management 'provides the strategy to manage the collected data to satisfy the user requirements. Data need to be integrated and stored from multiple sensors for future use'. According to [10], the purpose of WSN data management is to guarantee that data collected by the networks 'should be easily accessible to scientists at various levels of detail with low latency'. In addition, Gupchup defines data provenance as a number of transformation steps which are necessary to convert raw sensor values into final usable data: 'For scientific reasons, intermediate results need to be stored allowing the origins of every piece of data to be traced. This motivates the need for a unified system that manages the data gathered by WSNs and also stores various intermediate results' [10].

Moreover, there is no consensus in the literature about a common terminology for data management. Different terms are referring to similar or different concepts depending on the context of use. All the following terms are used: data sampling, data collection, data processing, data analysis, data compression, data aggregation, data fusion, data storage, data query, etc.

In this chapter, we do not aim to identify a common definition for data management in WSNs as it may be both difficult and inappropriate. We would rather discuss the main functions (or activities) of data management in the context of WSN-CPSs (Section 6.3) as well as the related opportunities and challenges (Section 6.4). Before doing so, and for the sake of clarity, we discuss in the next subsections the constraints under which WSNs are operating, and we compare them with WSN-CPSs.

6.2.2   Constraints of data management in WSNs

Before talking about constraints, it is important to understand how sensors work within their network. A WSN is usually composed of a large number of sensors (or sensor nodes) deployed in a large geographic environment. According to [11], sensor nodes perform three basic tasks: (a) take measurements about phenomena of interest from their surrounding environment, (b) process and/or store the collected measurements, and (c) transfer the collected and/or processed measurements to a sink node (also called base station) using the wireless network. The sink node is a device responsible for data storage and/or processing [11]. In addition to sensor and sink nodes, other node types can be found in the literature, such as the manager nodes presented in [12], which are used to provide managers with some data analysis and querying capabilities. So basically, each sensor node has to collect, process, and transfer data (optionally analyze data) while dealing with many constraints. Indeed, sensors operate with very limited resources in terms of energy, memory, and processing. In addition, they have to deal with uncertain large amounts of data in real time. We briefly present these constraints in what follows.

6.2.2.1   Energy

Energy is considered to be the most important design constraint in WSNs. Sensors are typically battery-powered, and in some scenarios they are deployed in harsh and inaccessible locations where it is difficult/impossible to get their batteries changed by a human. For this reason, the use of energy-efficient techniques is a critical issue in WSN design in order to maximize the lifetime of sensor nodes [13].

According to [13], sensor nodes basically spend energy for environment sensing, wireless communication and local processing functions. Energy-optimization strategies can thus be achieved by optimizing those three functions, especially the communication function which has been considered as the most expensive and energy-consuming. Therefore, the design of WSNs is based on a trade-off between these three functions, with a preference for more intelligent sensing and more local processing techniques whenever is possible in order to minimize communication costs. In addition, some research works have explored energy-efficient communication strategies, such as the work in [11] which demonstrated that short-range multi-hop transmission techniques are more energy-efficient than large-distance communications.

6.2.2.2   Memory and processing power

Sensor nodes have also limited processing power and memory. Indeed, sensors are usually equipped with limited processing and storage capabilities, which makes the use of large local data storage and processing strategies very difficult. Other strategies therefore need to be developed to overcome limited processing and memory capabilities. Distributed-data storage techniques and intelligent data-storage strategies (i.e. only store the relevant data for a limited period of time) can be used to reduce local-data processing. Similarly, distributed-computation strategies can be deployed, where different sensors collaborate to perform some complex computations. Dulman [13] presented the concept of 'network heterogeneity' as a solution for sensors' limited memory and processing capabilities. In this concept, a WSN is composed of heterogeneous devices, and the devices with limited capabilities can request certain computations from more powerful nodes. Other solutions will be presented in Section 6.3.

6.2.2.3   Data volume and real-time processing

WSNs usually deal with data streams where huge data volumes are continuously generated from multiple sensor nodes. Processing huge streaming data volumes is challenging, especially in real-time applications.

6.2.2.4   Data uncertainty

The data collected by sensors are typically imprecise and incomplete for several reasons, basically due to measurement noises and biases, transmission errors, or transmission interruption after batteries run out. Therefore, relevant methods need to be used in order to cope with data uncertainty. A substantial body of work treating WSNs' data uncertainty at different levels is proposed in the literature, such as coping with spatial location uncertainty [14], uncertainty in real-time applications [15, 16], and high-volume data streams [17].

6.2.3   WSN-CPSs compared to WSNs

Some researchers have studied the differences between WSNs and WSN-CPSs, such as Wu *et al.* [2], according to whom CPSs are featured by sensor cooperation across multiple domains (cross-domain sensing), heterogeneous information flows, and intelligent decision/actuation.

From a data management perspective, we summarize the main differences between WSNs and WSN-CPSs in three aspects: mobility, intelligence requirements, and real-time constraint (Table 6.1).

Table 6.1 Comparing data management in WSNs and CPS-WSNs

Criterion	WSNs	CPS-WSNs
Mobility	Limited controlled and uncontrolled mobility, mainly for network coverage purposes	Controlled and uncontrolled mobility, typically mission-oriented. Both sensing and actuating devices can be mobile
Intelligence	Data-mining and knowledge discovery techniques are used to monitor the physical system by detecting high-level events of interest (often using complex event processing techniques)	There is a need to extract high-level knowledge from heterogeneous sensing networks/applications Reasoning capabilities are required to decide which actions need to be assigned to the actuation component in order to control the physical world
Real-time constraint	Important only for critical applications, such as environmental monitoring systems (e.g. tsunami monitoring)	More important because the CPS needs to react timely and control the physical system. Delays are not commonly tolerated

6.2.3.1 Mobility

Historically, WSNs were static, i.e. with only stationary sensing devices. Mobile devices (sensors and sink nodes) have been introduced later with both controllable and uncontrollable mobility patterns, mainly for the purpose of improving network coverage. Considering the requirement of controlling the physical system, and in conjunction with technological advances in robotics, embedded systems and mobile applications, sensor and actuator mobility became more important in CPSs, compared to WSNs [2]. Mobility is not only used to improve the network coverage, but is considered as one of the means of achieving the CPS mission. Therefore, and as discussed further in Section 6.4, mobility carries more weight in CPSs' data management. However, it also leads to more uncertainty [2] caused by high network coverage as a result of change of network topology [18].

6.2.3.2 Intelligence requirements

In their early stage, researches into WSNs focused on efficiently collecting and managing sensing data. More recently, research trends have focused on using knowledge discovery and data-mining techniques for the detection of high-level events from raw sensorial data [19], especially using complex event processing techniques [20, 21]. In contrast, intelligent decision systems are an essential component of CPSs. For CPSs it is therefore not enough to collect the data or even to detect high-level events (as in WSNs), but it is fundamental to transform the sensed data into high-level knowledge that can be used for reasoning and decision-making.

CPSs require to extract new spatio-temporal knowledge from heterogeneous sensor networks and application scenarios, and to utilize intelligence properly in order to react timely and control the physical worlds [2].

6.2.3.3 Real-time constraint

The main objective of CPSs is to react to the dynamic happenings and control the physical world using the actuators. The real-time constraint is therefore more important in WSN-CPSs than in WSNs. The real-time decision-making component of a CPS must receive, process, and analyze inputs from a continuous stream of data. This constraint is stronger for some delay-sensitive applications such as healthcare and emergency applications.

Therefore, data management in WSN-CPSs is driven by three main constraints: real-time, mobility, and data heterogeneity. In the following section, we review the existing WSN-CPS data management solutions with a focus on the above-mentioned constraints.

6.3 Data management activities

The common perception of data management is derived from a database abstraction where data management in WSNs mainly refers to data storage ('how to store data efficiently') and query processing ('how to achieve fast and accurate information retrieval') [22]. In this chapter we adopt a broader scope, and we consider that data management comprises four typical tasks: data collection, data processing, data storage, and data querying. Since WSN-CPSs are usually dealing with a huge volume of data, it is particularly interesting to highlight other supplementary activities which may improve the data management process in WSN-CPSs, namely, data compression and data analysis.

Before we present these data management activities, it is important to mention that according to the mobility perspective, we may distinguish between two types of WSN-CPSs: static WSN-CPSs and mobile *WSN-CPSs*. Static WSN-CPSs are built on stationary WSNs. The evolution of stationary WSNs, in conjunction with advances in distributed robotics and embedded systems, has led to the emergence of mobile wireless sensor networks (MSNs), which in turn has led later to the emergence of mobile cyber-physical systems (MCPS) [23]. A mobile CPS refers to a system composed of physical and computation components that can move in space and can communicate over a network. Mobile cyber-physical systems have several applications and can be typically built using MANET and/or WSN networks. As the scope of this book primarily concerns WSN-CPS, and for the sake of simplicity, we do not address MANET networks. (More details about the differences and similarities between MANETs and WSNs can be found in [24].) Mobile WSN-CPSs are cyber-physical systems enabled by the use of mobile WSNs and are constrained by the same energy and processing limitations as in static WSN-CPSs.

In the literature, data management activities are quite similar in both static and mobile WSN-CPSs, except for data collection. Consequently, the data management activities presented in the following subsection apply for both static and mobile

WSN-CPS, except for data collection, which we discuss from a mobile WSN-CPS perspective.

6.3.1   Mobile data collection

This activity is also called data gathering, data acquisition or data sampling. The main purpose of data collection is to find out which data need to be collected, where (from which sensor/location) and when (time) while minimizing the energy consumption of the network. It is 'the task responsible for efficiently acquiring samples from the sensors in a sensor network. The primary objective of the sensor data acquisition task is to attain energy efficiency' [3].

In general, data collection methods and techniques used in static WSN-CPSs are the same as those used in traditional static WSNs. Details about these methods can be found, for instance, in [3]. As mentioned before, in this section we focus on data collection in mobile WSN-CPSs.

Mobile WSN-CPSs are built on mobile WSNs. In general a WSN is considered to be mobile if at least one of its components (sensor nodes, sinks, or support nodes) is mobile. In this context, Di Franceso *et al.* [25] identified three 'mobile elements' (MEs) in a mobile WSN:

1. Relocatable nodes: they are mobile sensor nodes that change their location either to enhance the coverage and connectivity of the networks or for deployment purposes [2].
2. Mobile data collectors: they are mobile devices dedicated to collect data generated by the sensor nodes. Mobile data collectors can be either mobile sinks or mobile relays. We talk about direct data collection if mobile sinks are used to move and directly collect data from sensor nodes. We talk about indirect data collection in the case where mobile relays (MR) are used to collect, store, and send data to the sinks [26].
3. Mobile peers: they are mobile sensor nodes that can both sense data and relay messages in the network. A mobile peer moves around the sensing area to collect data from sensor nodes. Once communication with a sink is established, a mobile peer transfers the data it collected by itself as well as data it received from other peers.

In general, most data collection solutions seek to minimize energy consumption by finding a trade-off between the environmental sensing, communication, and processing of the sensors. Moreover, when using mobile elements, the energy consumption of the routing process can also be further reduced by minimizing the distance from data source to data sink. The mobile data collector can move around the sensing area and collects data from the source nodes using short-range communications. Other advantages are the extension of the network lifetime, thanks to removing hotspots close to static sink nodes as well as the increase of network reliability by removing the dependency on static sink nodes [27, 28].

However, mobility in WSNs creates new data management challenges that have not been addressed before. In the literature, the sink's mobility is the most widely addressed in the context of mobile WSNs. It was first proposed by Luo *et al.*

[29] and since then several mobile data collection strategies have been proposed. The proposed solutions differ according to several aspects, where the most explored are mobility controllability and moving speed.

Controllability is considered to be the most significant mobility aspect for data collection. Two broad classes of controllability are identified in the literature: controlled and uncontrolled mobility.

Controlled mobility refers to the fact that mobile elements, which are often mobile sinks, are autonomous, can decide about their trajectory and speed, and are therefore able to move from one location to another. Sink mobility is generally controlled according to objective functions [30] or heuristics that try to maximize/minimize certain parameters of interest such as nodes' residual energy [31, 32] and Emax [31, 32]. Other solutions explored the planning of efficient moving tours for the mobile sinks in order to collect data with the minimum energy consumption [26, 33]. Other works have modeled the moving process as a Markov chain and determined the moving path using a Markov decision process [34].

By contrast, uncontrolled mobility refers to the case where mobile elements do not control their own movement. There are several uncontrolled mobility patterns. The first pattern is Fixed mobility [35], also called deterministic mobility [25] or predictable mobility [29], in which the sink is programmed to follow a fixed path according to a round-robin algorithm. This fixed path is predetermined and cannot be changed on-the-fly by the WSN. The mobility pattern is characterized by regularity, which is used to predict where and when mobile sinks can be in contact with sensor nodes. This can happen in case the mobile sink is placed on a public transportation vehicle such as in [36]. Several data collection strategies have taken benefit of the ability to predict spatio-temporal contacts between mobile sinks and sensor networks. 'Scheduled rendezvous' is one of those strategies, in which mobile sinks agree with sensor nodes on a specific time for their contact. Sensors will thus wake up during the selected time and wait for the mobile sink to transfer their data, which allows for efficient energy-saving policy. A similar strategy is discussed in [37, 38], where the mobile sink informs a mobile relay (a kind of mobile data collector) about its future location. The mobile relay collects the data from sensors deployed in its area, and when the mobile sink arrives, data can be transferred immediately and the sink does not have to wait.

In random mobility pattern, the sink moves randomly in the sensor field with no guarantee that it will visit all the sensor nodes and with no idea about how much time it will take to collect data [35]. Randomness is the opposite of regularity, and it is therefore difficult to predict when a mobile sink will be in contact with a sensor node and how long the contact will hold. Stochastic mobility has been modeled as Poisson arrivals in [28] while random direction has been addressed in [39]. In the worst case, sensor nodes should continuously perform discovery in order to increase the chance of detecting contacts with mobile sinks. However, in the case where some knowledge is available on nodes' mobility patterns, discovery activity can be performed only during the moments and locations where the probability of the co-existence of mobile elements is high. Approaches to mobile sinks discovery based on reinforcement learning in the field of artificial intelligence have been addressed in [25, 40, 41]. Reinforcement-based approaches provide the advantage of flexibility; they are adaptable to different mobility models and scenarios without the need of *a priori* knowledge.

Regarding the speed of mobile elements, Zeinalipour-Yazti and Chrysanthis [4] categorized mobile CPSs into three classes: (1) highly mobile, if the system contains mobile elements that move at high velocities, such as elements placed on cars, airplanes, and others; (2) mostly static, in cases where mobile elements move at low velocities, and (3) hybrid, in cases where the system contains both high- and low-speed mobile elements. Even though some works pointed out that sinks moving with high speed may not have enough time to stay in contact with sensor nodes and collect all the data from them, the effect of speed on data collection has not been fully explored in the literature [42, 43] and more work needs to be done.

The proposed strategies depend also on other aspects such as the mobility scale (indoor micro-scale mobility vs. outdoor large-scale mobility), the communication type (single-hop vs. multi-hop communication [44, 45]), and data collection strategy (pull or push strategy). For example, in a pull strategy, a node transfers its data only when it receives a request from the sink [46], whereas in a push strategy nodes initiate the data transfer toward the sink [47, 48]. It has been demonstrated that single-hop data collection achieves the lowest energy consumption because data relaying is not required, which alleviates the sensors' load [46]. However, with a random mobility pattern, single-hop data collection can lead to incomplete data collection from the WSN because, as we mentioned so far, randomness does not guarantee that the sink will reach all the sensor nodes in the network. Further details can be found in [35].

6.3.2   Data processing

As sensors are usually deployed in harsh and remote environments, the sensing process is prone to several failures (communication failure, battery failure, etc.). Consequently, the sensed data are often erroneous, missing, and/or incomplete [3]. Data processing consists of a set of transformations which are applied to the raw data collected by the sensor nodes and prior to its storage. The main transformations are: data cleaning and data fusion.

Data cleaning is the process of removing errors from the sensed data and addressing incomplete data (for example, by using regression or interpolation models to rebuild the missing data). Data fusion is a different process and refers to the reconstruction and combination of data obtained or derived from several sensors in order to provide comprehensive and accurate information.

While data cleaning is a clear and well-understood process, there is still much confusion around the concept of data fusion or rather around the terminology. Elmenreich [49] has reported different terms which are used in the literature to refer to techniques, technologies, and systems which derive information from multiple sources. These terms include 'sensor fusion', 'data fusion', 'information fusion', 'multi-sensor data fusion', and 'multi-sensor integration'.

Although the terms 'information fusion' and 'data fusion' are often used as synonyms, data fusion is mainly employed for data directly obtained from the sensors (raw data) while information fusion is used more for already processed data [49]. This divergence in terminology may be actually justified. In fact, and as stated by Castanedo [50], the fusion process can be applied to at least five levels of 'data' where raw data represent the lower level and derived representations represent the highest one.

Data fusion (lowest level according to [50]) is thus about aggregating raw data coming from different sensors into a single packet. This processing has of course a certain overhead. However, and since it prevents sensors from sending redundant data, it considerably reduces the communication traffic within the network. Knowing that the cost of transmission is higher than computation [22], data fusion thus represents a neat advantage for the network. Practically, data fusion also requires that the network is organized into clusters. Each cluster has at least a particular node called aggregator which is in charge of sending the aggregated data to the sink. The aggregators are more 'intelligent' nodes which can compute functions of modest complexity and their output (aggregated data) often depends on the corresponding aggregate query imposed by the central server [22].

6.3.3 Data storage

Since sensors are operating with limited resources (memory, bandwidth, energy), the large amount of collected data has to be efficiently stored [51]. This will facilitate user queries and data retrieval in general. To achieve this goal, two questions have to be answered: (1) Which data must be stored? (2) Where to store data?

The first question can be reformulated as follows: Do all the data collected by the sensor network need to be stored? The answer depends actually on the type of application. For example, in real-time applications where only recent measurements are important [52], collected data are often useful only within a small period of time after being collected. In such cases, a live (or streaming) approach is more adequate, where only recent data are stored. Nevertheless, in scientific applications for instance, all data are important regardless of the time where they have been collected [53]. In this case, old data cannot be discarded. This approach is called historical storage approach.

Combining both data storage approaches would make the system more flexible and adaptable to different applications. Nevertheless, in a large-scale context, which is a typical characteristic of many WSN-CPSs, this combination is very challenging [52].

Regarding the second question (where to store data), different approaches are also possible. In a centralized approach, all data collected by sensors are immediately transmitted to a base station. This approach, based on a communication-centric view of the network, suffers from two main problems. First, as networks become larger in size, nodes near the sink become congested and bottlenecks appear. Second, and since data transmission is the process which consumes the most energy, this centralized approach turns out to be not energy efficient.

As on-board storage is becoming cheaper and consuming less energy than before, sensor networks have slowly shifted from a communication-centric perspective to a storage-centric perspective where the primary objective of the network is to store the sensed data [53, 54]. This data can be retrieved later on or batched. Batching sensed data may considerably improve energy efficiency [54, 55]. However, since data is not immediately transmitted, any application running over a storage-centric sensor network must be delay-tolerant. Conversely, critical applications such as fire detection systems or industrial control systems are very sensitive to any data delay. Consequently, a communication-centric approach is more convenient in this case.

Data storage methods and techniques in sensor networks have been extensively studied [56], therefore for the sake of simplicity we do not discuss the further here.

6.3.4   Data querying

Optimizing data storing is a prerequisite for another process which is also crucial in sensor data management: data querying. One main objective of data querying is accessing/generating minimal amount of data [57, 58]. Many model-based techniques are available to deal with this objective. Some approaches use models for creating an abstraction layer over the sensor network [59, 60] while others rely on a hidden Markov model (HMM) [58] or dynamic probabilistic models (DPMs) for modeling spatio-temporal evolution of the sensor data [60]. More approaches are discussed in [3].

In the particular case of WSN-CPSs, data querying can be thought as an information flow processing system [61]. The term 'information flow processing' (IFP) refers to 'an application domain in which users need to collect information produced by multiple, distributed sources, to process it in a timely way, in order to extract new knowledge as soon as the relevant information is collected' [61]. In general, collected data in an IFP system are discarded after being processed. Except from complex applications, which require historical analysis, storing already processed data is not important. According to [61], there are two competing models today: the data stream processing model [62] and the complex event processing model [63].

Data stream management systems (DSMSs) derive from traditional database management systems (DBMSs) while being substantially different. Unlike DBMSs, which deal with persistent data and run queries which are only executed once, DSMSs deal with transient data that are continuously updated and run standing queries. Although some advanced and complete models of DSMSs have been proposed, DSMSs cannot entirely fulfill the IFP requirements [61]. Indeed, detection and notification of complex patterns of elements are not supported by these systems.

Conversely, the complex event processing (CEP) model, often seen as an extension to the traditional publish–subscribe domain [64], aims at detecting low-level patterns of events. These patterns have then to be filtered and combined in order to understand what is happening in terms of higher-level events and notify interested parties. CEP systems follow DAHP (database active, human passive) model, in which a system does continuous processing and notifies the user. Finally, CEP is regarded as an event-based system, which makes it more convenient for CPSs. Indeed, events are a 'natural way' to address reactive systems such as CPSs [65]. Events deal with interactions between components and observations rather than internal state, which allows specification and reasoning at higher levels while integrating easily with more detailed information [65]. Talcott *et al.* [66] enumerate other reasons for which CEPs would be more appropriate for CPSs, such as the causal partial order property [65].

6.3.5 Data compression

Sensors sometimes generate a huge quantity of data, which makes storage and query processing more demanding within the CPS. For example, sensors monitoring each single Boeing jet engine collect around 20 terabytes of data per hour of flight [67]. Although compressing data is a challenging process, it may considerably reduce this high volume. Many compressing techniques have been already proposed, such as techniques derived from signal processing [68, 69]. Other techniques use correlations within sensor data to compress data streams [70–72].

6.3.6 Data analysis

A cyber-physical system relies on sensor networks to monitor the real-world environment. While most recorded data are often normal, some may be abnormal (e.g. congestion in a traffic system or a very high temperature in a monitored forest). Such data, also called atypical data [1], are of high importance since they may indicate a change in the monitored environment. Analyzing atypical data is thus a crucial process within any CPS. However, many reported atypical data are erroneous. Many methods have therefore been proposed to assess atypical data in different CPSs [1, 73]. Once atypical data are identified as trustworthy data, they need to be analyzed with spatial, temporal, and other multidimensional information in order to infer the event behind this abnormality. Although this topic is still challenging, some efficient methods have already been proposed in some specific CPSs. For example, in traffic congestion management, a model of atypical cluster was proposed [74, 75] to retrieve traffic congestion events from massive data.

6.4 Cyber-physical cloud computing: opportunities and challenges

CPSs are often deployed for applications which are complex, large-scaled, and data-intensive. This makes data management processes deal with complex requirements and constraints. For example, CPS-based military applications need top-level security which may not be provided using normal storing and querying infrastructure. In addition, sensors collecting these data are constrained in terms of energy, processing, and storage capacity. The larger a WSN-CPS – which is often the case – the more these sensors become unable to store or process the resulting large amount of data. Nevertheless, these issues may be overcome by using cloud computing techniques [76]. In this section we discuss the potential of using cloud computing to support CPSs from the data management perspective. We also present the opportunities which may result from this integration as well as the challenges which have to be addressed.

Recent and promising research is slowly rendering cloud computing technologies as the computational backbone of CPSs [9]. Cloud computing is a service that offers computing, storage, networking, and software 'as a service' [77]. A new vision of cyber-physical cloud computing (CPCC) has recently emerged [78]. A CPCC is defined as 'a system environment that can rapidly build, modify and provision cyber-physical systems composed of a set of cloud computing based sensor, processing, control, and data services'. In other simpler words, a CPCC is a global intelligent system which collects data from the real world and understands the data according to its current context. CPCC aims at realizing a smart environment which is able to rapidly build and modify CPS composed of cloud services (see Figure 6.2). Examples of applications that CPCC can support include smart healthcare applications, smart disaster information systems, smart city applications, and social life networks.

CPSs are very complex systems in terms of structure and behavior, which may raise many challenges that researchers have to address so that applications can take full advantage of these systems. Despite the promise held by cloud computing, a lot of effort will be required in order to achieve the CPCC vision. Indeed, making the

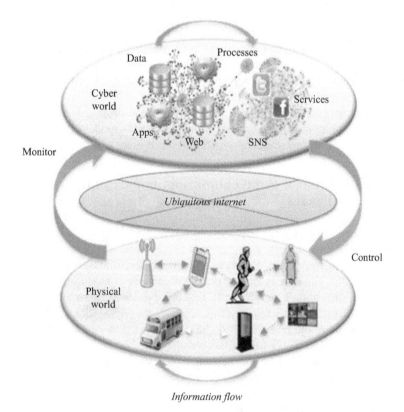

Figure 6.2   Cyber-physical cloud computing concept [77]

cloud support the real-time, safety, and reliability requirements of CPSs is not an easy task. In what follows, we briefly present the main challenges to be addressed.

6.4.1   Real time

Many mobile phone-based CPS services have been successful. Data from such applications can be perceived as large-volume continuous streaming data, which implies that they need to be stored and processed in a real-time manner [2].

Mobile-based CPS services, for instance, strongly rely on real-time operation. The CPS needs to process a constant stream of data from the sensors and often and at a very large volume [2]. Since accurate and fast actions are required in the real world, this data stream has to be efficiently associated to existing knowledge [77]. In particular, storing, processing, and data interpretation have to be achieved in real time. In a cloud-computing context, this is even more challenging. In order to allow a CPCC to operate in real time, delay between the sensing process and the actuating process must be optimized to the minimum possible duration [77].

6.4.2   Big Data

Thanks to low cost and ubiquitous sensors, the Big Data paradigm is clearly influencing CPS while bringing new challenges and opportunities [79]. When combined with cloud computing, Big Data has proven to be very successful in offloading most computation and data storage from both data centers and terminal devices [80]. As cloud computing looks very promising in a CPS context (i.e. CPCC), managing Big Data is inevitable [81]. Nevertheless, this is not an easy task. Indeed, while sensors are collecting data in the real world, the cloud shifts part of the data storage and computation to the edge of the network. This raises high requirements including real-time latency and response and mobility support [80]. In order to address these challenges, researchers have to focus on manipulating Big Data in ubiquitous WSN. In particular, they have to address many concerns about storage, data analysis and mining, and distributed algorithms for data placement [80].

6.4.3   Data mining

Since data in CPS may be uncertain and change dynamically, learning and data-mining tools will be very useful for retrieving useful knowledge by studying the temporal and spatial correlations of sensed data [2].

Moreover, when applied to CPS Big Data, data-mining and machine-learning techniques can help identify data patterns [79], learn context, and detect complex distributed events of interest. For instance, Das *et al.* [82] proposed a pattern-based approach to detect anomalies in the Big Data of commercial aircraft [82]. As CPS data can be easily converted to symbols (e.g. by using temporal abstractions [83]), different efficient algorithms [84] for learning patterns from symbolic data have been proposed by data-mining researchers. According to [79], when combined to traditional CPS modeling, pattern mining techniques are able to extract intelligence from monitoring data of CPS.

6.4.4   Data integration

When a CPS is supported by the cloud, it may solve issues related to data storage and computation but may raise other challenges. Indeed, it is not just a matter of data volume only. The CPCC will end up gathering, combining, and interpreting different and heterogeneous types of data (different units, uncertainties, structures, and semantics). Researchers have then to address several data management issues [77] including unifying data representation and processing models, ensuring intelligent data interpretation and semantic interoperability, and applying sampling and filtering techniques to improve data quality.

6.4.5   Load balancing

As cloud computing is becoming a mature technology, it is more likely that the concept will be widely adopted by CPS [2]. This will solve the storage, computing, and communication issues. Users can also interact with the part of CPS applications in which they are interested, instead of overloading the whole system. A new research direction would then be to study the load balancing for CPS over cloud platforms [2].

There are many other challenges that we do not discuss in this section, mainly because they are related to CPS in general and not to data management in particular. Privacy and security are perhaps two of the most important. Security and privacy are vital concerns especially for applications such as in healthcare or in military. Data can be confidential from a legal and/or ethical perspective. Consequently, special attention needs to be paid to ensure data security and confidentiality while designing CPS applications [9].

6.5   Conclusion

In this chapter we have reviewed what has been researched in WSN data management and how it drives the progress of CPS toward more intelligent applications. We have also highlighted important factors to add intelligence to the data management process, which may lead to the success of CPS technology.

Although CPS technology is still considered new, many opportunities are available to bring this technology to a more mature stage. CPCC is probably the framework to watch closely in the near future. Indeed, CPCC is supporting CPS with two successful paradigms: cloud computing and Big Data management. CPCC has also opened up new research directions in terms of real-time processing, data query, storage, and security.

References

[1]   L. A. Tang, X. Yu, S. Kim, J. Han, C. C. Hung and W. C. Peng. Tru-alarm: Trustworthiness analysis of sensor networks in cyber-physical systems. In *Proceedings of the 10th IEEE International Conference on Data Mining (ICDM '10)*, pp. 1079–1084, Sydney, Australia, December 2010.

[2]   F. J. Wu, Y. F. Kao and Y. C. Tseng. From wireless sensor networks towards cyber physical systems. *Pervasive and Mobile Computing*, 7(4): 397–413, August 2011.

[3]   S. Sathe, T. G. Papaioannou, H. Jeung and K. Aberer. A survey of model-based sensor data acquisition and management. In C. C. Aggarwaal (ed.) *Managing and Mining Sensor Data*. IBM T. J. Watson Research Center, Yorktown Heights, NY, USA, 2013.

[4]   D. Zeinalipour-Yazti and P. K. Chrysanthis. Mobile sensor network data management. In Ling Liu and M. Tamer Özsu (eds.) *Encyclopedia of Database Systems*. Springer, New York, NY, 2009, pp. 1755–1759.

[5]   M. Mokashi and A. S. Alvi. Data management in wireless sensor network: A survey. *International Journal of Advanced Research in Computer and Communication Engineering*, 2(3): March 2013.

[6]   S. Madden. Database abstractions for managing sensor network data. *Proceedings of the IEEE*, 98(11): 1879–1886, 2010.

[7]   B. Krishnamachari. Modeling data gathering in wireless sensor networks. In Y. Li, M. Thai and W. Wu (eds.) *Wireless Sensor Networks and Applications*, Section III Data Management, pp. 572–591, 2005.

[8]   A. Jacquot, J.-P. Chanet, K. M. Hou, G. De Sousa and A. Monier. A new management method for wireless sensor networks. In *Proceedings of the 9th IEEE IFIP Annual Mediterranean Ad Hoc Networking Workshop*, Juan les Pins, France, 2010.

[9]   S. A. Haque, S. M. Aziz and M. Rahman. Review of cyber-physical systems in healthcare. *International Journal of Distributed Sensor Networks*, Article ID 217415, 20 pp., 2014, doi: 10.1155/2014/217415.

[10]   J. Gupchup J. *Data Management in Environmental Monitoring Sensor Networks*. PhD dissertation, Johns Hopkins University, Baltimore, MD, USA, 2012.

[11]   I. Akyildiz, W. Su, Y. Sankarasubramaniam and E. Cayirci. A survey on sensor networks. *IEEE Communications Magazine*, 40(8): 102–114, August 2002.

[12]   Y.-W. Ma, J.-L. Chen, Y.-M. Huang and M.-Y. Lee. An efficient management system for wireless sensor networks. *Sensors*, 10: 11400–11413, 2010, doi: 10.3390/s101211400.

[13]   S. O. Dulman. *Data-Centric Architecture for Wireless Sensor Networks*. PhD dissertation, University of Twente, the Netherlands, 2005.

[14]   M. Kadkhoda and M.-R. Akbarzadeh-T. Uncertainty handling for sensor location estimation in wireless sensor networks using probabilistic fuzzy system. *International Journal of Artificial Intelligence and Applications for Smart Devices*, 1(1): 1–14, 2013.

[15]   M. A. Osborne, S. J. Roberts, A. Rogers and N. R. Jennings. Real-time information processing of environmental sensor network data using Bayesian Gaussian processes. *ACM Transactions on Sensor Networks (TOSN)*, 9 (1): November 2012.

[16]   T. Palpanas. Real-time data analytics in sensor networks. *Managing and Mining Sensor Data*, 173–210, 2013.

[17]   Y. Diao, B. Li, A. Liu, L. Peng and C. Sutton. Capturing data uncertainty in high-volume stream processing. In *Proceedings of the 4th Biennial Conference on Innovative Data Systems Research (CIDR),* Asilomar, CA, USA, 4–7 January 2009.

[18]   C. Y. Lin, S. Zeadally, T.-S. Chen and C.-Y. Chang. Enabling cyber physical systems with wireless sensor networking technologies. *International Journal of Distributed Sensor Networks,* 2012, Article ID 489794, 21 pp., doi: 10.1155/2012/489794.

[19]   J. Gupchup, R. Burns, A. Terzis and A. Szalay A. Model-based event detection in wireless sensor networks. In *Proceedings of the Workshop for Data Sharing and Interoperability on the World Wide Web (DSI 2007),* April 2007.

[20]   O. Saleh. Complex event processing in wireless sensor networks. In *Proceedings of the 25th GI-Workshop on Foundations of Databases (Grundlagen von Datenbanken),* Ilmenau, Germany, 28–31 May 2013.

[21]   M. Stocker, M. Ronkko and M. Kolehmainen. Abstractions from sensor data with complex event processing and machine learning. In *Proceedings of the 7th International Environmental Modelling and Software Society (iEMSs) Congress on Environmental Modelling and Software,* San Diego, CA, USA, D. P. Ames, N. W. T. Quinn and A. E. Rizzoli (eds.), 2014.

[22]   C. C. Aggarwal. *Managing and Mining Sensor Data.* Springer, 2013, ISBN 978-1-4614-6308-5.

[23]   X. Hu, T. H. S. Chu, H. Chan and V. Leung. Vita: A crowdsensing-oriented mobile cyber-physical system. *IEEE Transactions on Emergings Topics in Computing,* 1(1): June 2013.

[24]   J. Garcia-Macias and J. Gomez. Manet versus WSN. In *Sensor Networks and Configuration.* Springer, 2008, pp. 369–388.

[25]   M. Di Francesco, S. K. Das and G. Anastasi. Data collection in wireless sensor networks with mobile elements: A survey. *ACM Transactions on Sensor Networks (TOSN),* 8(1), August 2011.

[26]   S. Jain, R. C. Shah, W. Brunette, G. Borriello and S. Roy. Exploiting mobility for energy efficient data collection in wireless sensor networks. *Mobile Networks and Applications,* 11(3): 327–339, 2006.

[27]   R. C. Shah, S. Roy, S. Jain and W. Brunette. Data mules: Modeling a three-tier architecture for sparse sensor networks. In *Proceedings of the IEEE SNPA Workshop,* 2003.

[28]   A. A. Somasundara, A. Kansal, D. D. Jea, D. Estrin and M. B. Srivastava. Controllably mobile infrastructure for low energy embedded networks. *IEEE Transactions on Mobile Computing,* 5(8): 958–973, August 2006.

[29]   J. Luo, J. Panchard, M. Piorkowski, M. Grossglauser and J.-P. Hubaux. Mobiroute: Routing towards a mobile sink for improving lifetime in sensor networks. In *Proceedings of the IEEE International Conference on Distributed Computing in Sensor Networks (DCOSS),* pp. 480–497, 2006.

[30]   M. M. Mudigonda, T. Kanipakam, A. M. Dutko, M. Bathula, N. Sridhar, S. Seetharaman *et al.* A mobility management framework for optimizing the

trajectory of a mobile base-station. In *Proceedings of the 8th European Conference on Wireless Sensor Networks*, pp. 23–25, 2011.

[31] Y. Bi, L. Sun, J. Ma, N. Li, I. A. Khan and C. Chen. HUMS: An autonomous moving strategy for mobile sinks in data-gathering sensor networks. *EURASIP Journal on Wireless Communications and Networking*, 15: 2007.

[32] S. Basagni, A. Carosi, E. Melachrinoudis, C. Petrioli and Z. M. Wang. Controlled sink mobility for prolonging wireless sensor networks lifetime. *Journal of Wireless Networks*, 14: 831–858, 2008.

[33] G. Xing, M. Li, T. Wang, W. Jia and J. Huang. Efficient rendezvous algorithms for mobility-enabled wireless sensor networks. *IEEE Transactions on Mobile Computing*, 11(1): 47–60, 2012.

[34] X. Fei, A. Boukerche and R. Yu. An efficient Markov decision process based mobile data gathering protocol for wireless sensor networks. In *Proceedings of the IEEE Wireless Communications and Networking Conference (WCNC '11)*, pp. 1032–1037, March 2011.

[35] M. I. Khan, W. N. Gansterer and G. Haring. Static vs. mobile sink: The influence of basic parameters on energy efficiency in wireless sensor networks. *Computer Communications*, 36.9: 965–978, 2013. *PMC*. Web. 4 January 2015.

[36] A. Chakrabarti, A. Sabharwal and B. Aazhang. Using predictable observer mobility for power efficient design of sensor networks. In *Proceedings of the 2nd International Workshop on Information Processing in Sensor Networks (IPSN 2003)*, 129–145, 2003.

[37] A. Giannakos, G. Karagiorgos and I. Stavrakakis. A message-optimal sink mobility model for wireless sensor networks. In *Proceedings of the 8th International Conference on Networks*, pp. 287–291, 2009.

[38] Y. Wu, L. Zhang, Y. Wu and Z. Niu. Interest dissemination with directional antennas for wireless sensor networks with mobile sinks. In *Proceedings of the 4th International Conference on Embedded Networked Sensor Systems, SenSys '06*, Boulder, Colorado, USA, pp. 99–111, 2006.

[39] S. Poduri and G. S. Sukhatme. Achieving connectivity through coalescence in mobile robot networks. In *Proceedings of the 1st International Conference on Robot Communication and Coordination (RoboComm 2007)*, pp. 1–6.

[40] L. P. Kaelbling, M. L. Littman and A. W. Moore. Reinforcement learning: A survey. *Journal of Artificial Intelligence Research*, 4: 237–285, 1996.

[41] V. Dyo and C. Mascolo. Efficient node discovery in mobile wireless sensor networks. In *Proceedings of the 4th IEEE International Conference on Distributed Computing in Sensor Systems (DCOSS)*, pp. 478–485, 2008.

[42] A. Kansal, A. Somasundara, D. Jea, M. Srivastava and D. Estrin. Intelligent fluid infrastructure for embedded networks. In *Proceedings of the 2nd ACM International Conference on Mobile Systems, Applications, and Services (MobiSys 2004)*, pp. 111–124, 2004.

[43] G. Anastasi, M. Conti, E. Gregori, C. Spagoni and G. Valente. Motes sensor networks in dynamic scenarios: An experimental study for pervasive

applications in urban environments. *International Journal of Ubiquitous Computing and Intelligence*, 1, 1.

[44]   S. Basagni, A. Carosi, E. Melachrinoudis, C. Petrioli and M. Z. Wang. Protocols and model for sink mobility in wireless sensor networks. *ACM Mobile Computing and Communication Review*, 10(4): 28–30.

[45]   S. Basagni, A. Carosi and C. Petrioli C. Controlled vs. uncontrolled mobility in wireless sensor networks: Some performance insights. In *Proceedings of the 66th Vehicular Technology Conference*, pp. 269–273, 2007.

[46]   I. Chatzigiannakis, A. Kinalis and S. Nikoletseas. Sink mobility protocols for data collection in wireless sensor networks. In *Proceedings of the International Workshop on Mobility Management and Wireless Access, Mobi-Wac'06*, Terromolinos, Spain, pp. 52–59, 2006.

[47]   F. Yu, E. Lee, S. Park and S. Kim. A simple location propagation scheme for mobile sink in wireless sensor networks. *IEEE Communications Letters*, 14: 321–323, 2010.

[48]   Y. S. Yun and Y. Xia Y. Maximizing the lifetime of wireless sensor networks with mobile sink in delay-tolerant applications. *IEEE Transactions on Mobile Computing*, 9: 1308–1318, 2010.

[49]   W. Elmenreich, W. *Sensor Fusion in Time-Triggered Systems*. PhD thesis, Technische Universität Wien, Institut für Technische Informatik, Vienna, Austria, 2002.

[50]   F. Castanedo. A review of data fusion techniques. *The Scientific World Journal*, 2013, Article ID 704504, 19 pp., 2013.

[51]   K. Xing, X. Cheng and J. Li. Location-Centric Storage for Sensor Networks. *MASS* 2005.

[52]   L. Petit, A. Nafaa and R. Jurdak. Historical data storage for large scale sensor networks. In *Proceedings of the 5th French-Speaking Conference on Mobility and Ubiquity Computing (UbiMob '09)*, pp. 45–52, ACM, New York, NY, USA, 2009.

[53]   Y. Diao, D. Ganesan, G. Mathur and P. Shenoy. Rethinking data management for storagecentric sensor networks. In *Proceedings of the 3rd Biennial Conference on Innovative Data Systems Research (CIDR'07)*, Asilomar, CA, USA, pp. 22–31, 7–10 January 2007.

[54]   P. Dutta, D. Culler and S. Shenker. Procrastination might lead to a longer and more useful life. In *Proceedings of HotNets-VI*, Atlanta, GA, USA, November 2007.

[55]   G. Mathur, P. Desnoyers, D. Ganesan and P. Shenoy. Ultra-low power data storage for sensor networks. In *Proceedings of IPSN/SPOTS'06*, Nashville TN, USA, April 2006.

[56]   J. Yick, B. Mukherjee and D. Ghosal. Wireless sensor network survey. *Computer Networks*, 52(2008): 2292–2330, 2008.

[57]   A. Thiagarajan and S. Madden. Querying continuous functions in a database system. In *Proceedings of SIGMOD*, pp. 791–804, 2008.

[58]   A. Bhattacharya, A. Meka and A. Singh. MIST: Distributed indexing and querying in sensor networks using statistical models. In *VLDB*, pp. 854–865, 2007.

[59]   A. Deshpande and S. R. Madden. MauveDB: Supporting model-based user views in database systems. In *SIGMOD*, pp. 73–84, 2006.

[60]   B. Kanagal and A. Deshpande. Online filtering, smoothing and probabilistic modeling of streaming data. In *ICDE*, pp. 1160–1169, 2008.

[61]   G. Cugola and A. Margara. Complex event processing with T-REX. *Journal of Systems and Software*, 85(8): 1709–1728, 2012.

[62]   B. Babcock, S. Babu, M. Datar, R. Motwani and J. Widom. Models and issues in data stream systems. In *Proceedings of PODS*, pp. 1–16, ACM, New York, NY, USA.

[63]   D. C. Luckham. *The Power of Events: An Introduction to Complex Event Processing in Distributed Enterprise Systems.* Addison-Wesley Longman, Boston, MA, USA.

[64]   P. Eugster, P. Felber, R. Guerraoui and A.-M. Kermarrec. The many faces of publish/subscribe. *ACM Computing Surveys (CSUR)*, 35(2): 114–131.

[65]   C. Talcott. Cyber-physical systems and events. In *Software-Intensive Systems and New Computing Paradigms, LNCS*, 5380, pp. 101–115, Springer, 2008.

[66]   C. Talcott, M. Wirsing *et al.* Cyber-physical systems and events. In *Software-Intensive Systems*, pp. 101–115, Springer, 2008.

[67]   A. B. Zaslavsky, C. Perera and D. Georgakopoulos. Sensing as a service and Big Data. In *Proceedings of the International Conference on Advances in Cloud Computing (ACC)*, Bangalore, India, July 2012.

[68]   R. Agrawal, C. Faloutsos and A. Swami. Efficient similarity search in sequence databases. In *Foundations of Data Organization and Algorithms*, pp. 69–84, 1993.

[69]   Y. Zhu and D. Shasha. StatStream: Statistical monitoring of thousands of data streams in real time. In *VLDB*, pp. 358–369, 2002.

[70]   S. Gandhi, S. Nath, S. Suri and J. Liu. GAMPS: Compressing multi sensor data by grouping and amplitude scaling. In *SIGMOD*, pp. 771–784, 2009.

[71]   L. Wang and A. Deshpande. Predictive modeling-based data collection in wireless sensor networks. In *Proceedings of EWSN*, pp. 34–51, 2008.

[72]   A. Arion, H. Jeung and K. Aberer. Efficiently maintaining distributed model-based views on real-time data streams. In *Proceedings of GLOBE-COM*, pp. 1–6, 2011.

[73]   X. Yu, L.-A. Tang and J. Han J. Filtering and refinement: A two-stage approach for efficient and effective anomaly detection. In *Proceedings of the 9th IEEE International Conference on Data Mining*, Miami, FL, USA, 2009.

[74]   L. A. Tang, X. Yu, S. Kim and J. Han. Multidimensional Analysis of Atypical Events in Cyber-Physical Data, 2012.

[75]   L. A. Tang, X. Yu, S. Kim, J. Han, W. C. Peng, Y. Sun, A. Leung and T. La Porta. Multidimensional sensor data analysis in cyber-physical system: An atypical cube approach. *International Journal of Distributed Sensor Networks*, Article ID 724846, 19 pp., 2012.

[76]   S. K. Dash, S. Mohapatra and P. K. Pattnaik. A survey on applications of wireless sensor network using cloud computing. *International Journal of Computer Science & Emerging Technologies*, 1(4): December 2010.

[77] R. Buyya, C. S. Yeo, S. Venugopal, J. Broberg and I. Brandic. Cloud com-
      puting and emerging IT platforms: Vision, hype, and reality for delivering
      computing as the 5th utility. *Future Generation Computer Systems*, 25(6):
      599–616, 2009.

[78] E. D. Simmon, K. Kim, E. Subrahmanian, R. Lee, F. de Vaulx, Y. Mur-
      akami, K. Zettsu and R. D. Sriram. *A Vision of Cyber-Physical Cloud
      Computing for Smart Networked Systems*. NIST Interagency/Internal Report
      (NISTIR) 7951, August 2013.

[79] A. B. Sharma, A. Niculescu-Mizil, H. Chen and G. Jiang. Modeling and
      analytics for cyber-physical systems in the age of big data. *SIGMETRICS
      Performance Evaluation Review*, 41(4): 74–77, April 2014.

[80] F. Xiao, C. Zhang and Z. Han Z. Big Data in ubiquitous wireless sensor
      networks. *International Journal of Distributed Sensor Networks*, 2014,
      Article ID 781729, 2 pp., 2014, doi: 10.1155/2014/781729.

[81] E. Gelenbe and F.-J. Wu. Future research on cyber-physical emergency
      management systems. *Future Internet*, 5(3): 336–354.

[82] S. Das, B. L. Matthews, A. N. Srivastava and N. C. Oza. Multiple kernel
      learning for heterogeneous anomaly detection: Algorithm and aviation safety
      case study. In *Proceedings of the KDD*, 2010.

[83] I. Batal, D. Fradkin, J. Harrison, F. Moerchen and M. Hauskrecht M. Mining
      recent temporal patterns for event detection in multivariate time series data.
      In *Proceedings of the KDD*, 2012.

[84] J. Han, M. Kamber and J. Pei. *Data Mining: Concepts and Techniques*, 3rd
      edn. Morgan Kaufmann, 2011.

Chapter 7

Routing in wireless sensor networks for cyber-physical systems

Wenjia Li[1], Kewei Sha[2] and Sherali Zeadally[3]

Abstract

Cyber-physical systems (CPSs) and cyber infrastructures are key elements of national infrastructures, and they generally involve a tight coupling of cyber and physical components. CPSs can provide a variety of mission-critical services, such as pervasive health care, smart electricity grid, green cloud computing, and surveillance with unmanned aerial vehicles. CPSs generally include a set of sensors which are connected wirelessly. Thus, wireless sensor network (WSN) is an important enabling technology which supports the deployment of CPS. In WSN, sensor nodes have a limited transmission range, and their processing and storage capabilities as well as their energy resources are also limited. Therefore, routing protocols for WSN have played a key role in ensuring reliable multi-hop communication under these conditions. In this chapter, we present a survey of the state of the art on routing techniques in WSNs for CPS. We first outline the design challenges for routing protocols in WSNs followed by a comprehensive survey of different routing techniques. We also highlight the advantages and performance issues of each routing technique. Finally, we discuss some challenges and future research directions in the area of routing in WSNs for CPS.

7.1 Introduction

As the world has become more developed, industrialized, and globalized, its reliance on critical physical and cyber infrastructures continues to increase. This infrastructure includes many systems such as: (1) electrical power generation and distribution, (2) roads, bridges, and tunnels that make up our ground transportation

[1]Department of Computer Science, New York Institute of Technology, New York, NY, USA, e-mail: wli20@nyit.edu

[2]School of Science and Computer Engineering, University of Houston, TX, USA, e-mail: sha@uhcl.edu

[3]College of Communication and Information, University of Kentucky, Lexington, KY, USA, e-mail: szeadally@uky.edu

system, (3) airports and air traffic control supporting airline transportation, (4) communication networks including both wired and wireless, (5) systems for storing and distributing water and food supplies, (6) medical and healthcare delivery systems, and (7) financial, banking, and commercial transaction assets. A common theme in these systems is the role played by the underlying physical environment. The physical environment provides the information necessary for achieving many important functionalities of the above critical infrastructures, such as real-time road traffic status for avoiding traffic jams in road transportation systems. Systems that collect information from the physical environment, make decisions based on the collected information, and affect operations within the physical environment, are called cyber-physical systems (CPSs).

In general, a CPS, as its name suggests, is an integration of physical and computational elements to form a situation-aware system that responds intelligently to dynamic changes of real-world scenarios. In CPS, data is collected from various physical sensors, and then they are analyzed in a real-time manner. CPSs are widely used in both civilian and military scenarios, such as smart grid [1], traffic monitoring [2], and battlefield surveillance [3].

In CPSs, sensors, actuators, and other embedded devices are networked wirelessly to sense, monitor, and control the physical world. In contrast to traditional embedded systems, the CPS is a network of interacting appliances with physical inputs and outputs instead of standalone devices. In this configuration, wireless sensor network (WSN) is being considered as a key enabling technology for the successful deployment of CPS.

CPSs have a set of characteristics that make the already challenging tasks of data routing even more difficult. We summarize these characteristics as follows:

• **Unreliable and error-prone communication media:** Radio frequency (RF) signal is the only feasible transmission medium in many application scenarios of CPS, such as traffic monitoring and battlefield surveillance. The open nature of the RF signal makes the transmitted data highly susceptible to both transmission errors and intentional tampering. For instance, false alarms may be triggered by the sensors in battlefield CPSs if a rat happens to pass by the sensor and interferes with the communication channel or the data communication may be tampered with by adversaries [4, 5].

• **Rapid change of network topology:** In many CPS application scenarios, sensors may move constantly, which leads to rapid changes in the network topology. For example, wearable sensors are deployed around patients in body area networks to support various telehealth applications, such as remote patient monitoring and robotic surgery [6–8]. As a result, the network topology dynamically changes with the movements of patients.

• **Heterogeneity of sensor data:** In the CPS, data is sensed and reported from multiple types of devices. For example, there are multiple sensing and reporting devices in the advanced meter infrastructure (AMI) for smart grid application [9], such as electricity meters, gas meters, heat meters, and water meters. Since these meters sense and report different types of data, it is not feasible to process the data in one single dimension. For instance, we cannot simply apply the majority voting mechanism to find the truth from the sensor data because this comes from various types of sensors and has different formats.

- **Real-time constraint:** Some delay-sensitive CPS applications require data to be handled in a real-time manner. If this constraint cannot be met, then such delay-sensitive applications might become unreliable and unusable. For example, an intelligent transportation system is such a delay-sensitive CPS application [2].
- **Inconsistency of sensor data:** A well-deployed CPS should allow reasonable redundancies of sensors to ensure data availability. However, inconsistent data reports occur between reliable sensors and faulty/compromised sensors. Therefore, the truth needs to be inferred from inconsistent data, and the faulty or compromised sensors need to be identified so that the data from them can be better interpreted.

A WSN contains hundreds or thousands of sensor nodes. The sensors can communicate either among themselves or directly with an external base station (BS). In addition, sensors need to occasionally relay data to other sensors because of the limited transmission range, which naturally introduces the necessity of multi-hop communication. Figure 7.1 shows the schematic diagram of a WSN. We can clearly see from this figure that each sensor node is composed of several functional modules, such as sensors, micro-controller, transceiver, and power sources. Each of these scattered sensor nodes has the capability to collect and route data either to other sensors or back to an external BS. A BS can either be a fixed node or a mobile node which is capable of connecting the sensor nodes to an existing communication infrastructure or to the internet where a user can have access to the reported data.

It is well understood that WSN plays an important role in various military and civilian scenarios, such as situational awareness in the battlefield [10, 11], habitat monitoring for endangered animal species [12, 13], and volcano eruption monitoring [14, 15]. In these various applications, data is collected by multiple sensors and then processed to monitor specific events (such as enemy movement, animal migration trajectory, volcano activity). For instance, during the search and rescue following natural disasters, such as earthquake, hurricane, and wildfire, a huge amount of sensors can be scattered by helicopters or even unmanned aerial vehicles (UAVs) and then networked with others to help locate survivors, identify hazardous areas, and coordinate search/rescue efforts among different teams. As for the WSN application to a military scenario, the interconnected sensors can significantly reduce the personnel involvement in the normally dangerous reconnaissance and patrol missions. In addition, networked sensors can also help identify intruders to the military base.

Although sensors are used in a wide range of application domains, sensor nodes are usually constrained in various resources, such as battery supply, memory space, and communication bandwidth. Such resource constraints, combined with a typical deployment of large numbers of sensor nodes, have introduced many challenges to the design and management of sensor networks. These challenges make it necessary to ensure energy awareness at all layers of the networking protocol stack for WSNs. For instance, at the network layer, the main goal is to

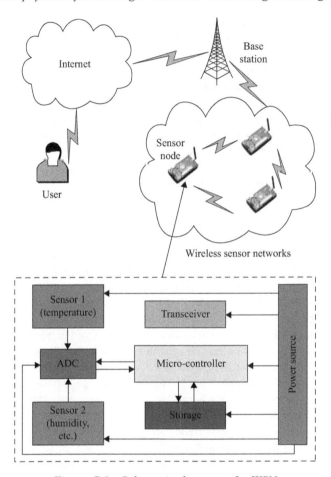

Figure 7.1   Schematic diagram of a WSN

achieve energy-efficient route discovery and relaying of data from the sensor nodes to the BS so that the lifetime of the network can be maximized.

Unlike other types of wireless networks, routing in WSNs is pretty challenging because of their unique features, which are summarized as follows.

1. The infeasibility of building a global addressing scheme for the deployment of large numbers of sensor nodes because the overhead of ID maintenance is high. Thus, the internet protocol (IP)-based protocols cannot be applied to WSNs. Sometimes it is also true for WSNs that getting the data is more important than knowing the IDs of the nodes which sent the data.
2. The requirement for multiple sensor nodes to send the sensed data to the same BS (or sink node). In this case, a data redundancy issue may naturally arise because multiple sensors may generate the same data within the vicinity of a phenomenon. Such redundancy needs to be considered by the routing protocols to improve energy and bandwidth utilization.

3. The tight constraints for sensor nodes in terms of transmission power, on-board energy, processing capacity, and storage, which consequently require careful resource management.
4. Position awareness of sensor nodes is essential because data collection is normally based on the location. For example, in transportation CPSs, it is important for a sensor to report both the traffic jam alert as well as the detailed location of the traffic jam – even the orientation of the road which is experiencing the traffic jam.
5. In many application scenarios, nodes in WSNs are generally stationary after deployment, which differs from other wireless networks, in which nodes are free to move and can lead to unpredictable and frequent topological changes.

As a result of the unique features listed above, many algorithms have been proposed for the routing problem in WSNs. These routing mechanisms have taken into consideration the inherent features of WSNs along with the application and network architecture requirements. The task of finding and maintaining routes in WSNs is non-trivial because energy restrictions and sudden changes in node status (e.g. failure) can cause frequent and unpredictable topological changes.

In this chapter, we present a survey of routing protocols in WSNs for CPSs that have been recently proposed in the literature. We classify the routing protocols into various categories. Our goal is to provide deeper understanding of the routing protocols in WSNs that support CPSs and identify some open research issues that need to be addressed in the future.

The rest of this chapter is organized as follows. In Section 7.2, we discuss challenges and issues when designing routing protocols for WSNs. A comprehensive survey of routing techniques in WSNs for CPSs is presented in Section 7.3. In Section 7.4, we discuss future research directions regarding routing in WSNs for CPSs. We conclude the chapter in Section 7.5.

7.2   Design challenges and issues for routing in WSN within the context of CPS

As we discussed earlier, WSNs have several features that make routing challenging. These features include limited power supply, limited computing capability, and limited bandwidth of the wireless communication links connecting sensor nodes. Various CPS applications impose additional constraints on the design of routing protocols as they require high reliability, time sensitivity, and security when the sensor data is transmitted wirelessly. Thus, the design of routing protocols in WSNs for CPSs is significantly influenced by all these features and restrictions. In this section, we describe some of the main challenges and issues that influence the design of routing protocols in WSNs for CPS.

*   **Energy consumption:** In the context of the wireless environment, each sensor node possesses only limited power supply. Thus, normal tasks such as processing or transmitting sensor data can easily exhaust the limited power of sensors. As a result, it is important to take energy consumption into

consideration when designing the routing protocols for WSNs. In general, the transmission power of a wireless radio is proportional to the square of the distance or at an even higher order in the presence of obstacles. In this case, multi-hop routing will consume less energy than direct communication. In a multi-hop routing scheme, each node acts as both data sender and data router. The malfunctioning of some sensor nodes due to power failure can lead to drastic topological changes and consequently require rerouting of data packets. In contrast, direct communication would perform well if all the nodes were very close to the sink node [16]. In practice, most of the time sensors are distributed over a large area, and thus multi-hop routing becomes inevitable.

- **Node deployment:** The topological deployment of sensor nodes significantly depends on the specific application scenario, and it also influences the performance of routing protocols. Sensor node deployment can be either deterministic or randomized in practice. In deterministic deployment, the sensors are placed in some fixed spots, and sensor data is routed through predetermined paths.

  In random node deployment, the sensor nodes are distributed randomly, which form the sensor network in an ad hoc manner. In this ad-hoc type of sensor network, clustering becomes necessary so that routing can be performed in an energy-efficient manner.

- **Node movement:** In many of the WSN applications, sensor nodes are supposed to be static. However, some WSN applications require sensor nodes to be able to move from time to time [17]. For example, it will be desirable if sensor nodes can move and collect data in various locations when they are scattered to support disaster rescue effort, in which case only a limited number of sensors may be available immediately after the natural disaster has occurred. In this case, the sensors may be deployed on various vehicles, such as UAVs, to help collect data in a timely fashion [18].

- **Reliability:** CPSs generally play an important role in supporting various critical infrastructures, such as electric grid, highway transportation, and air traffic control. In these critical infrastructures, it is important to ensure reliability and tolerate different types of faults and errors. Therefore, reliability and fault tolerance are other important factors that should be considered when designing routing protocols for WSNs in CPSs.

- **Security:** Security is always a key factor in the deployment of CPSs. Therefore, it is critical to take security into consideration when the routing protocols are designed. For instance, a wormhole attack can be launched against routing protocols for wireless sensor networks [19, 20]. In a wormhole attack, an attacker records packets at one location in the network, tunnels them (possibly selectively) to another location, and replays them there into the network. The replay of the information causes confusion to the routing protocols in wireless networks because the nodes that get the replayed packets cannot distinguish them from the genuine routing packets. Moreover, for tunneled distances longer than the normal wireless transmission range of a single hop, it is simple for the attacker to make the tunneled packet arrive with better performance than a normal route, which makes the victim node more likely to accept the tunneled packets instead of the genuine routing packets. As a result, the routing functionality in a wireless network will be severely interfered by the wormhole attack. In addition, there are other types of attacks aimed at disrupting the routing procedure of WSNs, such as Sybil attack [21] and rushing attack [22].

- **Data-sensing and collection model:** It is well understood that various types of CPS may have different data-sensing and collection models because of the unique characteristics of different application domains. According to the nature of CPS applications, data sensing and collection can be categorized as time-driven, event-driven, query-driven, or hybrid [23]. For example, in the smart grid application, phasor measurement units (PMUs) are used to measure the alternating current (AC) waveforms and send them to the control center at a particular rate per second. Thus, the data-sensing and collection model for PMUs is time-driven. In contrast, in the intelligent transportation system (ITS), traffic alerts will only be issued if a traffic jam is sensed, and it is appropriate to utilize the event-driven model to sense and handle the data. In addition, there may be combinations of these models. The data-sensing and collection models have a strong impact on routing protocols in terms of both energy consumption and route stability.
- **Scalability:** In general, there are many sensors in different areas, and any routing protocol in WSNs for CPSs should be able to handle a large number of sensor nodes. Moreover, routing protocols should be sufficiently scalable to respond to events that are detected in the environment when the number of available sensors varies: some of the sensors will remain in the 'sleep' state to save energy until an event occurs, and routing protocols in WSNs should still be able to handle the data collected from the remaining 'awake' sensors.
- **Quality of service (QoS):** In some CPS applications, the data should be delivered within a certain period of time from the moment they are sensed, otherwise the data will be useless. For instance, in ITS application, roadside sensors will sometimes detect that there is a significant reduction in the speed of a large number of vehicles, which indicates that there may be a traffic jam in this segment of the highway. In this case, the sensor data needs to be sent out as soon as possible because the real-time traffic status is highly time-sensitive and it may change even after several minutes of delay. Therefore, bounded latency for sensor data delivery is another major challenge for such time-critical CPS applications.

7.3 Routing protocols in WSNs for CPSs

In this section, we review the state of the art on routing protocols for WSNs that would support CPSs. First, we present a taxonomy of these protocols. Then, we highlight several types of widely used routing protocols in WSNs for CPS.

7.3.1 Taxonomy of routing protocols in WSNs for CPSs

In general, there are three approaches to classify routing protocols in WSNs for CPSs: classification according to the network addressing, classification according to

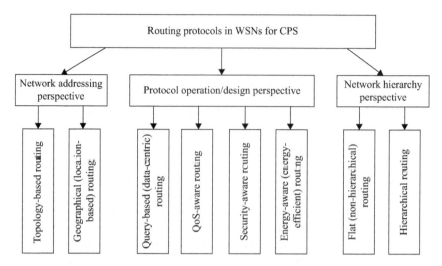

Figure 7.2   Taxonomy of routing protocols in WSNs for CPSs © IEEE 2004. Reprinted with permissions from [24]

the operations and prioritized design goals of the routing protocols, and classification according to the network hierarchy. First, depending on how we identify each sensor node in WSN, routing protocols can be divided into topology-based routing and geographical (location-based) routing. From the perspective of network hierarchy, routing protocols can be classified as flat (non-hierarchical) routing and hierarchical routing. Finally, according to their operations and prioritized design goals, routing protocols can be classified as query-based (data-centric) routing, QoS-aware routing, security-aware routing, and energy-aware (energy-efficient) routing. The detailed taxonomy of routing protocols for CPSs is shown in Figure 7.2 [24].

In the rest of this section, we discuss some popular routing protocols in WSNs.

7.3.2   Location-based routing protocols

Location-based routing, which belongs to the 'network addressing perspective' category in Figure 7.2, is an attractive approach in WSN routing. A lot of efforts have been made and a set of location-based routing protocols have been proposed in the past decade. Location-based routing requires position information but it has the advantages of low routing overhead, high scalability, and excellent adaptability to the dynamics of sensor nodes.

The main idea of location-based routing can be described as follows. First, the locations of the nodes and their neighbors are obtained. Each node gets its location by using GPS or other localization algorithms such as triangulation [25, 26] and range-free localization [27, 28]. Then, the locations are shared with neighboring nodes. In location-based routing, the message-forwarding decision is based on the location of the source node, neighboring nodes and the destination node. When an intermediate data routing node receives the data packet, it forwards the packet to the neighbor node which is currently the closest to the destination among all nodes in the neighbor list. In other words, when a sensor node needs to forward a message to the destination, it looks for the next hop in its neighbor list and selects the best next hop on the basis of the neighbor location and its distance to the destination node. These protocols are mostly greedy, stateless and on-demand routing protocols, therefore paths are established only when they are required.

Location-based routing is attractive because its design has several benefits. A mostly greedy mechanism is commonly adopted in the design of location-based routing, so it is very efficient because it generally finds a shortest path to the destination. Second, location-based routing makes message-forwarding decisions based only on local information such as a neighbor's location information. Thus, only one-hop neighbor message exchange is sufficient, but no flooding is needed. Moreover, as a stateless protocol, each sensor needs to store very little information to make the routing decision. Thus, the scalability of the protocol is very good. In addition, location-based routing adapts to the dynamics of the sensor network very well because the neighbor information can be updated periodically as needed at relatively small cost. Finally, the design of location-based routing can easily integrate other attractive routing design concepts such as energy-aware routing and hole-avoided routing. All these features of location-based routing make it a very attractive routing scheme for CPS.

With all the aforementioned advantages, in the design of location-based routing, we need to address several issues to make the routing protocol more efficient. Location-based routing frequently adopts a greedy design concept. As a greedy algorithm, it can find an efficient shortest path in most cases, but it may also cause two problems, namely local minimum and overusing a certain path. To address the local minimum problem, greedy perimeter stateless routing (GPSR) [29] proposes two routing models to get the route out of local minimum when the routing reaches a hole. Another protocol, WEAR [30] is designed to broadcast the hole information so as to avoid routing to holes. To address the problem of overusing a certain path and achieve load balance, energy-aware approaches have been proposed in several earlier works, including geography- and energy-aware routing (GEAR) [31], balanced, fault-tolerant and energy-aware routing (WEAR) [30], and geographic routing with environmental energy supply (GREES) [32]. To be adaptive in a highly dynamic environment, sensor nodes have to exchange location information periodically with their neighbors using beacons. Thus, a lot of messages have to be changed to keep neighbor information fresh. Several efforts have been targeted on the problem of reducing the number of control messages. For example, beaconless on-demand strategy for geographic routing (BOSS) [33] and energy-efficient beaconless geographic routing (EBGRES) [34] have been proposed to get neighbor information only when it is needed in routing. To further save energy, geographic adaptive fidelity (GAF) [35] has been designed to schedule the active/sleep status of the sensors. Security is another issue in location-based routing. For example, a malicious node may behave as a blackhole node or selectively forward messages for others. Thus, secure geographic routing [36] has been proposed to improve the security of location-based routing. Next, we introduce the main ideas of the aforementioned protocols one by one.

- **GPSR [29]:** GPSR is one of the earliest and widely used location-based routing protocols, which defines two scenarios of forwarding data packets. Greedy forwarding is used when a sensor needs to forward a data packet and can find a neighbor node that is closer to the destination than itself. Using greedy forwarding, the sensor chooses a neighbor that is closest to the destination and forwards the message to the chosen sensor. When greedy forwarding is not available, perimeter routing is adopted by using a simple planar graph traversal. A right-hand rule which forwards the packet around the face is used in perimeter routing. The sensor returns to greedy forwarding when a sensor node which is closer to the destination than the sensor entering the perimeter forwarding is found. The GPSR protocol successfully addresses the local minimum issue in greedy routing protocol while maintaining the efficiency of greedy routing.

- **GEAR [31]:** GEAR is proposed in order not to overuse some sensor nodes on certain shortest paths, i.e. to avoid using sensor nodes with low remaining energy when the next hop node is selected in making packet forwarding decisions. It uses greedy forwarding as for GPSR but it uses recursive geographic forwarding or restricted flooding when greedy forwarding is not available. By considering the remaining energy of neighbor sensors, GEAR can successfully distribute the load of message forwarding to sensors with more remaining energy so that it avoids overusing a certain set of sensor nodes that have a higher probability to be on a greedy path than others because of their locations.

- **WEAR [30]:** WEAR protocol observes that the local minimum is caused by the holes in the sensing area which is defined as a large area without active sensor nodes. Thus, sensor nodes start by collaboratively detecting existing holes in the sensing area and broadcasting related information to neighboring nodes by using controlled flooding. The routing decision at each node is made based on four factors, namely the distance to the destination, the energy level of the sensor, the location information and the available information on the local hole. WEAR tries not to route the package to the hole area to avoid overusing the perimeter sensor nodes around the holes. WEAR achieves a better load balance performance and it tolerates sensor failures which create holes in sensor networks.

- **GAF [35]:** GAF targets to save energy by turning off unnecessary sensor nodes in the network. GAF builds a virtual grid for the whole sensing area. Each sensor will be a member of one virtual grid and works in one of the three status modes – sleep for radio turned off, discovery for determining the neighbors in the grid, and active for participating in routing. Each sensor that can find other equivalent sensor nodes in the same virtual grid goes to sleep. Connectivity is preserved by the granularity of the virtual grid. In addition, GAF adopts a location-based routing as described earlier. A lot of energy is saved by scheduling the sleep/active status of the sensors so that the lifetime of the sensor network can be significantly extended.

- **GREES [32]:** GREES combines energy-efficient routing with geographic routing. In building routing heuristics, several factors including realistic wireless channel condition, packet advancement to the destination, the residual battery energy level, and the environmental energy supply are considered. Compared to other routing protocols, GREES considers realistic wireless channel condition and uses environmental energy supply in the protocol design. The system model in GREES is closer to reality. Experiments show that GREES can achieve better load balancing and can extend the lifetime of the sensor networks.

- **BOSS [33] and EBGRES [34]:** BOSS is designed to reduce the number of beacons in exchanging neighbor information, and it does not require proactive transmission of control messages such as beacons. Instead, the neighbors collaboratively decide the next forwarder. The sensor that needs to forward the message decides the next forwarder among its neighbors based on their positions. Moreover, in BOSS, the protocol considers realistic link condition and it sends out a real data message to discover the neighbors and automatically filter a set of unqualified neighbors with poor link condition, which cannot accomplish the task of forwarding the data message. As a result, BOSS significantly reduces the number of control messages. Sharing a similar goal as BOSS, EBGRES makes use of three types of messages, namely, DATA, ACK, and SELECT, to search a relay region and decides a forward strategy. Both protocols help to reduce the number of control messages.

- **Secure geographic routing [36]:** A secure geographic routing protocol is proposed to improve the security of location-based routing. It aims to address two types of attack, namely, excessive number of packet attacks to the overflow message queue, and the blackhole and selective forwarding attack. This was one of the early works to propose a rate control mechanism. Each sensor node estimates the approximate maximum rate and drops excessive packets. Forwarding verification is conducted by querying a trusted peer such as an anchor node or a trusted node. Then a probabilistic approach is used to select the next-hop node considering neighbor node locations. Finally, a multi-path routing is adopted to avoid blackhole and selective forwarding nodes. All those mechanisms work together to mitigate attacks to the location-based routing protocol.

7.3.3 Data-centric routing protocols

In many CPS applications that involve WSNs as an enabling technology, it is infeasible to assign a unique identifier (ID) for each sensor node because of the huge number of sensors that are commonly deployed. Furthermore, sensors are generally deployed in a random manner, which means that the network topology is unknown. Both of these two restrictions make it hard to identify and query a specific subset of sensor nodes. Thus, to ensure the delivery of sensor data, the data is usually transmitted from every sensor node within the area with significant redundancy, which is very inefficient in terms of energy consumption. To address this issue, some routing protocols choose to only select a subset of sensor nodes and utilize data aggregation during the data relaying process. These routing protocols are categorized as the data-centric routing protocols, in which the sink node sends queries to certain regions and waits for data from the sensors located in the selected regions. Note that data-centric routing protocols belong to the 'protocol operation/ design perspective' category in Figure 7.2.

Numerous research efforts have been made to study data-centric routing approaches. One such related work is sensor protocols for information via negotiation (SPIN) [37], which is the first data-centric routing protocol that considers data negotiation between nodes in order to eliminate redundant data and save energy. Directed diffusion [38] further improved the prior work and it also acts as the foundation for other research efforts that are related to data-centric routing, such as the research efforts discussed in Yeo and Gehrke [23] and Chu *et al.* [39]. Next, some of these protocols are described in detail.

7.3.3.1   Flooding and gossiping-based data dissemination

It is well understood that flooding and gossiping are both traditional methods to relay data in sensor networks without the need for any fancy routing algorithms and topology maintenance. In the case of the flooding method, after each sensor receives a data packet from its neighbors, the sensor broadcasts the packet to all of its neighbors. This process continues until the packet arrives at the destination or the maximum number of hops for the packet has been reached. In contrast, gossiping is an improved version of flooding, because the receiving node only sends the packet to one randomly selected neighbor rather than all of its neighbors, which will then repeatedly pick another one random neighbor to forward the packet to and so on. By this means, the number of repeated packets will be significantly decreased in gossiping when compared to flooding.

Even though the flooding method is easy to implement and maintain, it has several major problems that prevent its wide deployment. First of all, there is a problem called implosion, which is caused by duplicated messages sent to the same node from different sources. Second, when two nodes sensing the same region send similar packets to the same neighbor, the overlap problem occurs. Finally, due to the nature of broadcast, it is expected that a large number of packets will be redundant in the entire sensor network, which will consume a huge amount of energy, and consequently lead to the energy exhaust problem. In contrast, the gossip method improves the flooding method because it can avoid implosion by simply selecting a random node to send the packet to rather than broadcasting. However, this may cause delays in propagation of data to the sink node through the set of intermediate nodes. Furthermore, because next hops are selected randomly, the data packets may be sent further away from the destination after each hop or in the worst case all the passing nodes may even create a loop, which fails the gossiping process.

Based on the observations above, it is concluded that even though both flooding and gossiping methods are easy to understand and deploy, they have a lot of disadvantages that are difficult to address in practice. Therefore, more sophisticated data-centric approaches should be studied to further improve them.

7.3.3.2   First trial: sensor protocols for information via negotiation

SPIN is one of the earliest research efforts that focused on data-centric routing for WSN [37]. The basic idea of SPIN is that instead of flooding the data to every neighbor of each node, the data is only shared among those who are interested in it. To achieve this goal, the data is named using high-level descriptors or metadata. Before the transmission of the actual sensor data, metadata is first exchanged among sensors via a data advertisement mechanism, which is one of the key features of SPIN. When a node receives any new data, it will initially advertise the new data to its interested neighbors, i.e. those who do not have the data yet. The interested neighbors can then retrieve the data by sending a request message.

SPIN solves the three problems that the flooding method faces – implosion, overlapping, and energy exhaust – because it sends the sensor data in a targeted manner rather than simply flooding them out. It is worth noting that there are generally three types of messages in the SPIN protocol:

- Advertisement (ADV) message: used to advertise a particular metadata by a sensor.
- Request (REQ) message: used to request the specific data from another sensor.
- Data (DATA) message: used to transport the actual data.

An example is shown in Figure 7.3 to demonstrate the operation of the SPIN protocol. From this figure we find that whenever there is any new data that a sensor node obtains, the ADV message is used to notify the sensor's neighbors of this new data. If the neighbors are interested in getting the data, they reply with the REQ message. Finally, the actual data will be transmitted using the DATA message.

On the one hand, the SPIN protocol has several desirable advantages over the traditional flooding method. First of all, topological changes are localized because each node only needs to know its single-hop (i.e. 'immediate') neighbors. Second, the data negotiation process can save a significant amount of energy as well as network bandwidth because now the data is shared only with those who want it.

On the other hand, the SPIN protocol has some restrictions. One limitation of SPIN is that it cannot ensure that the data is delivered to the desired destination. Indeed, if a node is interested in some data that is far away from itself (for instance, several hops away), then it cannot successfully get it unless all of the intermediate nodes are interested in the same data, which sometimes may not be realistic in practice. Therefore, despite the fact that SPIN has many good features, it may not be the best choice for some mission-critical applications in the CPS domain, such as traffic monitoring or battlefield surveillance, because these applications really require on-time deliveries of sensor data.

7.3.3.3   Directed diffusion

Directed diffusion was an important milestone in the area of data-centric routing protocols [38]. In this protocol, a naming scheme is used for the sensor data to

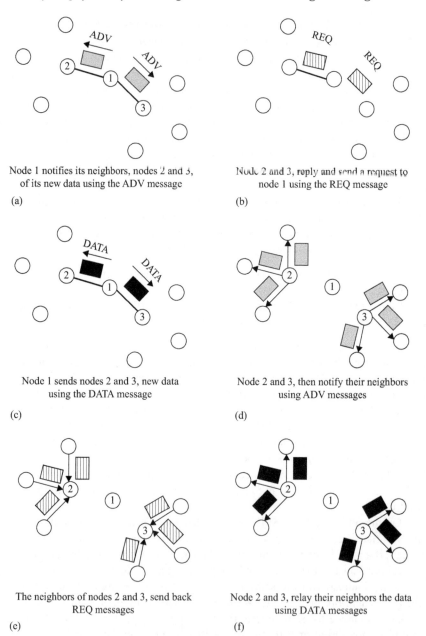

Node 1 notifies its neighbors, nodes 2 and 3,
of its new data using the ADV message

(a)

Node 2 and 3, reply and send a request to
node 1 using the REQ message

(b)

Node 1 sends nodes 2 and 3, new data
using the DATA message

(c)

Node 2 and 3, then notify their neighbors
using ADV messages

(d)

The neighbors of nodes 2 and 3, send back
REQ messages

(e)

Node 2 and 3, relay their neighbors the data
using DATA messages

(f)

Figure 7.3   An example of the sensor protocols for SPIN protocol

diffuse the data among sensor nodes, which can avoid the unnecessary operations
of the network layer routing and consequently save energy. More specifically,
direct diffusion uses attribute-value pairs to represent the data and queries the
sensors on an on-demand basis by using those pairs.

To create a query, an interest message is defined using a list of attribute-value pairs such as name of objects, interval, duration, and geographical area. The interest message is then broadcast by a sink through its neighbors, and the interest message is cached by these nodes for later use. The interest message also contains several gradient fields, which is a reply link to a neighbor from which the interest message was received. The gradient field is characterized by the data rate, duration and expiration time derived from the received message fields. Therefore, routes can be established between sink and sources by using both interest message and gradient field information. In addition, several routes can be established at the same time so that one of them is selected by reinforcement, in which the sink resends the original interest message through the selected path with a smaller interval and thus reinforces the source node on that path to send data more frequently.

Directed diffusion is also resilient to path failures. Whenever a path between a source node and the sink node fails, directed diffusion will reinitiate reinforcement by searching among other paths. As an alternative option, Ganesan *et al.* [40] proposed to employ multiple paths in advance so that in case of the failure of a path, one of the alternative paths is chosen without any cost for searching for another one. It is necessary to consume additional resources for keeping these alternative paths alive. However, more energy will be saved when a path fails and a new path becomes available immediately without any more reinforcement.

The major difference between directed diffusion and SPIN is the on-demand data-querying mechanism. In directed diffusion, the sink queries the sensor nodes if specific data is available using the flooding technique. In contrast, sensors in SPIN will actively advertise the availability of data, which allows the interested nodes to query that data.

The directed diffusion routing protocol has many advantages. First, there is no need to maintain a unique addressing system for sensor nodes because all communications happen locally among immediate neighbors. Second, the caching that each node supports is a big advantage in terms of minimizing delay. Third, it is highly energy-efficient because it is on-demand and there is no need for maintaining a global network topology.

Despite all of its advantages, directed diffusion cannot work well in some CPS applications because of the query-driven data delivery model. For example, in the CPS application that supports situational awareness in the battlefield, it is critical to continuously sense and report the real-time updates to the control center. In this case, the query-driven delivery model will become extremely inefficient. In addition, this type of routing cannot work well in other similar applications, including traffic surveillance as well as environmental monitoring.

7.3.4 Hierarchical routing protocols

As discussed earlier, scalability is a major challenge in the routing protocol design in wireless networks, particularly in WSN. Suppose that sensor nodes in WSN are deployed in a flat (or in other words, non-hierarchical) manner, then the sole gateway node will be gradually overloaded with the increase in sensor density. The overloaded gateway will lead to latency in communication and insufficient reporting of critical sensor events. Moreover, because sensors generally cannot support many hops of multi-hop communication due to the loss-rate of wireless channels, the flat sensor architecture is not able to scale well to a large set of sensors that covers a wider area of interest. To address the limitation in scalability, the concepts of network clustering and chaining have been applied to some routing protocols that help WSN better cope with additional load and serve a larger area. Protocols supporting clustering and chaining are generally classified as hierarchical routing protocols, which belong to the 'network hierarchy perspective' category in Figure 7.2.

The main goal of hierarchical routing protocols is to maximize the energy efficiency of sensor nodes by (1) grouping them into clusters, (2) involving them in multi-hop communication within a particular cluster, and (3) performing data aggregation and fusion in order to decrease the number of messages to the sink node. The first step of hierarchical routing is generally cluster formation, during which sensors in close proximity are grouped together and a cluster is formed with a designated cluster head [41]. Several well-known hierarchical routing protocols in WSN are summarized below.

7.3.4.1   Low-energy adaptive clustering hierarchy

Low-energy adaptive clustering hierarchy (LEACH) is one of the most well-known hierarchical routing algorithms for sensor networks [16]. In the LEACH method, sensor nodes are clustered according to the received signal strength, and local cluster heads act as routers and forward all the data packets within the cluster to the sink node. This significantly saves energy because the transmissions are done by cluster heads rather than all sensor nodes.

All the data processing such as data fusion and aggregation are achieved locally to the cluster. To ensure fairness among nodes in the same cluster, cluster heads change randomly over time in order to balance the energy dissipation of nodes. In addition, LEACH is completely distributed and thus requires no global knowledge of network. However, LEACH uses single-hop routing in which each node can transmit directly to the cluster head as well as to the sink. Therefore, LEACH is not applicable to networks deployed in large regions. Moreover, dynamic clustering introduces extra overheads (such as cluster head changes, advertisements), which may diminish the gain in energy consumption.

7.3.4.2   Power-efficient gathering in sensor information systems

Power-efficient gathering in sensor information systems (PEGASIS) is an improvement of the LEACH protocol [42]. Rather than forming multiple clusters, PEGASIS forms chains from sensor nodes so that each node transmits and receives data from a neighbor and only one node is selected from that chain to transmit data to the base station (sink). In the PEGASIS protocol, gathered data moves from node to node, gradually gets aggregated and is eventually sent to the base station. The chain construction is performed in a greedy way. Figure 7.4 demonstrates the token passing process in the PEGASIS protocol. In this figure, node 2 is the leader, and it will first pass the token along the chain to node 0. Upon receipt of the token, node 0 sends its

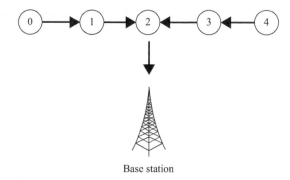

Figure 7.4   Token passing in the PEGASIS protocol

data toward node 2. After node 2 receives data from node 1, it passes the token to node 4, and node 4 sends its data toward node 2.

When compared to the LEACH protocol, PEGASIS uses multi-hop routing by forming chains and selecting only one node to transmit to the base station rather than using multiple nodes. Experimental results have shown that PEGASIS outperforms LEACH by about 100–300% in terms of the lifetime of the network depending on the network size and topology. The performance gain of PEGASIS over LEACH is made possible because of elimination of the overhead caused by dynamic cluster formation in LEACH and also decrease in the number of transmissions and receptions by using data aggregation.

As for its disadvantages, PEGASIS introduces excessive delay, especially for remote nodes on the chain that are far from the aggregating node. In addition, the fault tolerance ability of PEGASIS is questionable because one single leader in the chain can easily lead to the classic 'single point of failure' problem, in which any failure or unavailability of the single leader will make the data in the whole chain become unavailable.

To address the excessive delay in the PEGASIS protocol, an improved version of it, called hierarchical-PEGASIS, was proposed [43]. Hierarchical-PEGASIS aims at reducing the delay incurred for packets during transmission to the base station and proposes a solution for the data-gathering problem by considering energy and delay metrics. To reduce the delay in PEGASIS, simultaneous transmissions of data messages are pursued. However, to avoid collisions and possible signal interference among the sensors, two approaches have been investigated. The first approach incorporates signal coding (e.g. CDMA is used to avoid collisions and signal interference). Alternatively, in the second approach only spatially separated nodes are allowed to transmit at the same time.

Even if both the original and the hierarchical PEGASIS approaches avoid the clustering overhead of LEACH, they still require dynamic topology adjustment because the energy consumption of sensors is not tracked. For instance, every sensor needs to be aware of the status of its neighbor so that it knows where to route the data. Such topology adjustment generally introduces significant overhead to WSN especially if the WSN is heavily used.

7.3.4.3   Threshold-sensitive energy-efficient sensor network protocol

Threshold-sensitive energy-efficient sensor network (TEEN) protocol [44] is a hierarchical protocol which aims to be responsive to sudden changes in the sensed attributes such as temperature, humidity, or occupancy rate. In many CPS applications, such as smart grid and traffic monitoring, it is essential to sense, report, and respond to some events of interest in a timely fashion. In TEEN, the sensor nodes are organized in a hierarchical manner, and two layers of clusters are created based on the distance between a node and the cluster head. An example of the sensor node structure in the TEEN protocol is shown in Figure 7.5.

After the clusters are formed, every cluster head broadcasts two thresholds regarding the sensed attributes (namely hard threshold and soft threshold), to the members of the cluster. Hard threshold is the minimum possible value of an attribute to trigger a sensor node to switch on its transmitter and transmit the data to the cluster head. Thus, the hard threshold allows the nodes to transmit only when the sensed attribute is in the range of interest, thus reducing the number of transmissions significantly. Furthermore, once a node senses a value at or beyond the hard threshold, it transmits data only when the value of that attribute changes by an amount equal to or greater than the soft threshold. Therefore, soft threshold will further reduce the number of transmissions if there is little or no change in the value of sensed attribute. Both hard and soft threshold values can be tuned by the user so as to control the number of packet transmissions. However, TEEN is not

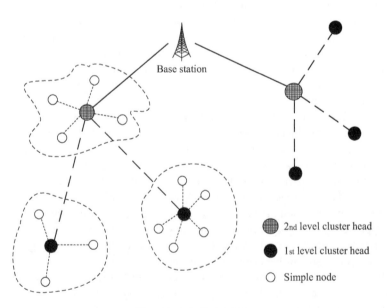

Hierarchical cluster in WSN

Figure 7.5   Hierarchical clustering for the threshold-sensitive energy-efficient sensor network (TEEN) protocol

suitable for applications where periodic reports are necessary because the user may not get any data at all if both of the thresholds are not properly set.

To meet the need from WSN applications where periodic reports are required, an improved version of TEEN was studied as well. Unlike TEEN, the adaptive threshold-sensitive energy-efficient sensor network (APTEEN) protocol [45] captures periodic data and reacts to time-critical events. The basic sensor node structure in APTEEN is similar to that in TEEN. However, APTEEN supports three different query types: (1) historical, to analyze past data values; (2) one-time, to take a snapshot view of the network; and (3) persistent, to monitor an event for a period of time.

7.3.5   *Summary of routing protocols in WSN for CPS*

Based on our earlier discussions, we compare and summarize these routing protocols and their features in Table 7.1. We also discuss whether or not the routing protocols in WSNs are suitable for the CPS applications.

From Table 7.1 we can observe that some of the location-based routing protocols (such as GPRS and GEAR) may not be easily applicable to CPS applications. This is because they generally introduce significant power consumption and they may have poor scalability. Some other location-based routing protocols, such as WEAR, GAF, GREES, BOSS, and EBGRES, are more suitable for CPS applications (such as the smart grid application), especially those without mobile nodes. Data-centric routing protocols can be used in CPS applications if they use homogeneous types of network infrastructure and nodes are not moving frequently. As for hierarchical routing protocols, they are generally suitable for CPS applications, because they are scalable and power supply is generally not a big concern in CPS applications.

7.4   Future directions of routing protocols in WSN for CPS

As WSN is an integral and important part of the CPS, the design of the routing protocols needs to be revisited to cater the new features in CPS, which are not present in WSN. This is especially because most CPS applications are more complicated than WSN applications, as they are dealing with additional heterogeneity issues in devices, sensing data types, communication channels, along with higher real-time requirements but coming with higher system mobility. In addition, CPS needs to make intelligent decisions based on the collected sensor data and actuators to take action based on the intelligent decisions from the system. To take into consideration these CPS requirements, a set of new routing protocols are expected and crucial. Next, we discuss the open issues and challenges in the routing protocol design when integrating WSN with the new features of the CPS.

- **Heterogeneous networking techniques and devices:** In CPS, it is common that different networking techniques and heterogeneous devices are used in one application. For example, several tiers of networking technologies can be all used in a smart grid system. Smart meters use ZigBee communication to connect smart devices and form a home-area network. Then, all smart meters in the same residential community and the nearby network gateway form a neighborhood-area

Table 7.1 Summary and comparison of routing protocols in WSNs for CPSs

Protocol	Category	Mobility	Powerusage	Dataaggregation	Scalability	QoS	CPS suitable
GPRS	Location-based	Limited	High	No	Limited	No	Very unlikely
GEAR	Location-based	Limited	Medium	No	Limited	No	Unlikely
WEAR	Location-based	Limited	Low	No	Limited	No	Partially
GAF	Location-based	Limited	Very low	No	Possible	No	Partially
GREES	Location-based	Limited	Very low	No	Limited	No	Partially
BOSSEBGRES	Location-based	Limited	Low	No	Limited	No	Partially
SPIN	Data-centric	Possible	Low	Yes	Limited	No	Partially
Directed diffusion	Data-centric	Limited	Low	Yes	Limited	No	Partially
LEACH	Hierarchical	Very limited	Maximum	Yes	Good	No	Likely
PEGASIS	Hierarchical	Very limited	Maximum	No	Good	No	Likely
TEENAPTEEN	Hierarchical	Very limited	Maximum	Yes	Good	No	Likely

network based on ZigBee communication. Furthermore, neighborhood gateways using cellular communication, WLAN or WiMAX techniques form a gateway network, which is further connected to the data centers of utility providers through traditional internet such as a fiber network. The same scenario exists in other CPS applications such as healthcare applications and vehicular network applications. In such a system, routing protocol should be adaptive to the heterogeneity of the networking techniques and devices because there is no common internet protocol (IP) layer as in traditional networked systems that cover all the heterogeneous devices and networking technologies in CPS. Cluster-based routing architecture could be an attractive solution here.

- **Higher real-time requirements:** One of the major differences between CPS and WSN is that actuators are added to control the physical world based on the decision from the data analysis of collected sensor data. As a result, more control messages to actuators will be delivered in CPS than those in WSN. For example, a temperature adjustment message could be sent from the data center to a smart meter and smart thermometers. In addition, a lot of service data could be sent from CPS to component devices of the system. For example, in a vehicular network scenario, vehicles upload collected city survey data to a roadside unit; meanwhile, they download video stream data from the roadside unit to support an online streaming application. Therefore, unlike in WSN where most messages are routed in one direction from sensor nodes to sinks, two-way traffic is common in CPS and could lead to high volumes of data. On the other hand, a lot of CPS applications have higher real-time requirements than WSN applications, as actuators need to take action on time or services need to be delivered with real-time requirements. Moreover, higher mobility of CPS nodes brings further challenges in routing protocol design. Thus, it is crucial that routing protocols for CPS need to be optimized to support efficient, two-way communications and to provide better real-time support. QoS routing protocols proposed for WSN could be revisited and modified in CPS.

- **Privacy and security:** With CPS applications widely deployed to improve the quality of life, a lot of sensitive data will be collected to support the decision-making function in CPS applications. For example, location-aware services have become pervasive in recent years. They are built based on the availability of sensitive location information. Electronic medical records which contain detailed private information are set up for each patient. Fine granularity electricity and water usage data are collected in a smart building so that building residents' daily activity could be transparent through the analysis of the utility data [46]. Lots of private photos and videos are taken and stored in smart phones. Security and privacy issues become critical factors to the success of CPS applications, but it becomes more and more difficult to secure this set of sensitive and private data because of the amount of information, the heterogeneity of devices and networking techniques, and so on. Routing protocol is an important attack point in CPS to get access to the sensitive data; therefore, privacy-preserving routing protocols and secure routing protocols should be examined further in CPS research.

- **Data aggregation:** In CPS, heterogeneous information flows will be common. Various sensors collect different types of data, which will be transmitted in the same network with lots of control messages. A significant volume of the data will be generated and routed every moment, and sometimes this set of data may contain a high level of redundancy. For example, in a vehicular network scenario, street view, city environmental data and accident and traffic data are collected from vehicles to roadside units, while service data is downloaded from the roadside units to vehicles. Considering the heterogeneity of the devices used in CPS and the heterogeneous capacities of networking techniques, in many cases, not all data can be successfully transmitted on time or it is not necessary to transmit all data because of their redundancy. Meanwhile, not all the data have the same level of delivery requirements and some data could be delay-tolerant. Routing techniques that support data aggregation to reduce the volume of the data and support traffic differentiation by considering the different real-time requirements of the data are essential in CPS.
- **Reliability of unreliable devices:** There are many heterogeneous devices in CPS. Some devices could be much more powerful and have more resources than others. As a result, those devices may have different levels of reliability. But most of the CPS critical applications need reliable routing services. How to build reliable routing protocols based on unreliable devices will be a significant challenge in CPS. We can assign different levels of responsibility to different devices in CPS by considering their available resources and reliability. Role-differentiated routing protocols and trust-based routing protocols would be interesting to explore in this context. Fault-tolerant mechanisms based on a certain level of redundancy could be used in routing protocol design for CPS. The trade-off among redundancy, performance and reliability needs to be studied further. Multipath routing and reliable routing protocols could also be explored to fulfill the routing requirements of CPS.
- **High system interaction and mobility:** Unlike most WSNs where the sensor nodes are either static or move very infrequently and there is only little interaction between the sensor nodes and humans, CPS will have a much higher level of system human–machine interaction and mobility. A typical vehicular network consists of vehicles that move much faster than sensor nodes. Sometimes mobility helps improve efficiency of the routing protocol, but meanwhile it brings more challenges if the receiver is located beyond the range of the communication. Generally speaking, high mobility brings more frequent information updates, thereby increasing the overhead of routing protocols. Prediction-based routing protocol could be attractive to address the high mobility in CPS and to achieve an efficient routing protocol design. In other scenarios, such as a typical healthcare application, there are a lot of human–machine interactions because humans are important players in the system. Therefore, we should also take human factors into consideration when the routing protocols are designed.

In summary, there are many new features in CPS that have not been seen in WSN. Several issues need to be addressed and examined in the routing protocol design in CPS. In addition, with so many CPS applications emerging in recent years, the lessons we learned from these applications are helping us to build customized, robust, high-performance, and energy-efficient routing protocols needed by different CPS applications.

7.5 Conclusion

Routing in sensor networks has attracted a lot of research attention during the past decade. This issue has introduced some unique challenges compared to traditional data routing in wired networks. Additional routing challenges are being faced with the increasing deployment of WSN in support of various CPS applications.

In this chapter, we reviewed and summarized some research efforts which aim to address the routing challenges for WSN as well as CPS. We also highlighted three major categories of routing protocols in WSN, namely location-based routing, data-centric routing, and hierarchical routing. In addition, we discussed some interesting open challenges and research issues that we believe are worthwhile investigating.

Acknowledgments

We thank the anonymous reviewers for their feedback and useful comments, which have helped us improve the quality and presentation of this chapter.

Appendix: List of Acronyms

ADV: Advertisement
APTEEN: Adaptive Threshold sensitive Energy Efficient sensor Network protocol
BOSS: Beacon-less On demand Strategy for geographic routing
CPS: Cyber Physical Systems
EBGRES: Energy efficient Beaconless Geographic Routing
GAF: Geographic Adaptive Fidelity
GEAR: Geography and Energy Aware Routing
GPSR: Greedy Perimeter Stateless Routing
GREES: Geographic Routing with Environmental Energy Supply
ID: IDentifier
IP: Internet Protocol
LEACH: Low-Energy Adaptive Clustering Hierarchy
PEGASIS: Power-Efficient GAthering in Sensor Information Systems
REQ: Request
SPIN: Sensor Protocols for Information via Negotiation
TEEN: Threshold sensitive Energy Efficient sensor Network protocol
WEAR: balanced, fault-tolerant and Energy-Aware Routing
WSN: Wireless Sensor Network

References

[1] M. Ilic, L. Xie, U. Khan and J. Moura. Modeling future cyber-physical energy systems. In *Proceedings of the 2008 IEEE Power and Energy Society General Meeting – Conversion and Delivery of Electrical Energy in the 21st Century*, pp. 1–9, IEEE, Pittsburgh, PA, USA, 20–24 July 2008.

[2] C.-H. Lo, W.-C. Peng, C.-W. Chen, T.-Y. Lin and C.-S. Lin. Carweb: A traffic data collection platform. In *Proceedings of the 9th International Conference on Mobile Data Management, 2008 (IEEE MDM '08)*, pp. 221–222, IEEE, Beijing, China, 27 30 April 2008.

[3] E. A. Lee. Cyber physical systems: Design challenges. In *Proceedings of the 2008 11th IEEE Symposium on Object Oriented Real-Time Distributed Computing*, Washington, DC, USA, pp. 363–369, IEEE Computer Society, Orlando, FL, USA, 5–7 May 2008.

[4] L.-A. Tang, X. Yu, S. Kim, J. Han, C.-C. Hung and W.-C. Peng. Tru-alarm: Trustworthiness analysis of sensor networks in cyber-physical systems. In *Proceeding of the 10th IEEE International Conference on Data Mining (ICDM '10)*, pp. 1079–1084, IEEE, Sydney, Australia, 13–17 December 2010.

[5] L.-A. Tang, X. Yu, Q. Gu, J. Han, A. Leung and T. La Porta. Mining lines in the sand: On trajectory discovery from untrustworthy data in cyber-physical system. In *Proceedings of the 19th ACM SIGKDD International Conference on Knowledge Discovery and Data Mining (KDD '13)*, pp. 410–418, ACM, Chigao, IL, USA, 11–14 August 2013.

[6] M. Chen, S. Gonzalez, A. Vasilakos, H. Cao and V. C. Leung. Body area networks: A survey. *Mobile Networks and Applications*, 16(2): 171–193, April 2011.

[7] K. Lorincz, D. Malan, T. Fulford-Jones, A. Nawoj, A. Clavel, V. Shnayder, G. Mainland, M. Welsh and S. Moulton. Sensor networks for emergency response: challenges and opportunities. *Pervasive Computing, IEEE*, 3(4): 16–23, October 2004.

[8] M. Hanson, H. Powell, A. Barth, K. Ringgenberg, B. Calhoun, J. Aylor and J. Lach. Body area sensor networks: Challenges and opportunities. *Computer*, 42: 58–65, January 2009.

[9] D. Hart. Using AMI to realize the smart grid. In *Proceedings of 2008 IEEE Power and Energy Society General Meeting – Conversion and Delivery of Electrical Energy in the 21st Century*, pp. 1–2, Pittsburgh, PA, USA, 20–24 July 2008.

[10] A. Arora, P. Dutta, S. Bapat, V. Kulathumani, H. Zhang, V. Naik, V. Mittal, H. Cao, M. Demirbas, M. Gouda *et al.* A line in the sand: A wireless sensor network for target detection, classification, and tracking. *Computer Networks*, 46(5): 605–634, 2004.

[11] T. Bokareva, W. Hu, S. Kanhere, B. Ristic, N. Gordon, T. Bessell, M. Rutten and S. Jha. Wireless sensor networks for battlefield surveillance. In *Proceedings of the Land Warfare Conference*, December 2006.

[12] A. Mainwaring, D. Culler, J. Polastre, R. Szewczyk and J. Anderson. Wireless sensor networks for habitat monitoring. In *Proceedings of the 1st ACM International Workshop on Wireless Sensor Networks and Applications (WSNA '02)*, pp. 88–97, ACM, Atlanta, GA, USA, 28 September 2002.

[13] D. Anthony, W. P. Bennett, M. C. Vuran, M. B. Dwyer, S. Elbaum, A. Lacy, M. Engels and W. Wehtje. Sensing through the continent: Towards monitoring migratory birds using cellular sensor networks. In *Proceedings of the 11th International Conference on Information Processing in Sensor Networks (IPSN '12)*, pp. 329–340, ACM, Beijing, China, 16–19 April 2012.

[14] G. Werner-Allen, K. Lorincz, M. Ruiz, O. Marcillo, J. Johnson, J. Lees and M. Welsh. Deploying a wireless sensor network on an active volcano. *IEEE Internet Computing*, 10 (2): 18–25, March 2006.

[15] G. Liu, R. Tan, R. Zhou, G. Xing, W.-Z. Song and J. M. Lees. Volcanic earthquake timing using wireless sensor networks. In *Proceedings of the 12th International Conference on Information Processing in Sensor Networks (IPSN '13)*, pp. 91–102, ACM, Philadelphia, PA, USA, 8–11 April 2013.

[16] W. Heinzelman, A. Chandrakasan and H. Balakrishnan. Energy-efficient communication protocol for wireless microsensor networks. In *Proceedings of the 33rd Annual Hawaii International Conference on System Sciences*, IEEE Computer Society, Maui, HI, USA, 4–7 January 2000.

[17] I. El Korbi and S. Zeadally. Energy-aware sensor node relocation in mobile sensor networks. *Ad Hoc Networks*, 16: 247–265, 2014.

[18] P. Rudol and P. Doherty. Human body detection and geolocalization for UAV search and rescue missions using color and thermal imagery. In *Proceedings of the 2008 IEEE Aerospace Conference*, pp. 1–8, Big Sky, MT, USA, 1–8 March 2008.

[19] Y.-C. Hu, A. Perrig and D. Johnson. Wormhole attacks in wireless networks. *IEEE Journal on Selected Areas in Communications*, 24(2): 370–380, February 2006.

[20] C. Karlof and D. Wagner. Secure routing in wireless sensor networks: Attacks and countermeasures. *Ad Hoc Networks*, 1: 293–315, September 2003.

[21] J. Douceur. The SYBIL attack. In *Peer-to-Peer Systems*, vol. 2429 of *Lecture Notes in Computer Science*, pp. 251–260. Springer, Berlin/Heidelberg, 2002.

[22] Y.-C. Hu, A. Perrig and D. B. Johnson. Rushing attacks and defense in wireless ad hoc network routing protocols. In *Proceedings of the 2nd ACM Workshop on Wireless Security (ACM WiSe 2003)*, pp. 30–40, ACM, San Diego, CA, USA, 19 September 2003.

[23] Y. Yao and J. Gehrke. The cougar approach to in-network query processing in sensor networks. *SIGMOD Record*, 31(3): 9–18, September 2002.

[24] J. N. Al-Karaki and A. E. Kamal. Routing techniques in wireless sensor networks: a survey. *Wireless Communications, IEEE*, 11(6): 6–28, 2004.

[25]   D. Moore, J. Leonard, D. Rus and S. Teller. Robust distributed network localization with noisy range measurements. In *Proceedings of the 2nd ACM International Conference on Embedded Networked Sensor Systems (ACM SenSys* 2004), pp. 50–61, 3–5 November 2004.

[26]   O. Tekdas and V. Isler. Sensor placement for triangulation based localization. *IEEE Transactions on Automation Science and Engineering*, 7(3): 681–685, 2010.

[27]   T. He, C. Huang, B. M. Blum, J. A. Stankovic and T. Abdelzaher. Range-free localization schemes for large scale sensor networks. In *Proceedings of the 9th Annual ACM/IEEE International Conference on Mobile Computing and Networking (MobiCom'03)*, pp. 81–95, San Diego, CA, USA, 14–19 September 2003.

[28]   L. C. S. Zhang and D. Chen. Accurate and energy-efficient range-free localization for mobile sensor networks. *IEEE Transactions on Mobile Computing*, 9(6): 897–910, June 2010.

[29]   B. Karp and H. T. Kung. GPSR: Greedy perimeter stateless routing for wireless networks. In *Proceedings of the 6th Annual International Conference on Mobile Computing and Networking (ACM MobiCom 2000)*, pp. 243–254, ACM, Boston, MA, USA, 6–11 August 2000.

[30]   K. Sha, J. Du and W. Shi. WEAR: A balanced, fault-tolerant, energy-efficient routing protocol for wireless sensor networks. *International Journal of Sensor Networks*, 1(3/4): 156–168, 2006.

[31]   Y. Yu, D. Estrin and R. Govindan. *Geographical and Energy-aware Routing: A Recursive Data Dissemination Protocol for Wireless Sensor Networks*. Computer Science Department, Technical Report UCLA/CSD-TR-01-0023, UCLA, Los Angeles, CA, USA, May 2002.

[32]   K. Zeng, K. Ren, W. Lou and P. J. Moran. Energy aware efficient geographic routing in lossy wireless sensor networks with environmental energy supply. *Wireless Networks*, 15(1): 39–51, January 2009.

[33]   J. A. Sanchez, R. Marin-Perez and P. M. Ruiz. BOSS: Beacon-less on demand strategy for geographic routing in wireless sensor networks. In *Proceedings of the IEEE International Conference on Mobile Ad Hoc and Sensor Systems 2007 (IEEE MASS 2007)*, pp. 1–10, IEEE, Pisa, Italy, 8–11 October 2007.

[34]   R. W. O. Jumira and S. Zeadally. Energy-efficient beaconless geographic routing in energy harvested wireless sensor networks. *Concurrency and Computation: Practice & Experience*, 25(1): 58–84, January 2013.

[35]   Y. Xu, J. Heidemann and D. Estrin. Geography-informed energy conservation for ad hoc routing. In *Proceedings of the 7th Annual International Conference on Mobile Computing and Networking (ACM MobiCom 2001)*, pp. 70–84, ACM, Rome, Italy, 16–21 July 2001.

[36]   K. L. K. Kang and N. Abu-Ghazaleh. Securing geographic routing in wireless sensor networks. In *Symposium on Information Assurance (SIA) 2006*, Albany, NY, USA, 14–15 June 2006.

[37] W. R. Heinzelman, J. Kulik and H. Balakrishnan. Adaptive protocols for information dissemination in wireless sensor networks. In *Proceedings of the 5th Annual ACM/IEEE International Conference on Mobile Computing and Networking (ACM MobiCom 1999)*, pp. 174–185, ACM, Seattle, WA, USA, 15–20 August 1999.

[38] C. Intanagonwiwat, R. Govindan and D. Estrin. Directed diffusion: A scalable and robust communication paradigm for sensor networks. In *Proceedings of the 6th Annual International Conference on Mobile Computing and Networking (MobiCom '00)*, New York, NY, USA, pp. 56–67, ACM, 2000.

[39] M. Chu, H. Haussecker, F. Zhao, M. Chu, H. Haussecker and F. Zhao. Scalable information-driven sensor querying and routing for ad hoc heterogeneous sensor networks. *International Journal of High Performance Computing Applications*, 16(3): August 2002.

[40] D. Ganesan, R. Govindan, S. Shenker and D. Estrin. Highly-resilient, energy-efficient multipath routing in wireless sensor network. *SIGMOBILE Mobile Computing and Communication Review*, 5(4): 11–25, October 2001.

[41] C. Lin and M. Gerla. Adaptive clustering for mobile wireless networks. *IEEE Journal on Selected Areas in Communications*, 15(7): 1265–1275, September 1997.

[42] S. Lindsey and C. Raghavendra. PEGASIS: Power-efficient gathering in sensor information systems. In *Proceedings of the 2002 IEEE Aerospace Conference*, vol. 3, pp. 1125–1130, Big Sky, MT, USA, 9–16 March 2002.

[43] S. Lindsey, C. Raghavendra, and K. Sivalingam. Data gathering algorithms in sensor networks using energy metrics. *IEEE Transactions on Parallel and Distributed Systems*, 13(9): 924–935, September 2002.

[44] A. Manjeshwar and D. Agrawal. TEEN: A routing protocol for enhanced efficiency in wireless sensor networks. In *Proceedings of the 15th International Parallel and Distributed Processing Symposium (IEEE IPDPS 2001)*, pp. 2009–2015, San Francisco, CA, USA, 23–27 April 2001.

[45] A. Manjeshwar and D. Agrawal. APTEEN: A hybrid protocol for efficient routing and comprehensive information retrieval in wireless. In *Proceedings of the 16th International Parallel and Distributed Processing Symposium (IEEE IPDPS 2002)*, Abstracts and CD-ROM, Fort Lauderdale, FL, USA, 15–19 April 2002.

[46] S. Zeadally, A.-S. Pathan, C. Alcaraz and M. Badra. Towards privacy protection in smart grid. *Wireless Personal Communications*, 73(1): 23–50, 2013.

Chapter 8

Resource management in cyber-physical systems

Ali Abedi[1], Fatemeh Afghah[2] and Abolfazl Razi[3]

Abstract

Cyber physical systems (CPSs) represent the new generation of systems as the integration of computational and physical resources with significant computation, sensing, communication, and control capabilities. Noting the limited battery energy, limited memory, and finite spectrum available in each wireless sensor node, development of more efficient resource allocation methods becomes critical in these networks. The question is whether the physical layer presents more opportunities for efficient resource allocation or whether medium access control is the place to implement such methods, or perhaps higher layers. This chapter attempts to answer this question by reviewing the state of the art in the literature and proposing some new ideas and future research directions to complement recent results. In particular, we review recent game theory approaches that not only optimize key performance metrics under system constraints but also make a balance between the system overall efficiency and fairness in the resource allocation policy. This fundamental study involves new policy design for scheduling, routing, and transmission for basic multiple access scenarios. It paves the way for designing and optimizing next-generation networks with an arbitrary number of mobile nodes having a wide range of capabilities, such that the overall performance of the system is kept as high as possible while the nodes with poor local conditions do not deprive from basic service requirements. A comprehensive study of recent results in the literature is provided to put this multidimensional optimization problem into the right context. Various examples and scenarios are introduced and some approaches to analyze them are discussed.

[1]Electrical and Computer Engineering Department, University of Maine, Orono, ME, USA, e-mail: ali.abedi@maine.edu
[2]Electrical Engineering and Computer Science Department, Northern Arizona University, Flagstaff, AZ, USA, e-mail: fatemeh.afghah@nau.edu
[3]Electrical Engineering and Computer Science Department, Northern Arizona University, Flagstaff, AZ, USA, e-mail: abolfazl.razi@nau.edu

8.1   Introduction

Cyber-physical systems (CPSs) represent the new generation of systems as the integration of computational and physical resources with significant computation, sensing, communication, and control capabilities [1–3]. CPSs have become a critical aspect in various applications including healthcare monitoring, smart grid [4], aerospace [5], smart transportation systems, infrastructure and environmental monitoring [6], and autonomous multi-agent systems. Wireless sensor networks (WSNs) are generally a key component of the CPSs given the considerable demand for reliable and real-time monitoring and control systems [7]. The implementation of WSN in CPS allows the data collection from various geographically distributed sensors as an input for the control system to make a real-time decision and activate the corresponding actuators.

Most engineering problems have to deal with limited resources when solving a real-world application. WSNs are not an exception and in fact have to be designed under very tight limitations and boundary conditions. Noting the limited battery energy, limited memory, and finite spectrum available in each wireless sensor node, development of more efficient resource allocation methods becomes critical in these networks. A desirable resource allocation scheme needs to have a very low overhead, while jointly managing power and spectrum. Other network resources, such as timing and quality of service (QoS), which is related to latency, need to be managed as well, making the resource allocation problem a multidimensional optimization challenge.

One question that comes up is where this optimization should take place. Does the physical layer present more opportunities for efficient resource allocation or is medium access control the place to implement such methods, or perhaps higher layers? There is no quick answer to this question and the solution strongly depends on the system conditions and the desired objective. In fact, changing a parameter in one layer may significantly impact the resource allocation in other layers. On the other hand, tuning parameters of one layer solely based on this particular layer and ignoring other layers may severely deteriorate the overall performance. For instance, we will show, in the following sections, that how adjusting packet lengths at higher layers based upon physical layer conditions may considerably reduce packet waiting time and hence better service utilizations. Therefore, cross-layer optimizations have become an important trend in the recent research area of resource allocation. When it comes to CPSs, the interactions between the network of sensor nodes and actuators in the physical world become an important factor in efficient resource allocation. Normally, the delay in changing the state of a physical device is orders of magnitude larger than latency and delays in a sensor node, i.e. an electronic component of the network. It is critical to take into account the characteristics of both types of network components when starting to think about resource allocation in CPSs.

This chapter attempts to answer some of the aforementioned questions by reviewing the state of the art in the literature and proposing some new ideas and future research directions to complement the recent results.

In order to have a general scenario applicable to a wide variety of applications, we do not limit this chapter to single, two-, or three-node systems in terms of research papers. Instead, we will consider multiple nodes transmitting sensory data and multiple nodes receiving the data and interacting with physical systems in parallel. This leads to the problem of scheduling and multiple access technique to limit the interference among the sensors in data transmission in order to save their limited available energy while increasing their throughputs.

In applications where the sensors are distributed over a vast geographical area, direct data transmission between the source node and the sink node may not be possible, in view of the limited transmission power at the sensors. Multi-hop communication is a promising technique to expand the communication range of the sensors and enhance the network connectivity [8, 9]. In this method, the source node requests the intermediate sensors in the network to forward their data to the corresponding destination node. This cooperative packet-forwarding by the nodes results in increasing the network throughput and extending the network coverage; however, it makes the resource management problem even more complicated. In this case, the sensors are not only required to optimize their own data transmission but also are supposed to provide a relaying service to the other nodes, which may result in exhausting their limited energy.

To represent the practical scenario of ad hoc multi-hop communication among the sensors, we consider the system model where the intermediate nodes in between the transmitter and receiver nodes can act as relay nodes creating a parallel relay channel. In this chapter, the problem of resource allocation in WSN is studied from the different aspects of (i) a single-sensor perspective to manage the limited power and bandwidth, (ii) a network perspective to find the optimum scheduling access to the shared wireless channel in order to reduce the interference, and (iii) design objectives related to the specific applications (e.g. latency). The optimum resource allocation solution for the sensors is obtained when accounting for the individual, network, and application-based preferences altogether. Indeed this results in a multi-faceted optimization problem. Solving a multidimensional optimization problem like this with conventional methods might be hard if not impossible. Game theoretical approaches seem to be more efficient ways of attacking this problem [10, 11].

In this chapter, we first review the different aspects of resource allocation problem in WSN-CPS. Then we provide a short overview of game theory as a powerful mathematical tool to investigate the problem of resource allocation in WSN-CPSs. Then we discuss the implementation of multi-hop communication and cooperative relaying in WSNs with more details. Some recent results are presented afterwards. We conclude with some future directions at the end of this chapter.

8.2 Resource allocation

Resource allocation can be viewed from various points of view. In this section, we present a few recent approaches with a diverse set of examples. A model called the Resource Space Model is proposed in [12] in order to manage flexible resources in a multidimensional manner. The proposed model with its probabilistic feature is capable of handling uncertainty in resource allocation when it comes to CPSs.

One example for resource allocation in WSNs cooperating with robots is presented in [13]. Robot task allocation and robot task fulfillment were considered in this work. Robots may cooperatively decide on the set of tasks to be individually carried out to achieve a desired goal. Fulfilling the assigned tasks through intelligent mobility scheduling is another aspect that was considered in this work.

Overlaying a cyber network on top of a physical network creates systems that are prone to attack due to the interdependence between the two networks. Failure in one network will affect the other network, creating an avalanche effect that might shut down the entire system. The robustness of these networks depends on the link allocation between the two networks. Optimum inter link allocation strategy against random attacks in the case where the topology of each individual network is unknown is considered in [14].

The problem of energy allocation over source acquisition/compression and transmission for energy-harvesting sensors is considered in [15]. Queue stability is considered as the main criteria when developing energy allocation strategies. Some suboptimal strategies are also studied where only the encoders' energy is allocated or there is no battery use. Meeting average distortion constraints in the sensor network posed by cyber–physical interactions using time division strategies in this work is another good example.

Another example is the electricity cost management for internet service providers [16]. Interesting challenges arise when physical power management is tied with internet dynamic demand. Instead of reducing power at each location independently, this work considers the total electricity cost for all locations jointly. This is a challenging problem with diverse spatio-temporal electricity cost. Both the center-level load balancing and the server-level power control are managed together.

Road safety and on-road infotainment systems are another set of examples in vehicular networks, which constitutes another class of CPSs. Limited network resources, possible pre-emption and contention between services for the display and non-negligible driver processing delay, make this an interesting example where driver-centric resource allocation is considered [17]. Heuristics solutions that are proposed in this work include wireless transmission failure, distributed implementation of the multi-sender systems, and utilizing real traces collected from taxis in the city of Shanghai.

Cross-layer optimization for hybrid crowd-sourcing in CPSs is considered in [18], where computing resource management, routing, and link scheduling are conducted jointly. Lower and upper performance bounds on the proposed algorithms provide an interesting insight into this problem.

Mobility is another aspect that needs to be considered before resource allocation is conducted. A novel mobile CPS for crowd-sensing applications is introduced in [19]. This scheme integrates service-oriented architecture with resource optimization.

Direct communication between the users in CPSs is not always possible due to the short communication range of the low power entities as well as the wireless channel situation. Therefore, cooperative multi-hop communications among the users can assure a reliable communication and extend the network throughput and connectivity. Cooperation among a large number of intelligent nodes brings up several questions for the users. The first problem is that each user needs to decide whether or not to participate in cooperation considering the amount of resources it requires to share with others during the cooperation and the benefits it may gain out of this [9, 20, 21]. The cooperative communication also brings some other issues to light, including the optimum routing path and best relay selection [22] and fair resource allocation [23]. Fairness can focus on energy usage, QoS, time sharing, or spectrum sharing [24]. Common properties of fairness and some management strategies are also presented in this work.

Mixing applications with different levels of importance makes the resource allocation a challenging task. Efficient utilization of hardware resources for worst-case blocking time computation used for scheduling analysis and resource sharing is presented in [25].

The resource allocation process can be performed in a central or distributed manner. In the central approach, a controller node needs to constantly monitor all network participants and capture their corresponding parameters in order to allocate the resources in an optimum manner. However, this approach is not always applicable to WSN-CPS systems, since it involves constant real-time communications between the central unit and the users. Noting the time-varying and random nature of these systems, the central control of the system imposes a heavy signaling overhead on the system.

In distributed approaches, the goal is to distribute the optimization algorithm and decision-making process among the users in a way that each node chooses its parameter based on its own observation of the network state. An important practical methodology to solve problems with distributed nature and multiple objectives is game theory, which provides a solution for the players in order to obtain the most possible benefits from both individual and overall point of views.

In continuation, we review the concept of game theory as a promising mathematical tool to study the interactions among the intelligent nodes in a WSN-CPS in order to select the optimum resource allocation, followed by a detailed discussion on cooperative communication and a review of the state-of-the-art advances in resource allocation in WSN-CPS.

8.3 Game theory

Game theory is capable of modeling the complex interactions among several rational agents. Agents that evaluate the network status and try to maximize a preset utility function are called rational. Therefore, game theory can describe the behavior of rational agents that compete with one another to take their shares from a common resource pool. It also models a set of cooperative players that cooperate with each other to achieve a common goal. Game theory analyzes the strategic thinking of players, which utilizes the available information to make the best decision at various situations [26, 27]. In recent years, game theory has been widely used in various fields, such as biology, economics, politics, and engineering [9, 28, 29].

In this section, a brief overview of game theory in the context of resource allocation in CPSs is presented. Noting the significant computation capability of the communication-enabled autonomous entities in CPSs, they are defined as smart agents since they are capable of observing the network situation and making the appropriate decisions to optimize their performance. We start this section with some general examples, followed by formulation and mathematical modeling, and continue with applications in power allocation and cooperative relaying in CPSs in the next sections. Materials presented in this section are adopted from [30].

To illustrate the operation of game theoretical solutions, we provide a famous game example, called *Prisoner's Dilemma*. In this game model, a policeman has arrested two criminals to charge them with a potential criminal action. The policeman is sure about the crime, but does not have enough evidence to prove it. The only way to convict is to urge the suspects to confess. If both deny the crime, they will only be convicted of a minor crime and spend a short time in the jail (one year).

Therefore, the policeman designs a scenario to make the suspects admit the crime. The officer interrogates the suspects in two separate rooms, where the suspects are not allowed to communicate with each other. However, they are both aware of the rules and consequences of their own and the other suspect's decisions.

In order to encourage the suspect to confess, the officer offers them a deal. If one of them confesses while the other one does not, the one who confesses will be released and the one who denies will be imprisoned for 10 years. If both confess, the verdict is five years for each of the suspects. If both deny the crime, one year of imprisonment is decided for each of the suspects. The possible options available to the suspects, along with their corresponding sentences are presented in Table 8.1.

Without considering the consequence of the other suspect, the logical decision might seem to deny the crime. However, knowing the game rules and the consequences of both players' decisions, and considering the fact that each suspect is answering the police independently and is not aware of the other suspect's action, they both may prefer to confess to the crime as to minimize their jail time. This is done by a hidden encouragement mechanism in the game rule that urges each of the suspects to confess in order to minimize the jail time.

Generally, a game is defined by a set of rules and parameters to simulate the interaction among rational agents. One commonly used notation to present a game model with N players is

$$G =< Q, \{A_i\}, \{U_i\} >, \{i \in Q\} \tag{8.1}$$

where the game parameters are defined as follows:

- $Q = \{1, 2, \ldots, N\}$ is the finite set of players.
- A_i is the action set of player i, where $a_i \in A_i$ denotes a non-empty subset of user i action.

Table 8.1   *Prisoner's dilemma as a classical example of game theoretical modeling of problems in WSN*

Suspect 1/Suspect 2	Confess	Deny
Confess	−5, −5	0, 10
Deny	−10, 0	−1, −1

- Action profile of the game is defined by $A = A_1 \times A_2 \times \cdots A_N$.
- U_i is the utility function of player i.

The utility function quantifies the objectives of players by mapping the action of users to real numbers. Maximizing the utility function models the motivation of users to achieve their goals. For instance, $U_i(a_i)$ represents the satisfaction that user i obtains by selecting the action set a_i. This means that the utility function assigns a number to every possible outcome of the game, where a higher number implies that the outcome is highly preferred. In game theory applications, it is assumed that the users are rational, meaning that they intend to maximize their utility functions and select the strategy that provides them with the highest possible benefits.

Games are divided into different categories based on various criteria [31]. One of these classifications is dividing the game into two groups of cooperative and non-cooperative games. In non-cooperative games, the players act selfishly and compete with each other to maximize their individual interests. In these games, the players are not able to negotiate with one another. Hence, finding the solution of the game is equivalent to a joint optimization problem, where each player tries to optimize his or her own utility. In contrast, in cooperative games, the players can communicate with one another to obtain a mutually beneficial solution. In these games, all players may have the same goal of optimizing a social welfare, or they may be divided into groups with the same benefits, called coalitions [32, 33].

In a non-cooperative case, the solution of the game is usually defined by Nash equilibrium (NE) strategy. In an NE strategy set, no player has any incentive to unilaterally change its strategy to increase its utility, provided that the other players keep their strategies unchanged. Solution of the game achieves the Nash equilibrium strategy set for all players when each rational agent (player) selects its best possible response to other players' strategies, provided that neither player can increase its utility by unilaterally changing its strategy. The following is a definition of NE.

A strategy profile, S^* achieves NE if [34],

$$\forall i \in Q, \forall s_i \in S_i, \quad U_i\left(s_i^*, S_{-i}^*\right) \geq U_i\left(s_i, S_{-i}^*\right) \tag{8.2}$$

where S_{-i} denotes the strategy of all players except player i. Every game with a finite number of players and a finite strategy profile has at least one NE in pure or mixed strategies [35].

Different classes of games including non-cooperative games, stochastic game, repeated game, and potential game [22] are widely used in the WSN-CPS to model the resource allocation [10, 11, 36], encourage cooperative relaying, and perform coordinative task management.

8.4 Cooperative relaying

With the growing demand for WSNs, and emerging CPSs, service providers are forced to increase network capacity and establish new base stations for data processing. Installing new radio base stations to provide network coverage for more users is time-consuming and costly and might be infeasible in some situations. Physical network expansion also changes the network topology and requires substantial changes in network parameter assignments. Hence, in some cases, utilizing new radio stations is not practical to expand the network coverage. The other issue gaining attention in CPSs with WSNs is the limited transmission power available at the nodes. This would limit the communication range of the sensors and restrict their direct accessibility to the base station.

Furthermore, even if a user is located in a network coverage area, the QoS may not be in an acceptable range. The quality of radio waves that propagate through communication channels might be affected by several undesired phenomena such as noise, interference, and fading. The fading effect could be due to signal reflection, diffraction, and scattering from fixed and mobile obstacles in the radio path. The fading may cause severe attenuation and distortion in the radio waves, which in turn causes problems in extracting data from the received signals at the destination node.

In recent years, different diversity techniques have been studied to overcome the fading effect [37]. The idea behind using diversity techniques is to transmit the signal through different radio channels and combine them in a constructive way at the receiver. Diversity techniques can be realized through different methods including time, frequency, and space diversity. In time/frequency diversity techniques, multiple versions of the original signal are transmitted at different time slots/frequencies. Another commonly used diversity technique is space diversity, where additional signal paths are created using multiple transmit or receive antennas. However, adding extra antennas at the transmitter or receiver is not always possible due to the size (relative to wavelength) and hardware limitations of WSN-CPS devices.

The aforementioned challenges regarding wireless network expansion as well as the QoS improvement have led researchers to employ the idea of cooperative relaying. In this technique, the existing users in the network provide a relaying service to the others [38, 39]. This model is general in the sense that it incorporates multiple sources, destinations, and relays to cover a wide variety of CPS applications.

Cooperative relaying can expand the network coverage and connectivity to provide service for the users, who do not have access to the network. Moreover, in this technique, parallel paths can be realized via relaying from other users in the network to obtain the diversity gain. Furthermore, cooperative relaying provides a number of advantages including higher network throughput, improved power, and spectral efficiency, as well as lower implementation cost; however, providing a relaying service to other users in the network is not an inherent characteristic of the users as it results in consuming their available network resources such as bandwidth, energy, and memory.

One important challenge in implementation of cooperative packet forwarding is providing enough incentive for users to encourage them to participate in cooperation via relaying other node packets. This encouragement could be in the form of service price (e.g. transferring money) or obtaining the chance to use cooperative relaying from others when needed. In this case, each user will decide whether it is beneficial to consume its resources and forward other users' message to the destination with the objective of receiving compensation from the source, or whether it is better to reject the cooperative relaying request from its neighbors. In WSNs, this issue becomes more significant, since each node has limited energy. Making a decision regarding participation in cooperative relaying depends on the role that each user plays in the network. In general, each node in the network may work as source or relay node in different phases.

While serving as source, each node needs to consider the quality of its direct link to the destination, the internal traffic of its neighbors, who are willing to relay the source node's packets, the energy costs of relaying as well as the signaling overhead. These considerations will assist in the decision process of asking the other nodes for relaying service or attempting to transmit the packet directly to the destination. On the other hand, to decide on providing relaying service for other nodes, each user should consider the potential gain it receives for cooperative service and whether this gain compensates the costs of consuming its limited resources, including bandwidth, energy, and memory, in addition to the extra signaling required to handle the cooperation.

Game theory is a proper approach to modeling a network, which consists of selfish nodes that try to maximize their own benefits. Therefore, it can appropriately model several resource allocation problems in wireless networks [40–43]. Due to the inherent trade-offs such as power allocation and packet forwarding in cooperative communication, a game theoretical approach can be used to study these networks.

Game theory is used to analyze the trade-off between nodes' interest to avoid forwarding others' packets in order to save transmission power versus providing relaying service to increase the system throughput [44]. Utilizing game theory maximizes the overall system performance in terms of different objectives such as maximum throughput and minimum delay requirements as well as low power consumption cost and implementation simplicity. Several game theoretical approaches have been studied in the literature to model the cooperative packet forwarding in wireless networks and encourage the selfish user to provide cooperative service [19]. Repeated games were considered in [45] to analyze cooperative packet forwarding. In these games, players interact with each other in consecutive stages considering others' actions in previous rounds of the game. Therefore, they can be encouraged to cooperate with each other by using incentive mechanisms such as reputation algorithms. A distributed self-learning repeated game is proposed in [46] to study the packet forwarding. In [47], a cooperation enforcement mechanism is proposed to prevent the users from misbehavior in ad hoc networks. Markovian games are utilized to address this issue. In these games, actions of players are determined based on the current state of the game, instead of considering the complete game history [18, 48].

8.5 Recent results

8.5.1 *Fairness versus efficiency*

An important aspect of optimal resource allocation is to address the fundamental inherent trade-off between efficiency and fairness of the solution. The optimal resource allocation is usually recast as an optimization problem under limiting constraints that are imposed by the system implementations and finiteness of the resource pool. The objective function quantifies one or more performance metrics of interest such as the information rate, QoS, and end to end latency. The objective function is usually characterized properly to reflect the overall system performance.

This optimization procedure aims at elevating the average QoS for the network users and the resulting solution is called the effective solution. This solution may settle in a totally unfair situation, causing some network users to be deprived of a basic service level [49].

A classic example of this situation is parallel Gaussian channel, where a number of source nodes communicate with their designated destination nodes via Gaussian channels under total power constraint. It is known that the optimal strategy for this problem that maximizes the total information rate follows the water-filling rule, where the most power is assigned to a user with the lowest noise level. This solution is depicted for a three-user system in Figure 8.1. It is seen that the highest amount of power is allocated to user 2, who benefits from the best channel signal-to-noise ratio (SNR), while user 1 with the worst channel conditions does not receive any share of the shared available power pool. The intuitive justification for this result is that it is desirable to invest more on the users that are more efficient in order to maximize the total system information rate.

The efficient solution may not be acceptable in real-world applications if it leads to unbearable difference between users' QoS. For instance, in a time division multiple access (TDMA) system, it is desired to ensure that a minimum number of time slots is granted for each user to facilitate the basic voice communications. In such scenarios the fair solution is desired, where all users are provided with similar QoS (e.g. information rate).

For instance, in the aforementioned parallel Gaussian example, the fairest solution is simply obtained by equaling the SNR for all users:

$$\frac{P_i}{N_i} = v, \quad \sum_{i=1}^{3} P_i = P \Rightarrow P_i = P \frac{N_i}{\sum_{i=1}^{3} N_i} \tag{8.3}$$

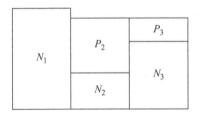

Figure 8.1   *Power allocation in parallel Gaussian channel. Three source nodes send their information to their target destinations via parallel Gaussian channels. N_i is the equivalent noise power at destination node i and P_i is the power assigned to user i, where the total power is constrained $\sum_{i=1}^{3} P_i = P$. The optimal solution is water-filling power allocation [50]*

which yields the same transmission rate for all users. In a practical system, we can make a balance between efficiency and fairness by incorporating various fairness parameters into the optimization objective function.

Various fairness indices including min-ratio, max-ratio, min-max fairness, proportional fairness, α-fairness, Atkinson's index, entropy measure, and Jain's index are proposed in the literature to characterize fairness level and leverage it into resource allocation problems. Recently, an axiomatic theory of fairness has been developed in [38], where five fairness axioms of continuity, homogeneity, saturation, partition, and starvation are defined to determine a legitimate fairness index. It has been shown that a unique family of globally decomposable functions satisfies these axioms. Fortunately, this family encompasses the most commonly used fairness indices. In an elegant formulation, a single generalized parametric function is proposed that specializes to various popular fairness indices by tuning parameters β as follows [51]:

$$f_\beta(x) = \text{sign}(1 - \beta) \left[\sum_{i=1}^{n} \left(\frac{x_i}{\sum_j x_j} \right)^{1-\beta} \right]^{\frac{1}{\beta}} \tag{8.4}$$

where x_i represents the portion of common resource allocated to user i, $i = 1, 2, \ldots, n$.

8.5.2   Unifying framework for multi-resource allocation

Another important dilemma arises when the resource allocation problem involves several heterogeneous infinitely divisible resource types. The two major trends to solve this problem can be listed as fairness on dominant shares (FDS) and generalized fairness on jobs (GFJ) [52]. In the FDS approach, each user is associated with a resource type with the highest share value for this user. The fairness index is then defined for all users to provide a fair resource allocation based on their dominant resource type. This elegant generalization extends the unified fairness index defined in [51] to a multi-resource scenario. Alternatively, GFJ takes a similar strategy and formulation, with only one difference that aims at balancing the completed jobs per users instead of pushing for the fair resource allocation.

To clarify more on the difference between these two approaches, note that in WSNs, the resource allocation is performed by assigning various types of shared resources such as time slots, frequency bands, transmission power to users in an optimal way in order to maximize the evaluation metric of interest such as average throughput, end to end delay or even application-specific QoS metric. For instance, peak signal to noise ratio (PSNR) and mean opinion score (MOS) metrics are used as application-specific evaluation metrics for image and speech reconstruction scenarios. Consequently, the fairness may be imposed either on the resource utilization or on the desired performance metric.

8.5.3   Advanced resource allocation in wireless sensor networks

Several resource allocation scenarios have been investigated for computer networks and wireless networks considering both efficiency and fairness, where a comprehensive survey can be found in [49, 53]. Another important variant of the resource allocation problem is optimal spectrum sharing in the context of cognitive radio networks (CRNs), where a shared frequency band is assigned to the users of coexisting networks with different level of priorities [54–56].

However, the resource allocation problem in wireless networks may be much more complex than simply partitioning a shared resource among users. For instance, a scenario of bundling sensor measurements into transmit packets with constant header sizes can be regarded as a resource allocation problem, where transmit packets as network resources are assigned to various sensor measurements. In [57], this problem is solved for a single-hop wireless channel equipped with first-come, first-served (FCFS) scheduling and automatic repeat request (ARQ) mechanism in order to minimize the end to end delay, while satisfying the queue stability conditions. This study suggests channel-adaptive framing policy as a cross-layer optimization, where the packet length as a higher layer protocol parameter is adjusted based on the underlying physical layer conditions. As a rule of thumb, it has been shown that for channels with extremely low error rate, longer packets are more efficient due to less overhead cost. However, as channel error increases, the optimal packet length becomes shorter to avoid frequent packet retransmissions due to high packet drop rate. If the channel conditions are slow-varying, the dynamic packet length policy based on the current channel status is highly recommended.

A similar study proposes the idea of local packet length adaptation based on the channel conditions in order to maximize throughput in wireless local area network (WLAN) channels [58].

Transmission parameter tuning in order to achieve different performance objectives has been investigated from different perspectives in WSNs. An interesting evaluation parameter is the age of information defined as the time difference between the symbol generation epoch and its delivery time to the destination. An optimum sampling rate is found in order to minimize the expected age of information at the destination for a single-hop queuing system in [59].

A novel resource allocation problem is discussed in [60], where a set of measurements' samples are queued and transmitted to the destination through packet erasure channels. Each packet is assigned with a strict delivery deadline and is discarded if received after the deadline. The optimization problem is defined as assigning a predefined total number of bits among the samples in order to minimize the expected distortion using an arbitrary convex distortion function. These problems demonstrate how various sampling and transmission parameters can be regarded as network resources in a complex wireless network equipped with queuing systems in order to optimize a desired performance metric.

An important extension to the resource allocation problem in wireless networks has recently emerged for the new generation of networks that consist of energy-harvesting nodes to prolong the operation of energy-constrained wireless networks. In these networks, the available energy at each node is not constant; rather it is an increasing function of time in general. The energy of each node can be constantly raised using an advanced energy-harvesting method or can even be exchanged among users under an energy communications framework [61].

Recently, the problem of optimal transmission policies for energy-harvesting nodes is tackled for various system set-ups, including queuing and blocking systems over wireless channels in order to maximize data throughput [62, 63]. The throughput optimization is considered for energy-harvesting nodes considering delay-constrained communications [64]. An optimal transmission policy is proposed for wireless sensor networks of energy-harvesting nodes communicating over slotted fading channels in order to minimize the average distortion level for delay-constrained communications in [65]. In [66], the main objective is delay minimizations. The most appealing recent attempt is to solve the optimization problem for cognitive radio networks where part of the users in one or more of the coexisting networks are energy-harvesting nodes [67].

8.6   Future directions

In this chapter, a general multi-node wireless sensor network with limited cyber-physical resources is considered. A comprehensive study of recent results in the literature is provided to put this multidimensional optimization problem into the right context. Various examples and scenarios are introduced and some approaches to analyze them are discussed.

The main challenge in future research into the resource allocation problem in CPSs is the scalability problem. Fundamental limits on latency, throughput, and fairness, when the number of users asymptotically approaches infinity, need to be derived. Other optimization methods, besides the game theoretical analysis proposed in this chapter, need to be developed. Even for game theoretical methods, there is a lot of room for research on how to define utility functions and the weights of the various parameters.

With emerging technologies in smart homes and smart grid areas, the number of CPS and WSN nodes supporting them is going to increase exponentially. This chapter is just a starting point towards development of robust, efficient, and fair networking protocols for the next generation of WSN and CPS.

Another important resource allocation is parallel processing and cloud computing. With the recent advances in the computational power of network nodes in wireless networks, the idea of in-network computation becomes more and more popular to divide the required processing task at network nodes rather than transforming huge amounts of data to the central data processing nodes. In most cases, a statistical function of data is desired as sufficient information for further processing. For instance, if the objective of a habitat monitoring sensor network is to periodically update the minimum, mean and maximum statistic of the temperature, sending the raw measurements from all sensor nodes to the central station is highly inefficient. In [68], the idea of in-network statistical function computation is used to achieve maximum throughput in WSNs. In [69], the theoretical limits for in-network computation is derived for directed and undirected graphs. In-network computation is realized in the context of cognitive radio networks by employing distributed source coding and broadcasting in wireless communication to increase spectral efficiency of sensor networks [70].

Another important direction of resource allocation is the application of energy harvesting. The general idea in energy harvesting is that the total power is available for the transmitter through a phonetically statistical time function with applications in remote powering, green communications and solar powers [71–73].

A comprehensive review on the recent advances in communication algorithms including scheduling, coding, retransmission policy, and multiple access techniques for energy-harvesting nodes is provided in [74].

Reference

[1]   E. A. Lee. Cyber physical systems: Design challenges. In *Proceedings of the 11th IEEE Symposium on Object/Component/Service-Oriented Real-Time Distributed Computing (ISORC '08)*, pp. 363–369, Orlando, FL, USA, May 2008.

[2]   J. Wan, H. Yan, H. Suo and F. Li. Advances in cyber-physical systems research. *KSII Transactions on Internet and Information Systems,* 5(11): 1891–1908, 2011.

[3]   E. A. Lee. Cyber-physical systems – are computing foundations adequate? In: *Position Paper for NSF Workshop on Cyber-Physical Systems: Research Motivation, Techniques and Roadmap.* Austin, TX, USA, October 2006.

[4]   A. Zaballos, A. Vallejo and J. M. Selga. Heterogeneous communication architecture for the smart grid. *Network, IEEE,* 25(5): 30–37, September–October, 2011.

[5]   W. Edmonson, S. Gebreyohannes, A. Dillion, R. Radhakrishnan, J. Chenou, A. Esterline and F. Afghah, Systems engineering of inter-satellite communications for distributed systems of small satellites. In *Proceedings of the IEEE Systems Conference (IEEE SysCon '15),* pp. 705–710, Vancouver, BC, Canada, April 2015.

[6]   R. A. Leon, V. Vittal and G. Manimaran. Application of sensor network for secure electric energy infrastructure. *IEEE Transactions on Power Delivery,* 22(2): 1021–1028, 2007.

[7]   C.-Y. Lin, S. Zeadally, T. Chen and C.-Y. Chang. Enabling cyber physical systems with wireless sensor networking technologies. *International Journal of Distributed Sensor Networks,* 12: 2012.

[8]   E. Yaacoub, and A. Abu-Dayya. Multihop routing for energy efficiency in wireless sensor networks. Ch. 8 in *Wireless Sensor Networks – Technology and Protocols* (Mohammad A. Matin (ed.)). INTECH, Rikeka, Croatia, 2012.

[9]   R. J. Aumann and S. Hart. *Handbook of Game Theory, with Economic Applications.* North-Holland, Amsterdam, 1992.

[10] H.-Y. Shi, W.-L. Wang, N.-M. Kwok and S.-Y. Chen. Game theory for wireless sensor networks: A survey. *Sensors,* 12(7): 9055–9097, 2012.

[11] S. Shen, G. Yue, Q. Cao and Y. Fei. A survey of game theory in wireless sensor networks security. *Journal of Networks,* 6(3), 521–532, March 2011.

[12] H. Zhuge and Y. Xing. Probabilistic resource space model for managing resources in cyber-physical society. *IEEE Transactions on Services Computing,* 5(3): 404–421, 2012.

[13]   X. Li, R. Falcon, A. Nayak, and I. Stojmenovic. Servicing wireless sensor networks by mobile robots. *Communications Magazine, IEEE,* 50(7): 147–154, July 2012.

[14]   O. Yagan, Q. Dajun, Z. Junshan and D. Cochran. Optimal allocation of interconnecting links in cyber-physical systems: interdependence, cascading failures, and robustness. *IEEE Transactions on Parallel and Distributed Systems,* 23(9): 1708–1720, September 2012.

[15]   P. Castiglione, O. Simeone, E. Erkip and T. Zemen. Energy management policies for energy-neutral source-channel coding. *IEEE Transactions on Communications,* 60(9): 2668–2678, September 2012.

[16]   L. Rao, X. Liu, M. D. Ilic, and J. Liu. Distributed coordination of internet data centers under multiregional electricity markets. *Proceedings of the IEEE,* 100(1): 269–282, January 2012.

[17]   X. Li, C. Qiao, A. Wagh, R. Sudhaakar, S. Addepalli, C. Wu and A. Sadek. A holistic approach to service delivery in driver-in-the-loop vehicular CPS. *IEEE Journal on Selected Areas in Communications,* 31(9): 513–522, September 2013.

[18]   M. Li and P. Li. Crowdsourcing in cyber-physical systems: Stochastic optimization with strong stability. *IEEE Transactions on Emerging Topics in Computing,* 1(2): 218–231, December 2013.

[19]   X. Hu, T. H. S. Chu, H. C. B. Chan, and V. C. M. Leung. Vita: A crowdsensing-oriented mobile cyber-physical system. *IEEE Transactions on Emerging Topics in Computing,* 1(1): 148–165, June 2013.

[20]   A. K. Sadek, W. Yu and K. J. R. Liu. On the energy efficiency of cooperative communications in wireless sensor networks. *ACM Transactions on Sensor Networks,* 6(1), 2009.

[21]   Z. Zhou, S. Zhou, S. Cui and J. Cui. Energy-efficient cooperative communication in clustered wireless sensor networks. In *Proceedings of the IEEE Military Communications Conference, MILCOM 2006,* pp. 1–7, Washington, DC, USA, October 2006.

[22]   Z. Zhou, S. Zhou, J. H. Cui and S. Cui. Energy-efficient cooperative communication based on power control and selective single-relay in wireless sensor networks. *IEEE Transactions on Wireless Communications,* 7(8): 3066–3079, 2008.

[23]   F. Afghah, A. Razi and A. Abedi. Power allocation in parallel relay channels using a near-potential game theoretical approach. In *Proceedings of the 48th Annual Conference on Information on Sciences and systems (CISS'14),* pp. 1–6, Princeton, NJ, USA, March 2014.

[24]   H. Shi, R. V. Prasad, E. Onur and I. G. M. M. Niemegeers. Fairness in wireless networks: Issues, measures and challenges. *Communications Surveys & Tutorials, IEEE,* 16(1): 5–24, 2014.

[25]   Q. Zhao, Z. Gu and H. Zeng. HLC-PCP: A resource synchronization protocol for certifiable mixed criticality scheduling. *Embedded Systems Letters, IEEE,* 6(1): 8–11, March 2014.

[26]   M. Osborne and A. Rubinstein. *A Course in Game Theory.* MIT Press, Cambridge, MA, USA, 1994.

[27]   R. B. Myerson. *Game Theory: Analysis of Conflict*. Harvard University Press, Cambridge, MA, USA, 1997.

[28]   L. Lambertini. *Game Theory in the Social Sciences*. Routledge, London, 2011.

[29]   J. von Neumann and O. Morgenstern. *Theory of Games and Economic Behavior*. Princeton University Press, Princeton, NJ, USA, 1944.

[30]   F. Afghah. Design and Analysis of Cooperative Communications Networks Using Game Theory. PhD thesis, University of Maine, Orono, ME, USA, 2013.

[31]   M. Osborne. *An Introduction to Game Theory*. Oxford University Press, Oxford, 2000.

[32]   R. J. Aumann and J. H. Droze. Cooperative games with coalition structures. *International Journal of Game Theory*, 3(4): 217–237, 1974.

[33]   C. J. Harsanyi. A simplified bargaining model for the n-person cooperative game. *International Economic Review*, 4: 194–220, 1963.

[34]   D. Fudenberg and J. Tirole. *Game Theory*. MIT Press, Cambridge, MA, USA, 1994.

[35]   J. Nash. Non-cooperative games. *The Annuals of Mathematics, Second Series*, 54(2): 286–295, 1951.

[36]   F. Afghah, M. Costa, A. Razi, A. Abedi and A. Ephremides. A reputation-based Stackelberg game approach for spectrum sharing with cooperative cognition. In *Proceedings of the IEEE Conference on Decision and Control (CDC'13)*, pp. 3287–3292, December 2013.

[37]   D. Tse and P. Viswanath. *Fundamentals of Wireless Communications*. Cambridge University Press, Cambridge, 2004.

[38]   A. Sendonaris, E. Erkip and B. Aazhang. User cooperation diversity. Part I. System description. *IEEE Transactions on Communications*, 51(11): 1939–1948, November 2003.

[39]   J. N. Laneman, D. Tse and G. Wornell. Cooperative diversity in wireless networks: Efficient protocols and outage behavior. *IEEE Transactions on Information Theory*, 50(12): 3062–3080, December 2004.

[40]   A. McKenzie and L. DaSilva. *Game Theory for Wireless Engineers*. Morgan and Claypool, San Rafael, CA, USA, 2006.

[41]   F. Afghah, A. Razi and A. Abedi. "Stochastic game theoretical model for packet forwarding in relay networks. *Springer Telecommunication Systems Journal*, Special Issue on Mobile Computing and Networking Technologies, 52(4): 1877–1893, 2013.

[42]   V. Srivastava, J. Neel, A. B. Mackenzie, R. Menon, L. A. Dasilva, J. E. Hicks, J. H. Reed and R. P. Gilles. Using game theory to analyze wireless ad hoc networks. *IEEE Communications Surveys Tutorials*, 7(4): 46–56, 2005.

[43]   F. Afghah and A. Abedi. Distributed fair-efficient power allocation in two-hop relay networks. In *Proceedings of the IEEE International Conference on Sensing, Communication, and Networking (SECON'13)*, pp. 255–257, New Orleans, LA, USA, June 2013.

[44]   M. Felegyhazi, J.-P. Hubaux and L. Buttyan. Nash equilibria of packet forwarding strategies in wireless ad hoc networks. *IEEE Transactions on Mobile Computing*, 5(5): 463–476, 2006.

[45] M. Felegyhazi, L. Buttyan and J.-P. Hubaux. Equilibrium analysis of packet forwarding strategies in wireless ad hoc networks – the dynamic case. *Personal Wireless Communications*, 2775: 23–25, 2003 (Lecture notes in computer science).

[46] Z. Han, C. Pandana and K. J. R. Liu. A self-learning repeated game framework for optimizing packet forwarding networks. In *Proceedings of the IEEE Wireless Communication & Networking Conference (WCNC)*, pp. 2131–2136, New Orleans, LA, USA, March 2005.

[47] P. Michiardiand R. Molva. A game theoretical approach to evaluate cooperation enforcement mechanisms in mobile ad hoc networks. In *Proceedings of the IEEE/ACM Workshop on Modeling and Optimization in Mobile, Ad Hoc and Wireless Networks (WiOpt 2003)*, pp. 107–121, Sofia Antipolis, France, March 2003.

[48] Y. E. Sagduyu and A. Ephremides. A game-theoretic look at simple relay channels. *ACM/Kluwer Journal of Wireless Networks*, 12(5): 545–560, 2006.

[49] H. Shi, R. V. Prasad, E. Onur and I. G. M. M. Niemegeers. Fairness in wireless networks: Issues, measures and challenges. *Communications Surveys & Tutorials, IEEE*, 16(1): 5–24, 2014.

[50] T. M. Cover and J. A. Thomas. *Elements of Information Theory*, 2nd edn. Wiley, Hoboken, NJ, USA, 2005.

[51] T. Lan, D. Kao, M. Chiang and A. Sabharwal. An axiomatic theory of fairness in network resource allocation. In *Proceedings of IEEE INFOCOM 2010*, pp. 1–9, San Diego, CA, USA, March 2010.

[52] C. Joe-Wong, S. Sen, T. Lan and M. Chiang. Multi-resource allocation: Fairness-efficiency tradeoffs in a unifying framework. In *Proceedings of IEEE INFOCOM*, pp. 1206–1214, Orlando, FL, USA, 25–30 March 2012.

[53] N. Katoh, A. Shioura and T. Ibaraki. Resource allocation problems. In *Handbook of Combinatorial Optimization*, vol. 3 (P. M. Pardalos, Ding-Zhu Du and R. L. Graham (eds.)). Springer, New York, 2013.

[54] G. Vijay, E. Ben Ali Bdira and M. Ibnkahla. Cognition in wireless sensor networks: A perspective. *Sensors Journal, IEEE*, 11(3): 582–592, March 2011.

[55] M. T. Masonta, M. Mzyece and N. Ntlatlapa. Spectrum decision in cognitive radio networks: A survey. *Communications Surveys & Tutorials, IEEE*, 15 (3): 1088–1107, 2013.

[56] M. Naeem, A. Anpalagan, M. Jaseemuddin and D. C. Lee. Resource allocation techniques in cooperative cognitive radio networks. *Communications Surveys & Tutorials, IEEE*, 16(2): 729–744, 2014.

[57] A. Razi, A. Abedi and A. Ephremides. Delay minimization with channel-adaptive packetization policy for random data traffic. In *Proceedings of the 48th Annual Conference on Information Sciences and Systems (CISS)*, pp. 1–6, 19–21, Princeton, NJ, USA, March 2014.

[58] M. N. Krishnan, E. Haghani and A. Zakhor. Packet length adaptation in WLANs with hidden nodes and time-varying channels. In *Proceedings of the IEEE Global Telecommunications Conference (GLOBECOM)*, pp. 1–6, Houston, TX, USA, December 2011.

[59] S. Kaul, R. Yates and M. Gruteser. Real-time status: How often should one update? In *INFOCOM, 2012 Proceedings IEEE*, pp. 2731–2735, Orlando, FL, USA, 25–30 March 2012.

[60] A. Faridi and A. Ephremides. Distortion control for delay-sensitive sources. *IEEE Transactions on Information Theory*, 54(8): 3399–3411, August 2008.

[61] B. Gurakan, O. Ozel, Jing Yang and S. Ulukus. Energy Cooperation in energy harvesting communications. *IEEE Transactions on Communications*, 61(12): 4884–4898, December 2013.

[62] C. Huang, R. Zhang and S. Cui. Throughput maximization for the Gaussian relay channel with energy harvesting constraints. *IEEE Journal on Selected Areas in Communications*, 31(8): 1469–1479, August 2013.

[63] J. Xu and R. Zhang. Throughput optimal policies for energy harvesting wireless transmitters with non-ideal circuit power. *IEEE Journal on Selected Areas in Communications*, 32(2): 322–332, February 2014.

[64] L. Liu, R. Zhang and K.-C. Chua. Wireless information transfer with opportunistic energy harvesting. *IEEE Transactions on Wireless Communications*, 12(1): 288–300, January 2013.

[65] C. Huang, R. Zhang and S. Cui. Delay-constrained Gaussian relay channel with energy harvesting nodes. In *Proceedings of IEEE International Conference on Communications (ICC)*, pp. 2433–2438, Ottawa, Canada, 10–15 June 2012.

[66] M. Khoshnevisan and J. N. Laneman. Minimum delay communication in energy harvesting systems over fading channels. In *Proceedings of the 45th Annual Conference on Information Sciences and Systems (CISS)*, pp. 1–5, 23–25 March 2011.

[67] A. E. Shafie and A. Sultan. Optimal random access for a cognitive radio terminal with energy harvesting capability. *Communications Letters, IEEE*, 17(6): pp. 1128–1131, June 2013.

[68] R. Sappidi, A. Girard and C. Rosenberg. Maximum achievable throughput in a wireless sensor network using in-network computation for statistical functions. *IEEE/ACM Transactions on Networking*, 21(5): 1581–1594, October 2013.

[69] H. Kowshik and P. R. Kumar. Optimal function computation in directed and undirected graphs. *IEEE Transactions on Information Theory*, 58(6): 3407–3418, June 2012.

[70] K.-C. Chen. Improving spectrum efficiency via in-network computations in cognitive radio sensor networks. *IEEE Transactions on Wireless Communications*, 13(3): 1222–1234, March 2014.

[71] M. Gorlatova, A. Wallwater and G. Zussman. Networking low-power energy harvesting devices: Measurements and algorithms. *IEEE Transactions on Mobile Computing*, 12(9): 1853–1865, September 2013.

[72] B. Gurakan, O. Ozel, Jing Yang and S. Ulukus, S. Energy cooperation in energy harvesting communications. *IEEE Transactions on Communications*, 61(12): 4884–4898, December 2013.

[73] Tao Han and N. Ansari. On optimizing green energy utilization for cellular networks with hybrid energy supplies. *IEEE Transactions on Wireless Communications*, 12(8): 3872–3882, August 2013.

[74] S. Ulukus, A. Yener, E. Erkip, O. Simeone, M. Zorzi, P. Grover and K. Huang. Energy harvesting wireless communications: A review of recent advances. *IEEE Journal on Selected Areas in Communications*, 33: 360–381, 2015.

Chapter 9

Mobile sensors in wireless sensor network cyber-physical systems

Faisal Karim Shaikh[1] and Sherali Zeadally[2]

Abstract

To overcome the problem of coverage and efficient data management, mobility in wireless sensor networks (WSNs) has been attracting a lot of attention in recent years. WSN are also an integral part of cyber-physical systems (CPSs) because of their strong sensing ability. CPSs are considered to be an emerging technology, and there are many challenges that need to be addressed to enable the flexible integration of the cyber and physical domains. We present a survey of mobile WSN-CPS approaches that have been proposed to date, along with a mobile WSN-CPS application classification. The mobile WSN-CPS applications are then compared to highlight the similarities in and differences between the state-of-the-art technologies used in mobile WSN-CPS. We have also identified the areas of mobile WSN-CPS that require further research in the future.

9.1 Introduction

Wireless sensor networks (WSNs) represent a key enabling technology for the emerging cyber-physical systems (CPSs). The fusion of sensing and wireless communication has led to the rapid and wide proliferation of WSN compared to other technologies. Accordingly, the commercial use of WSN has increased dramatically, and the WSN market is forecasted to reach $12 billion worldwide by 2020 [1].

Generally, a WSN comprises a large number of static sensor nodes with low processing, limited power capabilities, and often communicating over short-range unreliable radio links. Additionally, sensor nodes have limited storage capacity, battery life and multiple on-board sensors that can take readings such as temperature, level of humidity, and acceleration from the environment. Since the communication

[1]Science and Technology Unit, Umm Al-Qura University, Kingdom of Saudi Arabia, and Mehran University of Engineering and Technology, Jamshoro, 76062, Pakistan, fkshaikh@uqu.edu.sa
[2]College of Communication and Information, University of Kentucky, Lexington, KY 40506-0224, USA, szeadally@uky.edu

range of sensor nodes is limited, hop-by-hop communication is often adopted by sensor nodes to exchange data. Typically, a powerful base station, termed as *sink*, is also an integral part of a WSN. The sink mediates between the sensor nodes and the applications running on a WSN.

The traditional WSN model is based on the assumption that the network is dense so that sensor nodes and the sink can communicate with each other through multi-hop communication links. More recently, mobility has also been introduced to WSN [2–5], as shown in Figure 9.1. As nodes are mobile, in order to reach the sink, a dense deployment of sensor nodes may not be needed. In fact, mobile nodes can cope with the network coverage problem due to inherent perturbations such as link outages and node crashes. Mobile sensor nodes can also be used to connect isolated regions that occur with the sparse deployment of sensors or as a result of the failure of some sensor nodes. Moreover, mobile nodes can visit sensor nodes in the network and collect data directly through single-hop communication. This reduces contention, collisions, and message loss [4]. Mobility in WSN can be either inherent depending upon the application or introduced when required to maximize the energy efficiency [6].

Many WSN applications are deployed to interact with the physical environment and report on the phenomena of interest to the user via the sink. This leads to

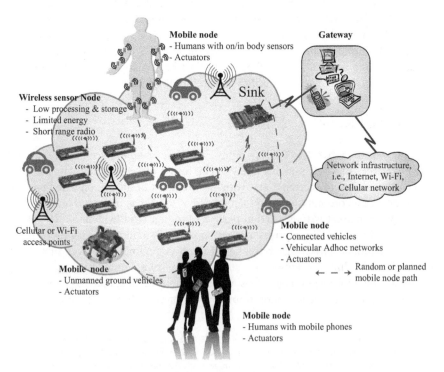

Figure 9.1   A mobile wireless sensor network system

an open-loop network where information is only flowing from the physical world to the user via a virtual world of networks. To close the loop, actuators are introduced in WSN, which led to CPS. In recent years, CPSs have emerged as a promising paradigm that can support human-to-human, human-to-machine, and machine-to-machine interactions in the physical and cyber worlds [7, 8]. CPSs provide the tight integration of physical and cyber systems which support abstract computations and physical processes [8–10], where sensors, actuators, and embedded devices are networked to sense, monitor, and control the physical world, and receive feedback on how physical processes affect the computations. With the advent of cellular technologies and the wide adoption of mobile devices such as smartphones, humans-in-the-loop (with mobile devices) have become an essential part of mobile WSN, enabling mobile WSN-CPS. A mobile WSN-CPS has inherent mobility [11, 12] and could be built from mobile devices carried by their owners. These mobile devices can take measurements of different phenomena at different places (e.g., road traffic situations) and may react in the physical world [13]. Thus, mobile CPS provides a convenient and an economical platform that facilitates sophisticated and ubiquitous mobile sensing applications between humans and their surrounding physical world.

Mobility in mobile WSN-CPS is very diverse due to various paradigms such as the dynamic physical environment. Generally, the mobility associated with it can be either controlled (e.g., robots) or uncontrolled (e.g., humans) [14]. In controlled mobility, the speed and direction of the mobile nodes can be controlled based on the physical environment. In the case of the uncontrolled mobility model, enforcing the speed and direction of mobile nodes is not possible. The two mobility patterns have different characteristics. For instance, the uncontrolled mobility model is suitable for delay-tolerant applications, whereas controlled mobility can be exploited to extend the lifetime of the network [15]. Furthermore, mobility can either be random in nature or have some regular pattern. For the random mobility model, the direction and speed of the mobile node are chosen randomly. However, for the regular mobility model, some defined direction and speeds can be derived, e.g., vehicles moving on the road in one direction within a certain speed limit. In some mobility models (e.g., controlled, regular), the mobility of the mobile node can be predicted using the historical record of the movement patterns [16]. With the emergence of new mobile WSN-CPS applications, new mobility models are also emerging, such as limited [17, 18] and group mobility [19, 20]. For example, in the hospital environment, a very limited and restricted mobility of patients can be observed. In addition, the body of a human/patient has very limited mobility patterns [21, 22]. Moreover, humans or cars tend to move in groups, giving rise to the group mobility model, where spatial redundancy is exploited for data transfer purposes.

Figure 9.2 depicts a typical mobile WSN-CPS architecture divided into cyber and physical domains. The physical domain provides interactions such as sensing with the physical environment and also dynamically impacts the environment via the actuators. The cyber domain seamlessly integrates with the physical domain and provides the necessary information needed for better operation of the physical domain, such as data processing, mobility management, and knowledge sharing on

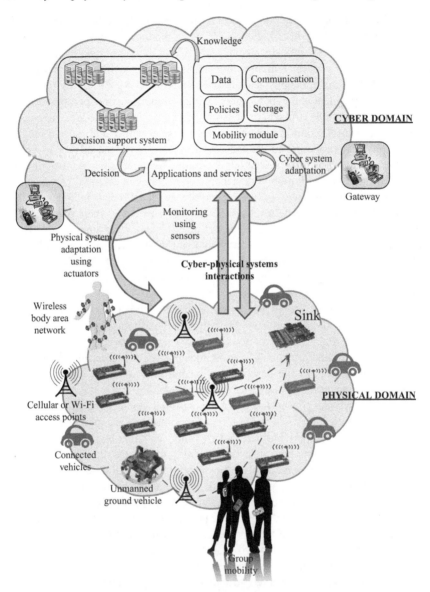

Figure 9.2   Mobile WSN-CPS architecture

the fly. Accordingly, in this chapter we focus on the classification of mobile WSN-CPS applications and show how they manage the physical and cyber domains.

The rest of the chapter is organized as follows. In Section 9.2, we present a classification of mobile WSN-CPS applications. We also compare existing applications based on how they utilize the cyber and physical domains. In Section 9.3, we discuss the challenges and opportunities for mobile WSN in the CPS environment. Finally, Section 9.4 concludes the chapter.

9.2   Mobile WSN-CPS Applications

Almost all WSN applications can be considered under CPS because of the pervasive nature of WSN, their tight integration with the environment, and their management by end-users using different types of web-based tools. For mobile WSN-CPS, we consider three main application classes which are going to have a significant impact on the lifestyles of people in the near future. These domains include smart transportation CPS, smart medical CPS, and smart social CPS (as shown in Figure 9.3).

Next we compare the different applications of each class based on their characteristics for the cyber and physical domains. In the cyber domain, we are interested in the type of services that can be supported by CPS: What type of infrastructure is used by the CPS? How does the decision support system assess the desired conditions efficiently? How is data management carried over to support real-time processing? In the physical domain, it is important to know the following: What type of sensors and actuators are involved to sense and to react in the real environment? What type of communication means are available? What type of mobility patterns are supported?

9.2.1   Smart transportation CPS

Transportation plays a vital role in our daily lives and represents one of the main applications of mobile CPS [8, 23]. To support smart transportation CPS, different types of heterogeneous sensors are being embedded in vehicles. Many types of intelligent transportation system (ITS) have been developed [24–26] using either the Global Positioning System (GPS) [27] or Wi-Fi to annotate maps for road and traffic conditions [28, 29] supporting ITS-CPS applications. Thus, an ITS-CPS is a system of collaborating computational elements gathering data from vehicles using Wi-Fi or cellular networks and processing the collected data over the cloud to

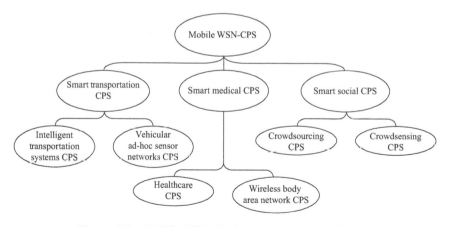

Figure 9.3   Mobile WSN-CPS application classification

provide passive feedback to users. ITS-CPS improves road safety and convenience, manages vehicle traffic, and provides alert to the users.

The growing adoption of cheap sensor hardware (such as tyre pressure and windscreen temperature sensors) and cyber infrastructures (such as economical computing services), for both inter-vehicle and intra-vehicle applications, is enabling vehicles to be "connected" with each other as well as with their surrounding infrastructures through wireless networks, leading to the development of vehicular *ad-hoc* sensor network cyber-physical systems (VASNet-CPS) [30, 31]. VASNet-CPSs are distributed and self-organizing in nature, comprising of moving vehicles, and are characterized by high speed and large-scale vehicular networks. VASNet-CPS can coordinate among vehicles, predict and avoid collisions and traffic congestion, provide real-time optimization of routes, and assist vehicle drivers. Smart parking using GPS, ultrasonic, laser, and magnetometer sensors [32–35] to efficiently find parking places is one VASNet-CPS application.

To enable ITS-CPS and VASNet-CPS, many inter- and intra-vehicle communication standards have been developed. OBD2 [36] and CAN Bus [37] are two examples of the intra-vehicle standards, whereas DSRC/WAVE [38] and 802.11p [39] are inter-vehicle standards. There are also various activities in progress for its standardization, such as CEN/TC 278 [40], CALM (ISO TC204 WG16) [41], and ETSI TC ITS [42]. Many challenging research issues resulting from the high mobility of vehicles, the wide range of relative speeds between vehicular nodes, and the real-time nature of the services have emerged.

A query-centric cyber-physical system (QCPS) for transportation has been proposed [43] to facilitate the various services. The proposed approach arranges similar types of sensors in the form of a grid and allocates a grid coordinator, which interacts with the cloud to process the query. Various sensors (such as vision, speed, and environmental) are used to estimate the road conditions and the surrounding environment. In QCPS, the query is sent by the vehicle coordinator to the roadside unit (RSU). The RSU forwards the query to the cloud for further processing. The communication is carried over Wi-Fi. The system has been evaluated using a small-scale testbed, and it was shown that QCPS requires less communication and computation power for query-centric processing than other related approaches.

The design and evaluation of the emerging application of smart transportation CPS using holistic approaches has been described in [44]. The authors have considered human driver behavior, studied the transportation and communication systems to improve driver safety, and provided support for on-road 'infotainment'. They have introduced several connected vehicle applications, such as video delivery on the move using heterogeneous networks, on-road advertisement delivery, taxi dispatch, and carpool sharing. The authors have suggested utilizing the collected historical GPS data to predict the mobility of vehicles and to provide customized services to drivers. Server-based infrastructure is used to process and store the data from vehicles over Wi-Fi and cellular networks. The data requirement is variable for different kinds of services. The processing of data is performed off-line on the servers. The decision support system (DSS) mainly consists of manual calculation based on predefined thresholds. The enforcement of decision by the DSS is carried out via human drivers. For different services, the authors have used a combination of different simulators such as PARAMICS [45], TRANSIMS [46], and NS2, and have emphasized the acceptability of cyber transportation applications by users.

The need to include the human factor in smart transportation CPS is highlighted in [47]. The authors' emphasis was on human-based VASNet-CPS which monitors/analyzes the physical movements of people (both individually and in groups) in the cyber domain and efficiently enables the mobility of people in the physical environment. A highway monitoring scenario is used as a case study, where 100 vehicles were equipped with mobile phones to collect the location and speed of the vehicle using GPS. The collected data is transferred through a cellular network to a web server, from where it is available to all users.

Real-time and reliable information dissemination in vehicle-to-infrastructure (V2I) communication networks is discussed in [48] from a CPS perspective. Different physical (i.e. speed of vehicle, radio range, etc.) and cyber (i.e. behavior of protocols) properties are discussed which hinder the dissemination of data from the vehicle to the cloud and vice versa. A CPS-based solution is proposed where the cyber world is adapted based on the data received from the physical world. The performance of the proposed approach is evaluated by manually tuning the data received from the physical world using simulations. It is shown that the reliability achieved using their approach is higher than for other approaches where each vehicle locally processes the data and performs information dissemination reliably using Wi-Fi.

V-Cloud [49] presents a three-layer architecture for vehicular CPS and a cloud-enabled system. The cloud computing component is responsible for real-time management of data and its processing. Different types of sensors and mobile phones are used to track the driver's behavior, and vehicle-to-vehicle (V2V) and V2I communications (over Wi-Fi and cellular networks) to provide safety and driver comfort.

A robust system architecture is proposed in [50] for controlling mobility and wireless communication in automotive CPS. The authors designed a controller to ensure the availability of communication resources for an autonomous robot team to optimize network performance as robots move to accomplish their assigned tasks. The DSS utilizes a stochastic process to optimize the routing parameters that maximize the probability of having a connected network for given positions. Robot trajectories are calculated locally for simple environments, and global algorithms are used for complex scenarios. The proposed controller design was evaluated on a testbed of five robots, and the results showed that end to end communication survives even if individual point-to-point links fail with high probability. In addition, the global trajectory planning for complex environments produces good results for end to end communication.

A VASNet-CPS for smart road networks has been described in [51]. The authors have proposed a delay model of a vehicle's travel across a city to provide smart road services. Commercial off-the-shelf (COTS) mobile cloud computing facilities were utilized for data management. Various types of heterogeneous sensors and communication technologies are assumed by the proposed mobile cloud-based VASNet-CPS.

In [52], a testbed for mobile CPS called Pharos is proposed. Pharos comprises wirelessly connected autonomous mobile devices that can coordinate with each other and with embedded sensors and actuators. The design architecture of Pharos is modular and flexible, comprising COTS-based nodes which make it possible to build an inexpensive and scalable testbed for mobile WSN-CPS. Using Pharos, it is easy to establish the empirical relation between cyber and physical properties of the mobile CPS. Despite the unpredictability and uncontrollability of mobile CPS, Pharos shows that the node movement can be repeated accurately. The authors compared their approach with the simulation of a small-scale mobile CPS and showed that the simulator results were not able to capture the complexity and proposed fine tuning for OMNeT++ to match the real-world mobile CPS behavior. The hardware components used for the testbed comprised mainly the Proteus platform [53] along with various COTS sensors. For communication, the 802.11 b/g standard is used. Pharos uses a three-layered architecture. The client layer controls the testbed components wirelessly. The Pharos Player server layer runs on a Proteus x86 computer to control the mobility and navigation of the nodes in the testbed. The micro-controller layer comprises the sensor/actuator drivers that reside within the Proteus micro-controller.

Table 9.1 compares the cyber domain and the physical domain of smart transportation CPS. To reap the benefits of smart transportation CPS, we need to address several challenges. Generally, the transportation system software is complex, requiring a tight integration among the software components of the cyber domain and the physical domain. We note that currently, although the DSS is an important component of CPS, it is not properly designed and implemented. Only simple and naive methods for decision making are utilized. Accordingly, we further observe that the actuators in CPS are not fully utilized, and applications focus more on passive actuation in the physical world in the form of alerts and warnings. There is a growing trend to move from a typical client server architecture to cloud computing, and this trend is also penetrating into smart transportation CPS as well. Although there is a strong interest in cloud computing, most of the existing smart transportation CPS applications still prefer to utilize central storage mechanisms and rely on traditional mechanisms for processing the data. Although new standards have become available for the vehicle communication paradigm, current smart transportation CPS still uses traditional communication technologies which emphasize the need to revisit the communication aspects in smart transportation CPS. Furthermore, the state space of such systems is huge. As a result, the current systems are not able to fully comply, and all the major components have not been fully considered for mobile CPS. In addition to the discussion of the related works here, there are many international efforts for connected vehicles and ITS testbeds in progress, as reported in [54]. Recently, Local Motors [55] have also joined efforts with IBM and Intel to work on a prototype of connected vehicles [56].

9.2.2 Smart medical CPS

In smart medical CPS, the combination of patient input such as smart feedback, the patient's digital records data (DRD), and real-time sensor input such as biosensors and/or smart devices can support data acquisition for efficient DSS [57]. Mobile

Table 9.1 Comparison of cyber domain and physical domain for smart transportation CPS

Approach	Cyber domain						Physical domain				Evaluation
	Service	Infrastructure	DSS	Data management			Sensors	Communication	Mobility	Actuators	
				Storage	Processing	Requirement					
QCPS [43]	various	cloud	n/a	central	RSU	n/a	heterogeneous	Wi-Fi	regular	n/a	testbed
[44]	various	server	manual	central	server	variable	n/a	Wi-Fi, cellular	predictable	humans	simulators
[47]	road traffic	server	n/a	central	server	low	homogeneous	cellular	random	n/a	testbed
[48]	reliable msg. transfer	server	manual	n/a	local	low	n/a	Wi-Fi	random	n/a	OMNeT++
V-Cloud [49]	various	cloud	n/a	central	cloud	variable	heterogeneous	Wi-Fi, cellular	random	n/a	theoretical framework
[50]	end to end connectivity	server	stochastic	n/a	local, global	low	n/a	Wi-Fi	controllable	n/a	testbed
[51]	road networks	cloud	n/a	central	cloud	low	heterogeneous	all	predictable	n/a	theoretical framework
Pharos [52]	various	server	n/a	central	local	variable	heterogeneous	Wi-Fi	uncontrollable	n/a	testbed

CPS in healthcare includes the coordinated inter-operation of autonomous devices with patients, the management and operation of medical physical systems using actuators, nano-sized implantable smart devices, programmable materials, and wireless body area networks (WBANs). Furthermore, enabling cloud services for healthcare CPS yields numerous improvements for massive data collection, storage, and management [58].

For monitoring the patient and providing information to care-givers, a medical CPS is described in [59]. As a case study, the design of a pacemaker is used to define the medical CPS. The detailed software development of medical CPS includes medical device coordination, publish/subscribe messaging services, and context-aware DSS. The patient's physiological data is collected and stored in a distributed manner, and the processing is carried out on the base station attached to the patient. From there the information is forwarded to the care-givers.

A secured healthcare architecture for CPS called $CPeSC^3$ is proposed in [60]. The $CPeSC^3$ architecture is composed of communication, computation, resource scheduling and management, cloud integration, and security components. A medical healthcare scenario is presented as a case study. Sensors are deployed on the patient's body to collect the health data, which is forwarded (using 3G services) to the cloud, where the computation is carried out in a real-time manner. The mobility of the patients is controlled within the hospital environment.

A social interaction-based power control game for inter-WBAN interference mitigation for WBAN-CPS is proposed in [61]. The authors developed the interference model based on social interaction information, e.g., how many times patient A met with patient B. To find the interaction type, a probability density function (PDF) model of social interaction is presented. Monte Carlo simulations are used to verify the correctness of the proposed model. To improve power efficiency of WBAN, a game theory approach is used. A typical WBAN scenario is considered where the people wear heterogeneous sensors and the data is collected by the base station, which forwards it to the server for further processing. The random mobility model for the movement of people is considered for the simulation of the scenario of inter-WBAN. It is shown that the inter-WBAN interference is significantly reduced using social data.

In [62] a secure and scalable architecture for huge amounts of data collection and processing of medical sensor networks is proposed. The proposed architecture includes sensors to collect patient data, a software module to monitor the patient's health using the collected data, a healthcare authority (HA), which is responsible for specifying and auditing of the security policies, and cloud servers to store and manage the data.

In [63] a cloud-based patient data collection scheme is presented. In that paper, sensors are integrated with cloud services to achieve better healthcare. The storing and processing of the collected data is done by open-source software such as Open Nebula and ANEKA as the middleware for cloud computing infrastructures. The sensors are fixed across the hospital bed and are deployed on the patient as needed. The sensors collect the patient's data and send it to the cloud using Wi-Fi. The authors do not describe how the data is manipulated to support DSS.

A modeling and analysis framework for medial CPS is presented in [64] for safety verification. The proposed framework models and analyzes the dynamic, nonlinear, time-delayed, and spatio-temporal complex interaction of the sensors with the human body. It also considers the non-trivial aggregation of interactions during networked operations. A generic construct for the formal specification of the CPS is considered. The safety requirements of the medical CPS are specified as constraints on the model's properties. The usage of the proposed framework is shown by two case studies, i.e., analgesic infusion pump drug overdose safety analysis and aggregate effect modeling for multichannel chemotherapeutic infusion.

A WSN-integrated cloud approach is discussed in [65] for healthcare applications. The authors discussed WSN–cloud integration and a dynamic collaboration mechanism. The experiments were conducted with commercially available motes. The testbed includes a medical server which can be utilized by a medical specialist, care-givers, and patients for accessing the collected medical data. The electrocardiogram (ECG), heart rate, and temperature sensors were interfaced with motes to monitor the patients physiological parameters. Motes transfer the collected physiological data to the base station using mesh networking.

Table 9.2 compares the cyber domain and the physical domain for smart medical CPS. We note that, similar to smart transportation CPS, there is a lack of proper DSS and actuators to interact with the physical environment. The existing medical CPS are not generic in nature and provide only specific services for the patients. The majority of smart medical CPS rely on the cloud computing infrastructure because they store data in the cloud. Generally the data requirement is high because medical sensors generate a large amount of data to be stored and processed. There are different types of sensors involved in order to monitor the patient, leading to a heterogeneous environment around the patient within the hospital or on the patient's body. A comprehensive report on medical devices and their usage in smart medical CPS is presented in [66]. That report highlights the heterogeneity of medical devices. Furthermore, for smart medical CPS, we see a lack of testbeds mainly because of patient privacy and the involvement of human lives.

9.2.3   Smart social CPS

Human interactions have been significantly influenced by social networks such as Twitter or Facebook. Social networks are now frequently used to report news, to advertise, to entertain, and even to organize social events [67]. If something happens anywhere in the world, people start to tune in to the social networks rather than the newspaper websites because the reporters of the event are an inherent part of the event that is occurring. This relates to the idea of crowdsourcing, where the group of people on the spot start sharing information about the phenomenon of interest [68]. Always-on smartphones have tended to drive the rise of crowdsource-based social networking CPS, where recording, maintaining, and sharing of rapidly changing environments becomes a rather simple task. This has also led to the generation of a large amount of data. One of the major challenges in such an environment is to determine the freshness of information along with the content.

Table 9.2 Comparison of cyber domain and physical domain for smart medical CPS

Approach	Cyber domain						Sensors	Physical domain			Evaluation
	Service	Infrastructure	DSS	Data management				Communication	Mobility	Actuators	
				Storage	Processing	Requirement					
[59]	monitoring	n/a	closed loop	distributed	base station	high	homogeneous	n/a	controlled	n/a	n/a
CPeSC [60]	healthcare	cloud	n/a	central	cloud	high	heterogeneous	3G	controlled	n/a	case study
[61]	healthcare	server	n/a	n/a	distributed	high	heterogeneous	any	random	n/a	simulation
[62]	healthcare	cloud	n/a	central	cloud	high	heterogeneous	n/a	controlled	n/a	n/a
[63]	monitoring	cloud	n/a	central	cloud	moderate	homogeneous	Wi-Fi	limited	n/a	n/a
[64]	healthcare	server	prediction	central	server	moderate	homogeneous	n/a	limited	inject	empirical
[65]	vital sign	cloud	n/a	central	base station	moderate	heterogeneous	Zigbee	limited	n/a	testbed

The current generation of smartphones is equipped with a number of sensors, such as camera, accelerometer, gyroscope, temperature, and light. With the help of these sensors, the collected data can be transformed into more meaningful information. This leads to the crowdsensing environment, where people can be allowed to sense the environment on the move and update information about their surroundings in an efficient manner [69].

In [70] the SocMessaging approach is proposed for distributed message transmission using a social-based CPS. SocMessaging utilizes the mobility patterns and the social network of the users to transmit messages from the source to the destination. To select a node for relaying messages among the encountered nodes, SocMessaging checks the likelihood of the node's reachability to the destination. The likelihood is based on how many times the node has relayed messages to the destination in the past, and whether the node is within the social network of the destination. The experimental results show that SocMessaging achieves high efficiency of message transmission among users.

Vita, a mobile CPS for crowdsensing, provides a systematic approach to facilitate different services by combining social networks, cloud computing, and various open-source tools [12]. Vita simplifies the development of multiple applications with lower communication overhead, consumes less battery power, and supports real-time communication. It also supports the automatic allocation of human tasks among members in the physical world and computing tasks among mobiles and cloud in the cyber world in an efficient manner. The data is stored and processed on the cloud to fulfill service requirements. As a case study, a SmartCity application developed on Vita was demonstrated.

Social Drive [71] supports a crowdsourcing-based vehicular *ad-hoc* network (VANET) social networking paradigm. Social Drive integrates cloud computing, a social network (Facebook), and an automatic rating system to inform users about the fuel economy of their trips through a mobile application. Using an on-board diagnosis device, the mobile application collects different car parameters, uses a rating mechanism for fuel efficiency, and posts related information (such as how much fuel was saved during a trip to Washington) on the social media cloud.

In [72], a CrowdHelp system is introduced for real-time patient assessment using mobile triage via crowdsourcing and sensors worn by people during a mass casualty incident. During the incident, data from the sensors and crowd is transferred (using the cellular network) to a server where the clustering is done for the different categories of patients.

ContriSenseCloud [73] is designed and implemented to reap the benefits of both CPS and social networks for public transportation. To make ContriSenseCloud generic in order to support other applications, it has two platforms, i.e., the service exchange platform and the application-specific exchange platform. The client server architecture is implemented to allow people to share their traveling experiences and needs. Commuters use the data for transportation route planning. Near-real-time algorithms are proposed and GPS-based new bus stops are mapped according to user needs. The authors developed an Android application for end-users and use a third-party Java-based server application to demonstrate the system.

A hybrid crowdsourcing architecture is proposed in [74] for providing computing resources for resource-limited CPS nodes to satisfy their workloads. The architecture supports various services by jointly considering resource allocation, routing, and link scheduling. The problem of providing computing resources is modeled as a finite queuing problem to be solved centrally at the cloud. An approximation algorithm is used to find the upper and lower bounds to guarantee strong network stability.

A comparison of the cyber and physical domains for smart social CPS is presented in Table 9.3. We note that the social CPS can be integrated with any other classes of mobile CPS easily to help and provide more efficient services to users. In this case, too, we need an appropriate DSS to make real-time decisions. Existing social networks generally rely on the use of cloud computing for storage and processing of data. Generally, humans are considered as the actuators in the real environment, and they have to physically interact with the environment to perform the actuation tasks. We also note from Table 9.3 that the majority of projects have utilized real testbeds.

9.3   Challenges and opportunities

In this section, we present a few challenges and opportunities for mobile WSN-CPS which require further investigation by researchers in industry and academia.

9.3.1   *Real-time mobility-aware scalable information dissemination*

The mobility of sensors must be considered along with the real-time deadlines of information dissemination to a large number of nodes in the physical environment. Real-time information dissemination is critical for applications such as vehicle collision avoidance, timely route replanning, and patient vital signs. A real-time publish/subscribe model can be worth investigating to solve this problem. Probabilistic gossiping can be another alternative where information is disseminated with certain probability to cover the area of interest.

9.3.2   *Heterogeneous communication*

Mobile nodes may communicate either with each other or with the surrounding infrastructure via different wireless standards. In contrast, the infrastructure may be connected at the back end to the database over the cloud via some dedicated wired network. Furthermore, the infrastructure may also process the information locally. In such a mixed mode environment, service differentiation and maintaining quality of service are of primary concern, and service requirements should be maintained accordingly. As the number of mobile nodes increases in the physical domain, traditional networks may fail because of contention and congestion. In such environments, a cognitive radio approach may provide some solution, but at the cost of more complexity.

9.3.3   *Capacity planning*

To help mobile devices, it is necessary to plan the infrastructure (both wired and wireless). The infrastructure location plays an important role in order to facilitate the flow of information and the scalability of the network.

Table 9.3 Comparison of cyber domain and physical domain for smart social CPS

| Approach | Cyber domain | | | | | | Physical domain | | | | Evaluation |
| | Service | Infrastructure | DSS | Data management | | | Sensors | Communication | Mobility | Actuators | |
				Storage	Processing	Requirement					
SocMessaging [70]	message transmission	cloud	n/a	distributed	local	less	mobile device	Wi-Fi, cellular	random, group	n/a	simulation
Vita [12]	various	cloud	n/a	server	cloud	moderate	mobile device	Wi-Fi, cellular	random, group	humans	testbed
Social Drive [71]	fuel economy	cloud	rating mechanism	local	local	moderate	mobile device, car sensors	Wi-Fi, cellular	predictable	humans	testbed
CrowdHelp [72]	emergency response	n/a	n/a	server	central	high	heterogeneous	cellular	random	n/a	theoretical framework
ContriSense-Cloud [73]	public transport	server	n/a	server	central	moderate	mobile device	cellular	random	n/a	testbed
[74]	various	cloud	queuing theory	cloud	central	high	heterogeneous	Wi-Fi, cellular	random	n/a	simulation

9.3.4   Real-time distributed and global management

The real-time information must be fed to the DSS in order to optimize the distributed and global management algorithms to balance the load on the overall system. Automated and intelligent DSS algorithms are needed in order to be able to deliver timely decisions for actuation in the physical domain. To ensure scalability, the region of interest can be divided into sectors and then the inter-sector and intra-sector management algorithms can be executed.

9.3.5   Software challenges

There are several open challenges for software development for mobile WSN-CPS which hinder the development and deployment of rapid applications [11]. It is difficult to forecast the power consumption of mobile CPS. Thus it is difficult for developers to consider this factor early on during software planning. Furthermore, diverse mobile CPS platforms make overall software development tricky and more complex for inter-operability. However, the different resource constraints and device capabilities make it harder to achieve seamless integration and collective evolution of heterogeneous sensor networks.

9.3.6   Humans-in-the-loop

Humans-in-the-loop is an essential part of the mobile WSN-CPS and needs to be taken into consideration during any analysis or design task right from the start. There are various challenges associated with humans-in-the-loop [75]. First, we need to fully understand the humans-in-the-loop control feedback. Second, modeling human behavior is very complex. We need to identify the different human behaviors and associated models. Third, formal analysis of these models is needed to support performance guarantees.

9.3.7   Aerial networks

Low-altitude unmanned aerial vehicles (also called drones) are getting a lot of attention for many applications [76]. There is a lot of focus on standardization and laws pertaining to the government, commercial, and home usage of such drones. Thus, we also need to consider future aerial networks where vertical mobility is going to play an important role in mobile WSN-CPS.

9.3.8   Simulators, testbeds, and real deployments

To validate mobile WSN-CPS as a whole, there is a strong need for scalable simulation tools, analytical mechanisms, and large-scale testbeds. Only then will it be possible to deploy mobile WSN-CPS in the real environment. Currently, there are only a few solutions [12, 52, 73] available, which are inadequate in helping us to better understand the behavior of mobile WSN-CPS.

9.4   Conclusions

We have presented a classification of mobile WSN-CPS applications based on various characteristics, including the supported infrastructure, decision support systems, data management strategies, types of sensors, communication techniques, mobility support, control/actuation, and system evaluation methods. The comparison presented in this chapter is aimed at understanding and visualizing the trends, techniques, and potential mobile WSN-CPS solutions around specific applications. It also highlights the gaps between the cyber domain and the physical domain where focus is much needed. From the survey we found that the use of decision support systems is the least explored area in mobile WSN-CPS. Furthermore, the control/actuation part of mobile WSN-CPS is almost absent or is often dependent on manual intervention of the humans-in-the-loop.

Acknowledgments

Faisal K. Shaikh was partially supported by Mehran University of Engineering and Technology, Jamshoro, Pakistan, and by grant number 10-Inf1236-10 from the Long-Term National Plan for Science, Technology and Innovation (LT-NPSTI), the King Abdul-Aziz City for Science and Technology (KACST), Kingdom of Saudi Arabia. We also thank the TCMCORE and STU at Umm Al-Qura University for their continued logistics support.

References

[1]   Semiconductor Wireless Sensor Internet of Things (IoT) Market 2014–2020 Forecasts. See http://cloud-computing.tmcnet.com/news/2014/02/20/7686049.htm. [Online; accessed 17 May 2015].

[2]   R. C. Shah, S. Roy, S. Jain and W. Brunette. Data mules: modeling and analysis of a three-tier architecture for sparse sensor networks. *Ad Hoc Networks*, 1(2–3): 215–233, 2003.

[3]   A. Kansal, A. A. Somasundara, D. D. Jea, M. B. Srivastava and D. Estrin. Intelligent fluid infrastructure for embedded networks. In *Proceedings of the 2nd International Conference on Mobile Systems, Applications, and Services (MobiSys)*, pp. 111–124, 2004.

[4]   A. Khelil, F. K. Shaikh, A. Ali and N. Suri. gMAP: An efficient construction of global maps for mobility-assisted wireless sensor networks. In *Proceedings of the Conference on Wireless On Demand Network Systems and Services (WONS)*, pp. 189–196, 2009.

[5]   I. El Korbi and S. Zeadally. Energy-aware sensor node relocation in mobile sensor networks. *Ad Hoc Networks*, 16: 247–265, 2014.

[6]   F. K. Shaikh, S. Zeadally and E. Exposito. Enabling technologies for green internet of things. *IEEE Systems Journal*, in press (12 pp.), 2015.

[7]   C.-Y. Lin, S. Zeadally, T.-S. Chen and C.-Y. Chang. Enabling cyber physical systems with wireless sensor networking technologies. *International Journal of Distributed Sensor Networks*, 489794 (21 pp.), 2012.

[8]   F.-J. Wu, Y.-F. Kao and Y.-C. Tseng. From wireless sensor networks towards cyber physical systems. *Pervasive and Mobile Computing*, 7(4): 397–413, 2011.

[9]   E. A. Lee. Cyber physical systems: Design challenges. In *Proceedings of the 11th IEEE International Symposium on Object Oriented Real-Time Distributed Computing (ISORC), 2008*, pp. 363–369. IEEE, 2008.

[10] J. Wan, H. Yan, H. Suo and F. Li. Advances in cyber-physical systems research. *KSII Transactions on Internet and Information Systems*, 5(11): 1891–1908, 2011.

[11] J. White, S. Clarke, C. Groba, B. Dougherty, C. Thompson and D. C. Schmidt. R&D challenges and solutions for mobile cyber-physical applications and supporting internet services. *Journal of Internet Services and Applications*, 1(1): 45–56, 2010.

[12] X. Hu, T. H. S. Chu, H. C. B. Chan and V. C. M. Leung. Vita: A crowd-sensing-oriented mobile cyber-physical system. *IEEE Transactions on Emerging Topics in Computing*, 1(1): 148–165, 2013.

[13] X. Li, C. Qiao, X. Yu, A. Wagh, R. Sudhaakar and S. Addepalli. Toward effective service scheduling for human drivers in vehicular cyber-physical systems. *IEEE Transactions on Parallel and Distributed Systems*, 23(9): 1775–1789, 2012.

[14] M. A. Alharthi and A.-E.M. Taha. Modeling mobility for networked mobile cyber-physical systems. In *Proceedings of the 4th ACM SIGBED International Workshop on Design, Modeling, and Evaluation of Cyber-Physical Systems*, pp. 1–6. ACM, 2014.

[15] S. Basagni, A. Carosi and C. Petrioli. Controlled vs. uncontrolled mobility in wireless sensor networks: some performance insights. In *Proceedings of the 66th Vehicular Technology Conference, VTC-2007, Fall 2007*, pp. 269–273. IEEE, 2007.

[16] Q. Dong and W. Dargie. A survey on mobility and mobility-aware MAC protocols in wireless sensor networks. *IEEE Communications Surveys & Tutorials*, 15(1): 88–100, 2013.

[17] M. M. Alam and E. Ben Hamida. Towards accurate mobility and radio link modeling for IEEE 802.15.6 wearable body sensor networks. In *Proceedings of the 10th International Conference on Wireless and Mobile Computing, Networking and Communications (WiMob)*, pp. 298–305. IEEE, 2014.

[18] S. S. Metcalf. Modeling social ties and household mobility. *Annals of the Association of American Geographers*, 104(1): 40–59, 2014.

[19] P. Traynor, J. S. Shin, B. Madan, S. Phoha and T. La Porta. Efficient group mobility for heterogeneous sensor networks. In *Proceedings of the 64th Vehicular Technology Conference, VTC-2006, Fall 2006*, pp. 1–5. IEEE, 2006.

[20] R. Chai, Y.-L. Zhao, Q.-B. Chen, T. Dong and W.-G. Zhou. Group mobility in 6LoWPAN-based WSN. In *Proceedings of the International Conference on Wireless Communications and Signal Processing (WCSP)*, pp. 1–5, 2010.

[21] A. Talpur, N. Baloch, N. Bohra, F. K. Shaikh and E. Felemban. Analyzing the impact of body postures and power on communication in WBAN. *Procedia Computer Science*, 32: 894–899, 2014.

[22]   Z. Aziz, U. M. Qureshi, F. K. Shaikh, N. Bohra, A. Khelil and E. Felemban. Experimental analysis for optimal separation between sensor and base station in WBANs. In *Proceedings of the 16th International Conference on e-Health Networking, Applications and Services (Healthcom)*, pp. 489–494. IEEE, 2014.

[23]   Intelligent Transportation Systems. See http://www.its.dot.gov/. [Online; accessed 23 March 2015].

[24]   California Center for Innovative Transportation. See http://www.calccit.org/. [Online; accessed 2 March 2015].

[25]   TCMCORE: Transportation and Crowd Management Center of Research Excellence. See http://www.tcmcore.org/en/. [Online; accessed 2 March 2015].

[26]   K. N. Qureshi and A. H. Abdullah. A survey on intelligent transportation systems. *Middle-East Journal of Scientific Research*, 15(5): 629–642, 2013.

[27]   P. Mohan, V. N. Padmanabhan and R. Ramjee. Nericell: Rich monitoring of road and traffic conditions using mobile smartphones. In *Proceedings of the 6th ACM Conference on Embedded Network Sensor Systems*, pp. 323–336. ACM, 2008.

[28]   A. Thiagarajan, L. Ravindranath, K. LaCurts, *et al.* VTrack: Accurate, energy-aware road traffic delay estimation using mobile phones. In *Proceedings of the 7th ACM Conference on Embedded Networked Sensor Systems*, pp. 85–98. ACM, 2009.

[29]   S. Coleri Ergen, H. S. Tetikol, M. Kontik, R. Sevlian, R. Rajagopal and P. Varaiya. RSSI-fingerprinting-based mobile phone localization with route constraints. *IEEE Transactions on Vehicular Technology*, 63(1): 423–428, 2014.

[30]   G. Karagiannis, O. Altintas, E. Ekici, *et al.* Vehicular networking: A survey and tutorial on requirements, architectures, challenges, standards and solutions. *IEEE Communications Surveys & Tutorials*, 13(4): 584–616, 2011.

[31]   M. Whaiduzzaman, M. Sookhak, A. Gani and R. Buyya. A survey on vehicular cloud computing. *Journal of Network and Computer Applications*, 40: 325–344, 2014.

[32]   S. Mathur, T. Jin, N. Kasturirangan, *et al.* Parknet: drive-by sensing of roadside parking statistics. In *Proceedings of the 8th International Conference on Mobile Systems, Applications, and Services*, pp. 123–136. ACM, 2010.

[33]   Y. Geng and C. G. Cassandras. New 'smart parking' system based on resource allocation and reservations. *IEEE Transactions on Intelligent Transportation Systems*, 14(3): 1129–1139, 2013.

[34]   X. Chen, E. Santos-Neto and M. Ripeanu. Crowd-based smart parking: a case study for mobile crowdsourcing. In *Mobile Wireless Middleware, Operating Systems, and Applications*, pp. 16–30. Springer, 2013.

[35]   VehicleSense. See http://www.vehiclesense.com/. [Online; accessed 2 April 2015].

[36]   On-Board Diagnostics (OBD). See http://www.epa.gov/otaq/regs/im/obd/index.htm. [Online; accessed 2 April 2015].

[37]   Controller Area Network (CAN-Bus). See http://www.gaw.ru/data/Interface/CAN_BUS.PDF. [Online; accessed 2 April 2015].

[38]   Wireless Access in Vehicular Environments (WAVE). See http://vii.path.
       berkeley.edu/1609_wave/. [Online; accessed 2 April 2015].
[39]   IEEE 802.11p Standard. See http://standards.ieee.org/findstds/standard/
       802.11p-2010.html. [Online; accessed 2 April 2015].
[40]   CEN/TC 278, ITS Standardization. See http://www.itsstandards.eu/.
       [Online; accessed 18 May 2015].
[41]   CALM, ISO TC204 WG16. See http://calm.its-standards.eu/. [Online;
       accessed 18 May 2015].
[42]   ETSI TC ITS, Standards on the Move. See https://portal.etsi.org. [Online;
       accessed 18 May 2015].
[43]   A. Mundra, G. Rathee, M. Chawla, N. Rakesh and A. Soni. Transport
       information system using query centric cyber physical systems (QCPS).
       International Journal of Computer Applications, 85(3): 12–16, 2014.
[44]   A. Wagh, Y. Hou, C. Qiao, *et al.* Emerging applications for cyber trans-
       portation systems. *Science and Technology*, 29(4): 562–575, 2014.
[45]   PARAMICS Simulator. See http://www.paramics-online.com/. [Online;
       accessed 22 April 2015].
[46]   The Transportation Analysis and Simulation System (TRANSIMS). See
       https://code.google.com/p/transims/. [Online; accessed 22 April 2015].
[47]   D. Work, A. Bayen and Q. Jacobson. Automotive cyber physical systems in
       the context of human mobility. In *National Workshop on High-Confidence
       Automotive Cyber-Physical Systems*, Troy, MI, 2008.
[48]   A. Gokhale, M. P. McDonald, S. Drager and W. McKeever. A cyber phy-
       sical systems perspective on the real-time and reliable dissemination of
       information in intelligent transportation systems. *Network Protocols and
       Algorithms*, 2(3), 2010.
[49]   H. Abid, L. T. T. Phuong, J. Wang, S. Lee and S. Qaisar. V-Cloud: vehicular
       cyber-physical systems and cloud computing. In *Proceedings of the 4th
       International Symposium on Applied Sciences in Biomedical and Commu-
       nication Technologies*, 165 (5 pp.), 2011.
[50]   J. Fink, A. Ribeiro and V. Kumar. Robust control for mobility and wireless
       communication in cyber-physical systems with application to robot teams.
       Proceedings of the IEEE, 100(1): 164–178, 2012.
[51]   J. P. Jeong and E. Lee. Vehicular cyber-physical systems for smart road
       networks. *KICS Information and Communications Magazine*, 31(3):
       103–116, 2014.
[52]   C. Fok, A. Petz, D. Stovall, N. Paine, C. Julien and S. Vishwanath. *Pharos:
       A Testbed for Mobile Cyber-Physical Systems*. University of Texas at Austin,
       Tech. Rep. TR-ARiSE-2011-001, 2011.
[53]   Proteus Mobile Nodes. See http://proteus.ece.utexas.edu/. [Online; accessed
       25 April 2015].
[54]   J. Cregger, R. Wallace and V. S. Brugeman. *International Survey of Best
       Practices in Connected Vehicle Technologies: 2012 Update*. Center for
       Automotive Research, Transportation Systems Analysis Group, 2012.
[55]   Local Motors. See https://localmotors.com/. [Online; accessed 19 May 2015].

[56]   Connected Car Project (IoT). See https://localmotors.com/awest/connected-car-project-internet-of-things/. [Online; accessed 23 April 2015].

[57]   S. A. Haque, S. M. Aziz and M. Rahman. Review of cyber-physical system in healthcare. *International Journal of Distributed Sensor Networks*, 217415 (20 pp.), 2014.

[58]   J. Wan, C. Zou, S. Ullah, C. F. Lai, M. Zhou and X. Wang. Cloud-enabled wireless body area networks for pervasive healthcare. *IEEE Network*, 27(5): 56–61, 2013.

[59]   I. Lee, O. Sokolsky, S. Chen, *et al.* Challenges and research directions in medical cyber-physical systems. *Proceedings of the IEEE*, 100(1): 75–90, 2012.

[60]   J. Wang, H. Abid, S. Lee, L. Shu and F. Xia. A secured health care application architecture for cyber-physical systems. *Preprint, arXiv*:1201.0213, 2011.

[61]   Z. Zhang, H. Wang, C. Wang and H. Fang. Interference mitigation for cyber-physical wireless body area network system using social networks. *IEEE Transactions on Emerging Topics in Computing*, 1(1): 121–132, 2013.

[62]   A. Lounis, A. Hadjidj, A. Bouabdallah and Y. Challal. Secure and scalable cloud-based architecture for e-health wireless sensor networks. In *Proceedings of the 21st International Conference on Computer Communications and Networks (ICCCN)*, pp. 1–7. IEEE, 2012.

[63]   C. O. Rolim, F. L. Koch, C. B. Westphall, J. Werner, A. Fracalossi and G. S. Salvador. A cloud computing solution for patient's data collection in health care institutions. In *Proceedings of the Second International Conference on eHealth, Telemedicine, and Social Medicine, ETELEMED'10*, pp. 95–99. IEEE, 2010.

[64]   A. Banerjee, S. K. S. Gupta, G. Fainekos and G. Varsamopoulos. Towards modeling and analysis of cyber-physical medical systems. In *Proceedings of the 4th International Symposium on Applied Sciences in Biomedical and Communication Technologies*, p. 154. ACM, 2011.

[65]   B. Perumal, M. Rajasekaran, H. M. Ramalingam, *et al.* WSN integrated cloud for automated telemedicine (ATM) based e-healthcare applications. *Proceedings of the 4th International Conference on Bioinformatics and Biomedical Technology; International Proceedings of Chemical, Biological and Environmental Engineering*, vol. 29, pp. 166–170. IACSIT, 2012.

[66]   NITRD. High Confidence Software and Systems Coordinating Group of the Networking and Information Technology Research and Development Program. *High-Confidence Medical Devices: Cyber-Physical Systems for 21st Century Health Care*. NITRD, 2009.

[67]   V. Kalogeraki and D. Gunopulos. Reliable mobile cyberphysical systems. In *Proceedings of the 1st International Workshop on Reliable CyberPhysical Systems*, 2012.

[68]   M. C. Yuen, I. King and K. S. Leung. A survey of crowdsourcing systems. In *Proceedings of the IEEE Third International Conference on Privacy, Security, Risk and Trust (PASSAT), and IEEE Third International Conference on Social Computing (SocialCom)*, pp. 766–773. IEEE, 2011.

[69]   W. Z. Khan, Y. Xiang, M. Y. Aalsalem and Q. Arshad. Mobile phone sensing systems: A survey. *IEEE Communications Surveys & Tutorials*, 15(1): 402–407, 2013.

[70]   K. Chen and H. Shen. A social-based cyber-physical system for distributed message transmission. In *Proc. SPIE 9091, Signal Processing, Sensor/ Information Fusion, and Target Recognition XXIII,* 90911M. SPIE, 2014.

[71]   X. Hu, V. C. M. Leung, K. G. Li, *et al*. Social Drive: A crowdsourcing-based vehicular social networking system for green transportation. In *Proceedings of the Third ACM International Symposium on Design and Analysis of Intelligent Vehicular Networks and Applications*, pp. 85–92. ACM, 2013.

[72]   L. I. Besaleva and A. C. Weaver. CrowdHelp: Application for improved emergency response through crowdsourced information. In *Proceedings of the 2013 ACM Conference on Pervasive and Ubiquitous Computing, Adjunct Publication*, pp. 1437–1446. ACM, 2013.

[73]   J.K.-S. Lau, C.-K. Tham, and T. Luo. Participatory cyber physical system in public transport application. In *Proceedings of the Fourth IEEE International Conference on Utility and Cloud Computing (UCC)*, pp. 355–360. IEEE, 2011.

[74]   M. Li and P. Li. Crowdsourcing in cyber-physical systems: Stochastic optimization with strong stability. *IEEE Transactions on Emerging Topics in Computing*, 1(2): 218–231, 2013.

[75]   S. Munir, J. A. Stankovic, C.-J.M. Liang and S. Lin. Cyber physical system challenges for human-in-the-loop control. In *Proceedings of the 8th International Workshop on Feedback Computing*, 4 pp., 2013.

[76]   Y. Saleem, M. H. Rehmani and S. Zeadally. Integration of cognitive radio technology with unmanned aerial vehicles: Issues, opportunities, and future research challenges. *Journal of Network and Computer Applications*, 50: 15–31, 2015.

Chapter 10

Intelligent wireless sensor networks (iWSNs) in cyber-physical systems

Farhad Mehdipour[1], Mirza Ferdous Rahman[2] and Kazuaki J. Murakami[3]

Abstract

Wireless sensor networks (WSNs) receive considerable attention as one of the key and enabling technologies of cyber-physical systems (CPSs) that, in turn, present a new computing paradigm now and into the future. WSNs are open to a wide range of potential applications, such as environmental monitoring, human health monitoring, target detection and tracking, and industrial process control. To be smart and highly efficient services are the present and future demand of WSNs. In this chapter, we aim to introduce the fundamentals of an intelligent wireless sensor network (iWSN) and its requisites, characteristics, and applications. Also, we distinguish between the traditional WSN and the iWSN. First, the definition and properties of an intelligent system are described. Then we explain how a WSN can be deployed in a cyber-physical system and provide the facility for communications among different components of the system. In the next sections, various characteristics of an iWSN, its applications, and comparison to WSN are highlighted. Two applications of WSN, comprising smart grid and smart field monitoring system for pest control, will be introduced and compared in terms of the features of intelligence they require.

10.1 Intelligent systems

Intelligent systems (ISs) can learn and take actions based on experience. On the one hand, they show some superior characteristics/functionalities such as adaptability,

[1]E-JUST Center, Graduate School of Information Science and Electrical Engineering, Kyushu University, Fukuoka, Japan, and WaioMio Ltd, New Zealand, e-mail: farhad@ejust.kyushu-u.ac.jp
[2]Department of Advanced Informatics, Graduate School of Information Science and Electrical Engineering, Kyushu University, Fukuoka, Japan, e-mail: ferdous@soc.ait.kyushu-u.ac.jp
[3]Department of Advanced Informatics, Graduate School of Information Science and Electrical Engineering, Kyushu University, Fukuoka, Japan, e-mail: murakami@soc.ait.kyushu-u.ac.jp

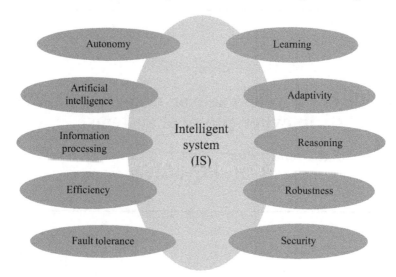

Figure 10.1   Intelligent system's conceptual diagram

robustness, improving efficiency over time, information processing, and reasoning to complete the task [1]. Additionally an IS provides a methodological approach to perform more important and relatively complex tasks to obtain consistent results over a particular time frame [1, 2]. On the other hand, depending on the intelligence level of the system, perceptions are different and debatable. An intelligent system can differ considerably from the traditional system although the intelligent system is constructed on the basis of the standard system concept. Figure 10.1 shows the essential features required for the system to be intelligent [3].

Intelligence level is a parameter that determines how much a system exploits various properties, as listed in Figure 10.1. For example, in terms of a computer system, in the sense of the capability of such a system to implement multiple and innovative applications, the system is intelligent. But at the same time, because of the limited capacities of the computer components (processor speed, memory through-put) and ability to run only predefined programs, the intelligence level is trivial. Moreover, the capabilities of an intelligent system might change and habitually increase over time. However, on the other hand, it is not mandatory for an intelligent system to be more capable than its normal counterpart system. In addition, among all the properties, some may be mandatory and some may be optional as added values to the system. The main characteristics of an intelligent system are described below.

- **Autonomy:** An autonomous system can achieve operational goals in complex domains by adding self-governing behavior and decision-making abilities [2, 4]. Such a system makes decisions on its own using high-level policies, regularly checks and optimizes its status, and automatically adapts itself to changing conditions. An autonomous system consists of autonomic components interacting with each other.

- **Learning:** The ability to learn is one of the major criteria for a truly intelligent system. With respect to time, the system gathers knowledge and learns autonomously from the present condition and interacts with the new environment or process and explores it for the future [2].
- **Artificial intelligence:** Artificial intelligence is the human-like intelligence exhibited by the system. It is classified into two categories: humanistic artificial intelligence and rationalistic artificial intelligence. The former studies machines that think and act like humans. The latter examines machines that can be built on the understanding of intelligent human behavior. There is little doubt that artificial intelligence is an essential basis for building intelligent systems [2].
- **Adaptivity:** This is about the ability for rapid self-adjustment when faced with changes in the surrounding environment [5]. Intelligent systems should dynamically evolve and be able to adapt to any normal and critical conditions. A major task of developing an intelligent system is to create methodologies, concepts, algorithms, and techniques that lead toward a higher level of adaptivity and flexibility, so that the system can improve its structure and knowledge of the environment, and autonomously learn and evolve its intelligence [2, 6].
- **Reasoning:** Reasoning is associated with thinking, cognition, and intellect. Reasoning about changes over time plays an important role in building an intelligent system [6, 7].
- **Information processing:** This is a process that describes everything that is happening in the system and its ability to manage uncertain and imprecise events. Data storage, data analysis, and data validation capabilities are necessary for an intelligent system to be efficient.
- **Efficiency:** Efficiency has changeable meanings in widely differing scenarios, but it is mainly concerned with achieving goals when it is performing in an expected manner. Fundamentally, efficiency is a measurable concept, quantitatively determined by the ratio of input to output in respect of time and effort. For an intelligent system, efficiency describes the capability to produce a specific outcome effectively with a minimum amount of waste, expense, or unnecessary effort.
- **Robustness:** Robustness is the ability of a system to cope with errors during operation or the ability of a procedure to continue to operate despite abnormalities. Abnormality or error happens in the event of inconsistent data input. A robust system must be able to respond to this erroneous input. An intelligent system must meet a high level of robustness [8].
- **Fault tolerance:** Fault tolerance means the capability for continuation of system operation in a normal manner when any fault occurs due to an abnormality in the environment or system itself (e.g. a faulty unit). It is vital for an intelligent system to exhibit a strong mechanism for handling fault and system recovery. However, the level of fault tolerance might vary from system to system [8].
- **Security:** Nowadays most systems are interconnected through intranet or internet. The importance of security components is increasing as intelligent systems become a source and repository of key business information and Big

Data. In addition, as intelligent systems take on critical roles, the physical impact of cyber attacks is becoming a crucial issue [9]. Consequently, an intelligent system should be protected against any unauthorized login, threat, information sharing, and data and program manipulation.

10.2   Wireless sensor networks

A wireless sensor network (WSN) consists of spatially distributed autonomous sensor nodes connected via a (wireless) communications infrastructure. It cooperatively monitors, records, and stores physical or environmental conditions, such as temperature, sound, vibration, pressure, motion, or pollutants. A sensor network differs considerably from contemporary wireless networks in terms of communication protocol, throughput and power constraints, communication bandwidth, etc. The network may consist of a small or a large number of sensor nodes depending on the target application [10]. A sensor node is a small battery-powered device capable of sensing, processing and communicating with other connected nodes along the network. The WSN can be divided into three parts: (1) data acquisition unit, (2) processing unit, and (3) communication/transmission unit. Depending on the application, homogeneous or heterogeneous sensor nodes may perform the main task of the sensor network. WSN architecture is constructed on the basis of various sensor network topologies, such as star, tree, and mesh [5, 11]. There are three basic technologies involved in smart sensor networks [12]:

- Ultra-low power sensors that are usually fabricated as micro-electromechanical systems (MEMS), using silicon-based integrated circuits.
- Embedded silicon chips, wireless transceivers, and firmware for point-to-point (P2P) communications and self-organizing systems. While the individual nodes are relatively fragile and communicate over only small distances, the complete network is robust, with communication through multiple redundant paths.
- Software for communications, control, and optimization purposes for thousands of nodes.

10.3   Deployment of WSN in cyber-physical systems

Cyber-physical systems (CPSs) are being introduced as a solution for future smart infrastructures. A CPS is like an extension to well-known embedded systems that integrate computing and communication systems to the physical world being sensed and controlled by sensors and actuators (Figure 10.2) [13]. In order to realize a CPS-based smart solution, three basic elements should exist, including sensors and actuators (physical world), (wireless) sensor network, and computing system for data processing, storage, etc. The sensor network, together with a wideband network backbone, provides the facility for transmitting data (that may be Big Data in terms of volume and velocity, etc.) from the physical world to the computing center, and vice versa.

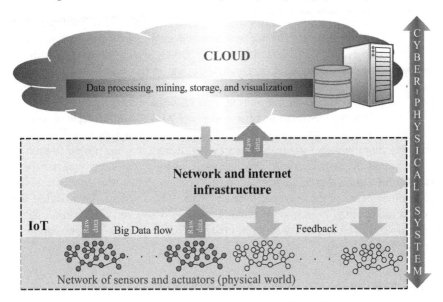

Figure 10.2   Components of a cyber-physical system

A large number of sensors deployed in the field can produce a lot of data that should be converted into meaningful information. To achieve this, one needs to answer a number of questions, such as when and where a sensor should sense, whom the data must be passed on to, and how the data is routed to destination. In addition, to answer the above questions, several constraints, such as the limited power of each sensor node, its low processing power and bandwidth, and the dynamic nature of the sensing phenomena (nodes may move or die due to energy depletion), should be taken into account. Finally, the answers to these questions need not be stationary since conditions may change dynamically. Using various techniques, such as aggregating data and reducing the communication cost from sensing nodes to the central computing system, may have a significant impact on reducing the total energy consumed by the WSN [14].

As mentioned above, multiple factors such as equipment capabilities, battery life limits, processor power, types of available sensors, sensor accuracy, and transmission range can affect system development and operation. Further, application needs determine the requirement on the frequency of sensing phenomena. Consequently, these factors raise several questions on the deployment and operation of sensors and sensor networks, such as the appropriate number of sensors, their arrangement, efficient rate of sensing, and amount of data to be transmitted. Most of these questions may be answered through a learning process employed by an intelligent WSN.

A possible approach for dealing with the complexity of a huge sensor network and enhancing efficiency in terms of power consumption and coverage is to develop a sensor network in a hierarchical manner [5, 11]. Data-centric or distributed processing can be performed depending on the conditions and processing capabilities of the sensor nodes. For example, using sensor nodes with simplified functionalities necessitates more intensive centralized computation, which likely results in less power consumption on the sensors. One way to realize this idea is to group several nodes into clusters, where nodes within a cluster will transmit the data through cluster heads that are responsible for transmitting data to the base stations.

In an intelligent WSN, the number of hierarchy levels can be decided based on the size of network and clusters. Given that the communication consumes significantly more power compared to processing, it may not be feasible to send the data acquired by each sensor to the sink node using a multi-hop sensor network [10]. In this case, an intelligent mechanism for routing data through intermediate sensors may enhance the energy-efficiency of the WSN, whereas intermediate data aggregation or data fusion may be avoided in nodes. However, cluster heads should perform such complementary tasks in order to reduce the energy consumption in the sensor nodes [14].

10.4   Intelligent wireless sensor networks

A WSN with intelligence properties offers a solution for overcoming issues such as limited energy source, network lifetime, topology and protocol maintenance, scalability, adaptivity, quality-of-service, computational power, and security, while also enhancing the network performance [15]. An intelligent wireless sensor network (iWSN) relies on the ordinary WSN concept and ensures the superior features of the system as well. On the basis of traditional WSN, the system has to develop and introduce new features and functionalities such as efficiency, programmability, maintainability, etc. in order to make the WSN intelligent. At the same time, it is not compulsory to embed all functionalities into WSNs for a particular target application. Since an iWSN is an extension of the traditional WSN, it may target similar applications with WSN, such as:

- large geographical area monitoring
- health monitoring
- infrastructure monitoring
- internet protocol (IP)-based sensor network
- ubiquitous computing
- web-based sensor monitoring service
- flight test monitoring
- wildlife habitats monitoring
- disaster management
- emergency response
- asset tracking
- manufacturing process flows
- home energy conservation
- traffic control
- smart grid

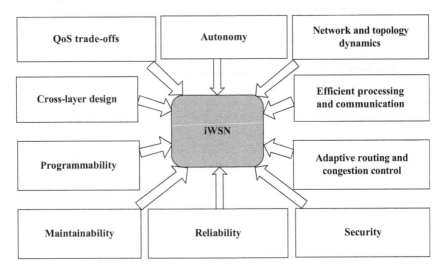

Figure 10.3 Requirements of an iWSN

It should be noted that some of the features mentioned in Section 10.1, such as reasoning, learning, and artificial intelligence, may not directly apply to the iWSN. However, features specific to the iWSN should be carefully recognized and incorporated within it. Furthermore, prerequisites of the iWSN may differ from the ordinary WSN with respect to the smart behavior of the application. In particular, they might be more or less satisfied with the basis of application type. The following (depicted in Figure 10.3) are the essential features for an iWSN:

(a)   **Autonomy:** Similar to a conventional WSN, an iWSN system should be autonomous in the sense that it could be able to operate over its full lifetime without maintenance. In such a system, sensors capture data autonomously, and data is sent to the base stations and computing systems for further processing and decision-making. Ensuring a long lifetime by applying energy-efficient or energy-harvesting techniques [16] can help render a WSN autonomous and could thus enable widespread use of these systems in many applications.

(b)   **Quality of service (QoS) trade-offs:** Of all the challenges facing the WSN, the energy problem is the most crucial. A sensor network is a battery-powered system and charging and recharging of the battery is cumbersome. Thus, a WSN has a limited lifetime. A WSN may suffer from energy waste problems due to collision, ideal listening, and overhearing on medium-access control (MAC), inefficient routing protocol, large control packet overhead, inappropriate network structure and sensor arrangement, etc. Furthermore, data transmission consumes a high amount of energy. Thus, for an intelligent WSN, employing efficient mechanisms in various levels of the network stack, such as MAC layer and routing, in order to reduce unnecessary energy usage and ensure autonomous operation, is crucial. Needless to say, QoS parameters such as the minimum throughput and maximum latency should not be compromised for the sake of energy efficiency. This is possible by making necessary trade-offs and decisions in favor of better QoS.

(c)   **Dynamics:** There are two types of dynamic facilities that are required to truly make an iWSN: network dynamics and topology dynamics.

- *Network dynamics*: Although mobility of the nodes is a factor that influences energy consumption and routing stability, in some applications both base station and sensor nodes may be mobile. Further, the sensing phenomena may be dynamic (e.g. target detection/tracking applications) or stationary (e.g. forest monitoring). Consequently, the sensor node either operates as a stationary node or as a mobile node [5].

- *Topology dynamics*: A WSN may support various network topologies such as star, tree, and mesh. The topology control of WSN is to save energy, reduce interference among nodes, and extend the lifetime of the network. During the architecture design for WSN, a dynamic topology for the intelligent network architecture must be ensured [17]. The strategy of dynamic topology control is to build a system in a distributed self-organizing manner against any changes in the whole network architecture.

(d)   **Cross-layer design:** Most of the work on the network layer design has focused on the individual network layer issues, and ignored the importance of cooperation among different layers in a sensor network. The aim of cross-layer design is to establish efficient and reliable inter-cooperation of transport, routing, medium access functionalities with the physical layer (wireless channel). The cross-layer design principle relies on a complete and unified cross-layering such that both the information and the functionalities of traditional communication layers are melded in a single protocol. The objective of the cross-layer protocol is to establish a highly reliable communication with minimal energy consumption, adaptive communication decisions, and local congestion avoidance [18, 19].

(e)   **Efficient processing and communication:** The purpose of data processing is to avoid the duplication of data and reduce the amount of data and the total number of transmissions in favor of energy saving and throughput enhancement. Data-processing operations include integration, fusion, filtering, storage, aggregation, etc. For an iWSN, efficient data processing can improve data transmission and communication quality. Some of the essential requirements for more efficient data processing are: (i) separate dedicated central processing units (CPUs) both for data computing and data communication; (ii) efficient data integration and filtering mechanisms relying on appropriate techniques such as machine learning algorithms; and (iii) sufficient storage facility for data, etc.

(f)   **Programmability:** A WSN is a self-configured autonomous system. To enhance the WSN's flexibility in every aspect, programmability can play an important role. In this regard, three types of software programs may be considered including (i) application software, (ii) communication software, and (iii) development software. The application software controls the sensor and actuator devices and forwards processed data to the wireless communication software. In turn, the communication software provides APIs (application programming interfaces) that bridge application with the CPU and the wireless RF (radio frequency) modules. Moreover, the objective of the development software is to implement special or additional functionalities for the network system, e.g. a program loader that helps users to upload the application programs into the network system [20]. All the above software parts can be modified and extended depending upon the variant conditions of the system.

(g) **Adaptive routing:** Adaptive routing ensures data delivery from source to sink nodes in any circumstances and reduces the energy consumption of the whole network while at the same time guaranteeing demanded data throughput. For the general routing method, first of all the source node calculates the number of hops to destination and chooses the shortest path for data transmission. But in the case of adaptive routing, the route from source to sink node may change at various conditions due to the presence of dead nodes (power shortage may cause the death of the sensor node). Therefore, it should be done in such a way that the redundant data paths are assured in favor of decreasing the probability of node death. In this case, the routing algorithm must be aware of the energy level of all clusters and of the entire network in order to avoid node death [21, 22].

(h) **Congestion control:** WSN belongs to the category of networks operating based on packet data transmission. In such networks, congestion is a major problem in the sense of packet loss. At the same time, congestion may create other problems such as an increase in the complexity of network operation, higher energy waste, and degradation of the overall performance of the network. For a high-density iWSN network, appropriate congestion control mechanisms should be exploited [23].

(i) **Maintainability:** Maintainability indicates the overall management of the WSN system including scalability, fault tolerance, redundancy approach, and remote maintenance [17]. These are discussed in turn below.

- *Scalability*: In some cases, WSN architecture might be extended to meet additional conditions or to increase coverage area. Scalability is the ability of a WSN system to handle the growing amount of work in a capable manner or its ability to be enlarged to accommodate network growth. Adapting to new conditions such as expansion of the existing network and ensuring performance of the network at a predefined level, are examples of the range and extent of intelligence required for WSN.

- *Fault tolerance*: A WSN has the capability to continue operation in a normal manner in the event of errors. In an iWSN, the system should have a strong mechanism for handling failures and error recovery. However, the level of fault tolerance might vary for different WSN applications.

- *Redundancy approach*: When there is a failure in any active node or active connectivity, a standby node and logical connectivity must be ensured for highly critical WSN applications such as human health monitoring, and disaster management. Maintaining system operation is performed depending on the failure of the node or connectivity: (i) node redundancy: a standby node takes place instead of the active node in failure, hence no interruption occurs in the service; or (ii) multi-string connectivity: a standby logical link takes place instead of the active logical link in failure. Hence, there would not be any connection failure along the whole network.

- *Remote maintenance*: It is not always feasible to observe the network condition from a close distance to the field (e.g. battlefield surveillance, forest monitoring). Remote login and configuration management (when necessary) can add to the value of an intelligent WSN.

(j)   **Reliability:** The principle behind reliable data communication is to prevent packet loss across the monitoring of critical events (e.g. earthquake, health monitoring) due to the significant value of the information. Data packets may be lost because of electromagnetic interference, multi-hop routing, and collision of the packets along the network. In many WSN applications, it is not critical if a small percentage of packets are lost, but for an iWSN application it is vital to have a protocol that guarantees reliable transmission of data. This is because, in an intelligent system, information loss may affect the quality of the decisions and the outcome determined by the system in critical conditions. An iWSN should guarantee a high percentage of packets correctly routed in any circumstances from source to destination, even if packets are routed through multi-hop paths [24].

(k)   **Security:** To connect with a wide network domain and for convenient communication, the WSN is mostly connected to an intranet or the internet. In the very near future, IP communication might be used in WSNs [25]. Thus, connection and bilateral cooperation between WSN and internet/intranet must be authentic and reliable. Ensuring proper security is a matter of high priority to render the WSN a truly intelligent system. Consequently, the iWSN must be protected against any unauthenticated access, threat, information sharing, data, and program manipulation [26].

10.5   WSN versus iWSN

There are some common functionalities among traditional WSN and iWSN, but incorporating additional features into WSN can raise the degree of intelligence in the system. The CPS can be constructed based on a WSN or an iWSN. The impact of intelligence is highlighted when human beings' or society's role comes into play in a CPS. Such a system is referred to as a cyber-physical social system (CPSS) [13] and involves human beings/society in the making of decisions. Thus, every component of the system should introduce sufficient degree of smartness. Consequently, the iWSN may be a better match for the CPSS than its regular counterpart WSN. For example, in a smart grid (which is explained in Section 10.6), the system can provide the user with information to make choices on their own energy usage. The user, as a smart

Table 10.1   Comparison of WSN versus iWSN

Functionality	WSN	iWSN
Autonomy	○	○
QoS trade-offs	○	○
Network dynamics	△	◎
Topology dynamics	△	◎
Hierarchical network	△	△
Congestion control	△	◇
Adaptive routing	△	◇
Cross-layer design	△	◇
Programmability	△	◎
Efficient data processing	○	◎
Efficient data communication	○	◎
Maintainability	◇	◇
Reliability	◇	◇
Network security	◇	◇

△: not essential, but may be incorporated
○: needed by default
◇: may be incorporated as an essential feature
◎: essential

element, is a part of such a system. This requires the WSN to be employed for the system supply data to enable analysis and customer usage in a secure and reliable way. Furthermore, the WSN should incorporate a sufficient degree of smartness to avoid any imbalances in total system configuration. Table 10.1 compares the WSN and iWSN in terms of specific parameters.

10.6   Example applications with distinct intelligence level

10.6.1   Smart grid

The grid traditionally refers to the electricity grid, a network of transmission lines, substations, transformers, and more that deliver electricity from the power plant to users (Figure 10.4). The smart grid combines a conventional power grid with the information technology that allows for two-way communication between the utility and its customers, and the sensing along the transmission lines is what makes the grid smart. In addition, major innovations such as renewable energy developments and plug-in electric vehicles ought to be accommodated by the smart grid. There are several benefits associated with the smart grid including: (i) more efficient transmission of electricity; (ii) quicker restoration of electricity after power disturbances; (iii) reduced operations and management costs for utilities, and ultimately lower power costs for consumers; (iv) reduced peak demand, which will also help lower electricity rates; (v) increased integration of large-scale renewable energy systems; (vi) better integration of customer–owner power-generation systems, including renewable energy systems; and (vii) improved security [27–29].

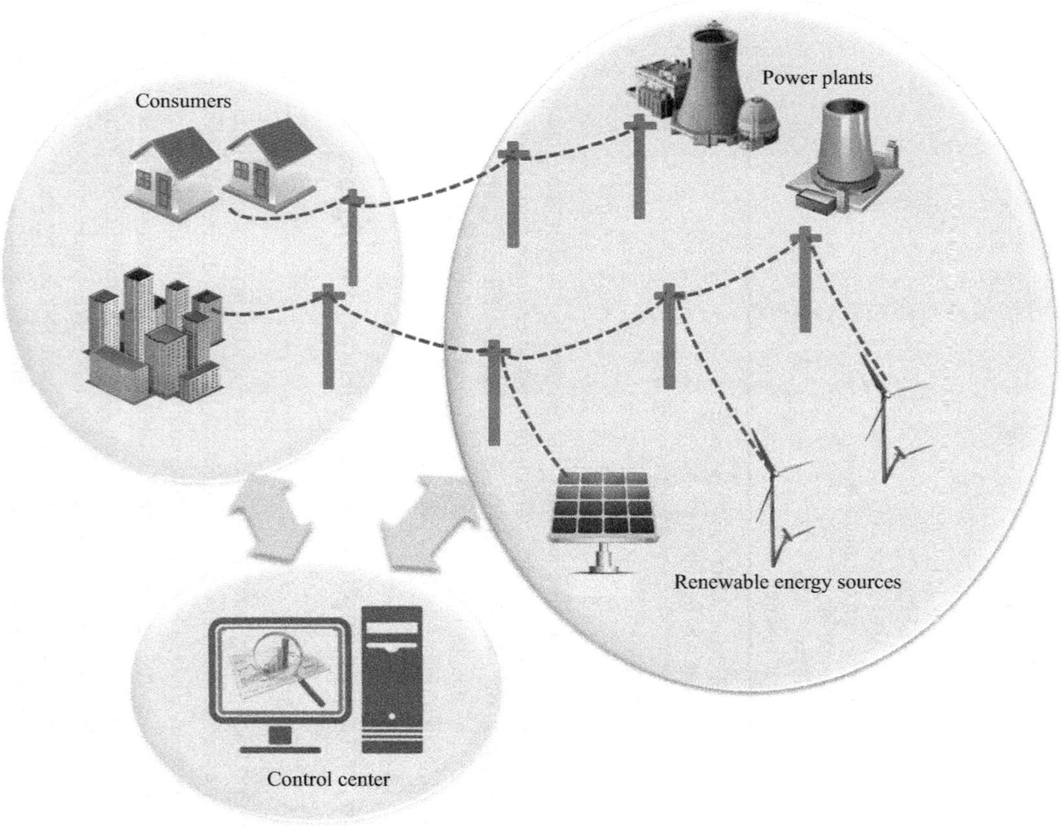

Figure 10.4 Smart grid

The smart grid gives the information and tools a user needs to make choices about their own energy use. It enables an unprecedented level of consumer participation. Smart meters and other mechanisms will allow the user to see how much electricity is used, when it is used, and its cost. Combined with real-time pricing, this will allow the user to save money by using less power when electricity is most expensive.

10.6.2 Smart field monitoring

Over the past decade, a number of environmental and habitat monitoring projects have emerged which have introduced attractive applications of WSNs [30–32]. Forest fire detection, flood detection, complexity mapping of the environment are all examples of environmental applications of WSNs. Animal monitoring is performed for various purposes such as research and wildlife protection and management [33–37]. As shown in Figure 10.5, a smart animal monitoring system can sophisticatedly analyze and mine the data acquired from sensors and cameras based on multiple biological features. This system can help biologists and pest management experts to extract their required information from the large amounts of data, and index the events of interests for future use.

The monitoring system can be implemented using technologies such as sensors and cameras. However, extracting knowledge using conventional techniques from a huge amount of data (particularly video) collected from the environment might be tedious and very time-consuming. For example, tracking a rare animal species in videos recorded by cameras may generate a huge amount of data which requires significant effort to extract the desired information. A number of methods related to animal species classification, individual animal recognition, and animal-related events have been introduced [35, 37] which mostly rely on detecting motion, color, texture, or the face of animal. Also, thermal imaging has been widely used by biologists for animal study [33]. However, existing solutions for processing wildlife data are semi-automated, need human interaction, or are limited to detection of a

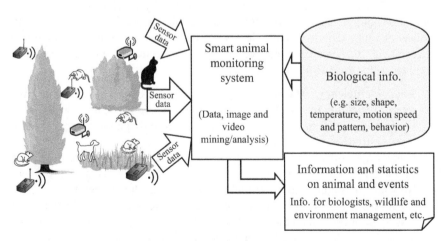

Figure 10.5 How a smart animal monitoring system works

specific single feature of the animal in fixed or predefined conditions. This means that more features need to be incorporated into existing solutions to build a sufficiently sophisticated and smart system.

As a multiplier effect, the smart animal monitoring system, for example, can provide pest control experts with information for monitoring agricultural fields. Hence it provides facilities for applying control techniques efficiently, cost-effectively, and with less risk to the animals and the environment. In [14] and [34] a smart pest monitoring system has been proposed for (a) reducing the environmental contamination caused by pesticides and (b) more efficient pest (particularly rat) control.

The introduced solution, referred to as SPeC (smart pest control), is based on the concept of cyber-physical systems. It is assumed to be intelligent, autonomous, energy- and cost-efficient, free from side effects on the environment, and a unique solution for monitoring information on the rats living in agricultural fields. SPeC supplies the infrastructure for constant tracking of rats' nests and activities and facilitates pest control measures. It generates a map of the field including various information such as the estimated number of rats, their nest locations, and the time and location in which they are mostly active. Consequently, various functions are developed for SPeC to increase the precision of the system in detecting rats. The underlying WSN exploited for this system should be able to operate in a very energy-efficient manner.

10.6.3　Applications need variant smartness

The function of the smart grid is to supply electricity with less failure, thus minimizing the risk of minor or major blackout; factors such as reliability, latency, maintainability, and security are also crucial. On the other hand, SPeC provides the image of an agricultural field to pest control experts, and there is no strict requirement on the real-timeliness or security of the system. Energy efficiency is what SPeC strongly needs, which can be provided by efficient protocol design, congestion control, network dynamics, adaptive routing, etc. Each of the above two applications may require different levels of intelligence due to their distinct properties. Table 10.2 compares the two applications, indicating different aspects of intelligence for each of them.

- **Energy efficiency:** Although both systems should be energy-efficient, this is more crucial for the SPeC due to the lack of energy sources in the field. Conversely, in the smart grid, sensors can be fed through the electric power of the system when needed.
- **Congestion control:** In a smart grid, packet loss due to congestion may give rise to a critical situation. However, the impact of packet loss in SPeC may not be crucial.
- **Network dynamic and hierarchical network architecture:** Both systems should be able to support hierarchical architecture and dynamic network formation in order to increase efficiency in terms of communication and energy consumption. In the case of a dynamic network, communication to the base station can be established by relying on the dynamic behavior of the network

Table 10.2   Various requirements of intelligence for smart grid versus smart pest control system

Functionality	Smart grid	SPeC
Energy efficiency	○	◎
Network dynamics	◎	◎
Dynamics of topology	◎	◎
Hierarchical network	△	△
Congestion control	△	◎
Adaptive routing	◎	◎
Throughput	○	○
Latency	◎	△
Cross-layer design	△	◎
Programmability	△	◎
Efficient data processing and data communication	○	◎
Reliability	◎	△
Maintainability	◎	△
Redundancy approach	◎	△
Network security	◎	△

△: less priority
○: default
◎: essential

with less latency and packet loss. The hierarchical network architecture is significant for the scalability measures as well.

- **Adaptive routing and efficient hardware:** The smart grid and SPeC both need to support adaptive routing protocol and employ efficient hardware, though for two different purposes, namely, reliable and secure service in the smart grid and enhanced energy efficiency in the SPeC.
- **Dynamics of topology:** A variant topology is crucial for the smart grid for critical situations where some nodes fail to avoid blackout and power system failure in the whole network. In the SPeC, to keep the network more energy-efficient, a fixed topology may be employed due to reduced criticality of network failures.
- **Throughput and latency:** Any of these systems need high throughput for data transmission; however, the smart grid needs less latency to ensure minimum response time to any failure in the system.
- **Cross-layer design:** Due to the severe energy constraints of SPeC, these features can be significant in designing an efficient network protocol, while for the smart grid they can be an added value.
- **Programmability:** Both systems need higher degrees of programmability to be able to act in various conditions, and more specifically for distinct purposes such as to prevent system failure in the smart grid, and to reduce energy consumption in the SPeC.
- **Efficient data processing and data communication:** Higher efficiency in data processing and communication results in more energy saving as well as better decision making. Thus, for the SPeC, such features require high priority; in the

smart grid, although these features can help improve the quality of user–system interaction, it is not a strict requirement.

- **Reliability, maintainability, and redundancy approach:** In a smart grid, any failure should be quickly and appropriately tolerated and faulty parts should be bypassed or replaced to prevent failure from spreading within the system as well as any blackout. These constraints on the SPeC are much more relaxed due to the lower criticality of the system.
- **Network security:** The smart grid provides a two-way communication between utility and users, and the data generated by the smart meters is analyzed for extracting knowledge. Any physical or cyber threat and attack to the system may cause loss of privacy for the users or physical failure and massive blackout. Therefore, preserving security is of significantly high priority. This requirement is much more relaxed for the SPeC, which does not provide critical information, although it can impose high costs on the system.

In summary, the smart grid needs intelligence features mainly in favor of providing highly secure, reliable, and scalable service for users, while a field monitoring system requires adaptability and programmability for the sake of reducing energy consumption.

10.7   Summary

Different applications may require various levels of intelligence. Depending on the target application, system characteristics, requirements, and functionalities may vary. However, it must be emphasized that to make a WSN system intelligent, some of the particular functionalities must be incorporated into the system as high priorities.

References

[1]   T. Poggio and L. Sringa. Project for an intelligent system: Vision and learning. *International Journal of Quantum Chemistry*, 42(4): 727–739, 1992.

[2]   A. Bouchachia. Adaptive and intelligent systems. In *Proceedings of the 2nd International Conference (ICAIS 2011)*, Klagenfurt, Austria, 6–8 September 2011.

[3]   I. J. Rudas and J. Fodor. Intelligent systems. *International Journal of Computers, Communications & Control*, III (Suppl. issue): 132–138, 2008.

[4]   R. Fjellheim, E. Landre, R. Nilssen, T. O. Steine and A. A. Transeth. Autonomous systems: Opportunities and challenges for the oil & gas industry. See http://nfaplassen.sitegen.no/customers/nfa/files/NYAutonomyOilGas01803.pdf.

[5]   T. Al-Mahmud. A survey on how dynamically changes topology in wireless sensor network. *ABC Journal of Advanced Research*, 1(1), 28–34, 2012.

[6]   B. Bouchon-Meunier, L. Magdalena, M. Ojeda-Aciego, J. L. Verdegay and R. R. Yager (eds.). *Foundations of Reasoning Under Uncertainty*. Springer-Verlag, Berlin and Heidelberg, 2010.

[7]   J. Cassens and A. Kofod-Petersen. Explanations and case-based reasoning in ambient intelligent systems. See http://www.idi.ntnu.no/~anderpe/publications/Context-07-AKP-JC.pdf. Accessed: 10 January 2016.

[8]   A. Schuster. *Robust Intelligent Systems*. Springer, 2008.

[9]   M. Mariana. Intelligent system for information security management: Architecture and design issues. *Informing Science: International Journal of an Emerging Transdiscipline*, 4.1: 29–43, 2007.

[10] I. F. Akyildiz, W. Su, Y. Sankarasubramaniam and E. Cayirci. Wireless sensor networks: A survey. *Computer Networks*, 38(2002): 393–422, December 2011.

[11] Y. Al-Obaisat and R. Braun. On wireless sensor networks: Architectures, protocols, applications, and management. See http://epress.lib.uts.edu.au/research/handle/2100/160.

[12] Y. Kawahara, M. Minami, H. Morikawa and T. Aoyama. Design and implementation of a sensor network node for ubiquitous computing environment. In *Proceedings of the 58th IEEE Vehicular Technology Conference* (vol. 1), pp. 3005–3009, Orlando, FL, USA, 6–9 October 2003.

[13] J. Paek, O. Gnawali, K. Y. Jang, D. Nishimura and R. Govindan. A programmable wireless sensing system for structural monitoring. In *Proceedings of the 4th World Conference on Structural Control and Monitoring (4WCSCM)*, pp. 1–8, July 2006.

[14] F. Mehdipour, K. C. Nunna and K. Murakami. A smart cyber-physical systems-based solution for pest control (work in progress). In *Proceedings of the IEEE International Conference on Cyber, Physical and Social Computing*, pp. 1248–1253, Beijing, China, 20–22 August 2013.

[15] J. Pinto. Intelligent sensor networks. *Automation World*, 2004. See http://www.jimpinto.com/writings/sensornets.html.

[16] R. J. M. Vullers, R. van Schaijk, H. J. Visser, J. Penders and C. Van Hoof. Energy harvesting for autonomous wireless sensor networks. *IEEE Solid-State Circuits Magazine*, 29–38, Spring 2010.

[17] H. Karl and A. Willig. *Protocols and Architectures for Wireless Sensor Networks*. Wiley, 2007.

[18] F. Bouabdallah, N. Bouabdallah and R. Boutabaz. Cross-layer design for energy conservation in wireless sensor networks. In *Proceedings of the IEEE International Conference on Communications (ICC '09)*, pp. 1–6, IEEE, Dresden, Germany, 14–18 June 2009.

[19] S. Jagadeesan and V. Parthasarathy. Cross-layer design in wireless sensor networks. *Advances in Intelligent and Soft Computing*, 166: 283–295, 2012.

[20] J. Paek, O. Gnawali, K. Y. Jang, D. Nishimura and R. Govindan. A programmable wireless sensing system for structural monitoring. In *Proceedings of the 4th World Conference on Structural Control and Monitoring (4WCSCM)*, pp. 1–8, July 2006.

[21] O. Changsoo, P. Mitra, L. Seokche on and S. Kumara. Distributed energy-adaptive routing for wireless sensor networks. In *Proceedings of the IEEE International Conference on Automation Science and Engineering (CASE)*, pp. 905–910, IEEE, Scottsdale, New Zealand, 22–25 September 2007.

[22]   J. Sen. An adaptive and multi-service routing protocol for wireless sensor networks. In *Proceedings of the 16th Asia-Pacific Conference on Communications (APCC)*, pp. 273–278, IEEE, Auckland, New Zealand, 31 October–3 November 2010.

[23]   J. Zhao, L. Wang, S. Li and X. Liu. A survey of congestion control mechanisms in wireless sensor networks. In *Proceedings of the 6th International Conference on Intelligent Information Hiding and Multimedia Signal Processing (IIH-MSP)*, pp. 719–722, IEEE Computer Society, Darmstadt, Germany, 15–17 October 2010.

[24]   Z. Zinonos, V. Vassiliou, C. Ioannou and M. Koutroullos. Dynamic topology control for WSNs in critical environments. In *Proceedings of the 4th IFIP International Conference on New Technologies, Mobility and Security (NTMS)*, pp. 1–5, IEEE, Paris, France, 7–10 February 2011.

[25]   P. Thubert, IP in wireless sensor networks, Cisco Systems, 2006. https://docbox.etsi.org/workshop/2008/2008_06_M2MWORKSHOP/CISCO_THUBERT_M2MWORKSHOP.pdf. Accessed: 20 January 2016.

[26]   D. Liu and P. Ning. *Security for Wireless Sensor Networks*. Springer, 2007.

[27]   Foundations for Innovation. *Strategic R&D Opportunities for the Smart Grid*. Report of the Steering Committee for Innovation in Smart Grid Measurement Science and Standards, March 2013. See http://www.nist.gov/smartgrid/upload/Final-Version-22-Mar-2013-Strategic-R-D-Opportunitiesfor-the-Smart-Grid.pdf.

[28]   The smart grid: An introduction. See http://energy.gov/oe/downloads/smartgrid-introduction-0. P. Thubert. *IP in Wireless Sensor Networks*. Cisco Systems, 2006. See https://docbox.etsi.org/workshop/2008/2008_06_M2MWORKSHOP/.

[29]   What is the smart grid? See http://www.hitachi.com/environment/showcase/solution/energy/smartgrid.html.

[30]   A. Ailamaki, C. Faloutsos, P. Fischbeck, M. Small and J. Van Briesen. An environmental sensor network to determine drinking water quality and security. *SIGMOD Record*, 32(4): 47–52, December 2003.

[31]   C. Gries, A. Stephenson and K. Morrison. Sensors at north temperate lakes. Information Management Newsletter of the Long Term Ecological Research Network, 2012.

[32]   C. G. Panayiotou, D. Fatt and M. P. Michaelides. *Environmental Monitoring Using Wireless Sensor Networks*. University of Cyprus, Nicosia, Cyprus, 2005.

[33]   T. Burghardt *et al*. Analysing animal behaviour in wildlife videos using face detection and tracking. *Vision, Image and Signal Processing, IEEE Proceedings*, 153(3): 305–312, June 2006.

[34]   F. Mehdipour. Smart field monitoring: An application of cyber-physical systems in agriculture (work in progress). In *Proceedings of the 3rd IIAI International Conference on Advanced Applied Informatics*, pp. 181–184, Japan, 31 August–4 September 2014.

[35] K. A. Steen, A. Villa-Henriksen, O. R. Therkildsen and O. Green. Automatic detection of animals in mowing operations using thermal cameras. *Sensors*, 12(6): 7587–7597, 2012.

[36] R. Szewczyk, A. Mainwaring, J. Polastre, J. Anderson and D. Culler. An analysis of a large scale habitat monitoring application. In *Proceedings of the 2nd international Conference on Embedded Networked Sensor Systems (SenSys '04)*, pp. 214–226, Baltimore, MA, USA, 3–5 November 2004.

[37] M. Zeppelzauer. Automated detection of elephants in wildlife video. *EUR-ASIP Journal on Image and Video Processing*, 46: 1–23, December 2013.

Chapter 11

Resilient wireless sensor networks for cyber-physical systems

Waseem Abbas[1], Aron Laszka[2] and Xenofon Koutsoukos[3]

Owing to their low deployment costs, wireless sensor networks (WSN) may act as a key enabling technology for a variety of spatially distributed cyber-physical system (CPS) applications, ranging from intelligent traffic control to smart grids. However, besides providing tremendous benefits in terms of deployment costs, they also open up new possibilities for malicious attackers, who aim to cause financial losses or physical damage. Since perfectly securing these spatially distributed systems is either impossible or financially unattainable, we need to design them to be resilient to attacks: even if some parts of the system are compromised or unavailable due to the actions of an attacker, the system as a whole must continue to operate with minimal losses. In a CPS, control decisions affecting the physical process depend on the observed data from the sensor network. Any malicious activity in the sensor network can therefore severely impact the physical process, and consequently the overall CPS operations. These factors necessitate a deeper probe into the domain of resilient WSN for CPS. In this chapter, we provide an overview of various dimensions in this field, including objectives of WSN in CPS, attack scenarios and vulnerabilities, the notion of attack resilience in WSN for CPS, and solution approaches toward attaining resilience. We also highlight major challenges, recent developments, and future directions in this area.

11.1 Introduction

A wireless sensor network (WSN) is a collection of sensor devices organized into a wireless network. Traditionally, WSNs have been used as cost-effective means of monitoring spatially distributed processes and phenomena. Their potential uses include military applications, such as battlefield surveillance and chemical attack detection, environmental applications, such as forest-fire detection and precision

[1]Vanderbilt University, Nashville, TN 37212, USA, e-mail: waseem.abbas@vanderbilt.edu
[2]Vanderbilt University, Nashville, TN 37212, USA, e-mail: aron.laszka@vanderbilt.edu
[3]Vanderbilt University, Nashville, TN 37212, USA, e-mail: xenofon.koutsoukos@vanderbilt.edu

agriculture, and health applications, such as monitoring human physiological data [1].

A cyber-physical system (CPS) is an integrated system of *computational elements* and *physical processes*, in which the physical processes are controlled by the computational elements [2]. Since the computational elements must have reliable information about the evolving state of the physical processes in order to control them, every practical CPS has to include *sensor devices*. These sensor devices monitor the physical processes, providing the computational elements with information that can be used for various tasks, such as state estimation and fault identification. Finally, the output of the computational elements is fed into actuator devices that can influence the physical processes in the desired way, which closes the loop between the physical and cyber parts of the system.

In the case of *spatially distributed physical processes*, however, the sensing task can prove to be challenging, as the sensor devices may need to be deployed over a larger area. For example, in order to provide intelligent traffic control for a whole city, we must have reliable information about the current traffic situation in various parts of the city. In order to have such information, we must deploy a large number of traffic sensors over a vast area. With wired sensors, the cost of deployment could be prohibitively high and, in some cases, it may even be physically or legally impossible. Consequently, WSNs, whose deployment is much simpler and more cost-effective, may act as a key enabling technology for spatially distributed CPSs.

The rest of the chapter is organized as follows. In the remainder of this section, we illustrate the role of WSN in the context of CPS along with information-security goals in CPS. In Section 2, various applications of WSN for CPS are stated along with examples. An overview of different attack scenarios and vulnerabilities in WSN along with instances of such attacks in practical networks is provided in Section 3. In Section 4, the notion of attack resilience in WSN is discussed along with the modeling issues and related challenges. Different approaches toward making WSN resilient against attacks, as well as a couple of detailed examples, are presented in Section 5. Finally, some future directions in this field are outlined in Section 6.

11.1.1   Cyber-physical systems and sensor networks

Monitoring and surveillance applications. Traditional sensor network applications focus on acquiring, transmitting, and fusing data. In these applications, the physical and cyber parts do not form a closed loop, or, in some cases, form a closed loop which includes a human element. See Figure 11.1 for a simple illustration of the system architecture of such applications.

For example, in a typical habitat-monitoring application [3], sensors measure environmental properties, such as light, temperature, humidity, and barometric pressure, and transmit their data through the sensor network to a gateway. Then, the gateway transmits the data through a transit network to a base station, which

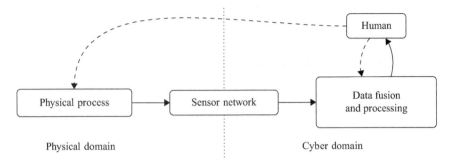

Figure 11.1   Wireless sensor networks for monitoring and surveillance applications

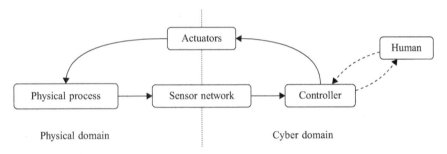

Figure 11.2   Wireless sensor networks for cyber-physical systems

provides wide-area network (WAN) connectivity. Finally, the processed data is displayed on a user-friendly interface to scientists.

As another example, in a forest-fire surveillance application [4], sensors collect temperature, humidity, and illumination data, and transmit it through the sensor network to a gateway node. The gateway node then forwards the data to some middleware, which stores the measurements in a database server and calculates forest-fire risk levels from real-time and historical data. Finally, the results are displayed in a web application, and, if a forest fire is detected, alarms are automatically sent to fire stations or nearby residents.

Cyber-physical systems. In CPSs, on the other hand, the physical processes and computational elements are tightly integrated: physical processes, sensors, controllers, and actuators form a *closed loop*. Note that CPSs can still be supervised by human operators; however, there is a closed, real-time control loop which does not contain a human element. See Figure 11.2 for a simple illustration of the architecture of CPSs using WSNs.

Since sensor networks in CPSs are part of closed, real-time control loops, ensuring their security is more critical than in traditional sensor network applications. In a CPS, malicious sensor data will result in incorrect control decisions, which are immediately executed by the actuators. Consequently, an attacker who has compromised a sensor network has some level of control over the physical process and may cause physical damage or financial losses using malicious control. For example, in a smart electric grid, an attacker who can tamper with real-time power consumption data may be able to cause physical damage by simulating a rapid increase in consumption.

Therefore, security is a crucial issue for WSNs in CPSs. In the following subsection, we summarize the traditional goals of information security and how they can be applied to CPSs. For a general overview of WSN in CPS, we refer readers to the other chapters in this book and to the survey of Wu *et al* [5].

11.1.2   *Information-security goals and cyber-physical systems*

Traditionally, the three key goals of information security are *confidentiality*, *integrity*, and *availability* (CIA). For CPSs, however, these properties are often listed in reverse order (AIC) to emphasize that, in many CPS, availability and integrity requirements have priority over the confidentiality objective [6, 7].

Availability. Availability means ensuring that a system remains in an operable state even if it is attacked. Providing availability for CPS requires ensuring not only that every element of the closed control loop remains operable, but also that the connections between them remain functional. Moreover, most CPS have strict timing constraints, which the system must satisfy even in the case of an attack.

Consequently, providing availability for a CPS entails defending against a variety of attacks. For example, the devices have to be protected against physical attacks, such as physical node destruction and wireless jamming, which can prove to be very challenging for spatially distributed systems. Furthermore, the system has to be protected against cyber attacks as well, in which an attacker exploits some software or protocol vulnerability in order to compromise the system. This can also be very challenging, since the lifetime of many CPS is measured in decades, and deploying software updates is usually a difficult process.

Integrity. Integrity means ensuring that information cannot be modified in an unauthorized and undetected manner. In a CPS, providing integrity means protecting the sensor and control data from modification attacks, which can be achieved by using a message authentication scheme. However, even though there are a number of message authentication schemes which are considered to be secure, providing integrity for a CPS remains challenging. Firstly, many CPS contain legacy devices, which were not designed for security. Consequently, in order to enable message authentication on these devices, their software has to be updated, which can be a difficult process in a CPS. Moreover, these legacy systems have limited computational capabilities and communication bandwidth, which may be insufficient for computing cryptographic functions and transmitting message authentication codes.

Confidentiality. Confidentiality means ensuring that information is accessible only to authorized entities. In CPS, confidentiality is usually viewed as the least important security goal, since sensor and control data rarely constitutes sensitive information. For example, in an intelligent traffic-control system, sensor data usually measures the current traffic situation, while control data adjusts the traffic signals; both of these are easily observable even without any access to the system.

However, in some CPS, sensor data can include sensitive personal information from end-users. In these systems, we must provide privacy, for which providing confidentiality is a necessary condition. For example, privacy issues are a very important concern for smart metering [8–10]. Nevertheless, for most WSN applications, confidentiality is not a major concern. For this reason, we will focus on availability and integrity in the remainder of this chapter.

11.2 Objectives of WSN for CPS

WSNs are used for a wide variety of applications in CPS. In this section, we give an overview of some of these applications, especially in the context of security and resilience of CPS, and present two specific applications in more detail.

11.2.1 Main objectives

Surveillance and monitoring. Continuous monitoring and surveillance of physical processes is one of the primary services offered by a WSN. In CPS, this sensory data is then utilized to regulate and control various system modalities in real time. For instance, traffic management through the use of traffic controls, route advisories, and road pricing has been made possible by the observations collected by the sensing devices installed at various locations in traffic-flow networks. Similarly, in the healthcare industry, WSNs are used to enable early detection of clinical deterioration through real-time patient monitoring [11]. The role of WSN in CPS is not just limited to monitoring physical processes by collecting data, but also impacts on the process by being part of the decision-making framework. Smart agriculture, intrusion detection in industrial systems, target tracking in security applications, regulation of necessary conditions in production processes, and energy management systems in buildings are a few instances that have greatly benefited from WSN deployment.

State estimation. In order to dynamically control a system, information about its evolving states is required, which can be acquired through state estimation techniques. Typically, states are estimated based on the noisy measurements collected from various sensing devices and the physical model of the process while minimizing (or maximizing) a certain criterion, such as the weighted least square, the maximum likelihood, and the minimum variance. For instance, in power systems, estimation of system variables such as voltages and relative phasor angles is required throughout the system nodes for power flow and contingency analysis.

In the case of CPSs that extend over a large area, reliable state information can be acquired by deploying sensor nodes at various locations within the system, and then processing their observations alongside the dynamical model of the system. For instance, assume that the dynamics or the evolution of states during a process are modeled by a linear time-varying system as

$$x_{k+1} = A_k x_k + B_k u_k + N_k w_k \tag{11.1}$$

Here, x_k is a state vector representing n state variables of the system at time instant k, $A_k \in \mathbb{R}^{n \times n}$ is a time-varying matrix that models state transitions, $u_k \in \mathbb{R}^m$ is the control input applied to the system, $B_k \in \mathbb{R}^{n \times m}$ is a control input matrix, and $w_k \in \mathbb{R}^o$ is process noise.

Sensors deployed at various points within the network make observations y_k, which can be represented in a compact way as

$$y_k = C_k x_k + v_k, \tag{11.2}$$

where $C_k \in \mathbb{R}^{p \times n}$ is a matrix that maps the true state space to the observed space, and $v_k \in \mathbb{R}^p$ is the measurement noise.

The state estimation problem is that, given a measurement sequence and an input sequence up to time k, i.e., $\{y_i, u_i : 0 \leq i \leq k\}$, what is the best possible estimate $\hat{x}_{k+t}$ of x_{k+t}? For $t = 0$, the problem is referred to as the *filtering* problem. *Kalman filtering* (KF) and its extensions are the most widely used tools for state estimation in physical processes using sensor observations (e.g., [12, 13]).

Fault detection and localization. Critical infrastructure networks, such as water distribution networks, provide good examples of CPS. These networks are prone to faults and failures, which, if not taken care of in a timely way, might cause tremendous losses. Thus, sensors measuring various system attributes are deployed throughout the network to observe system failures once they occur. For instance, in water networks, flow and pressure sensors are deployed to monitor any abrupt changes in these variables that might correspond to pipe bursts or leakages. If the objective is just to detect the failure without localizing the failure point, then the problem is the *detection* problem. However, if it is desired to uniquely identify each fault, the problem is referred to as the *localization* problem.

In the context of detection and identification of faults, efficient sensor placement is a crucial design problem and has been an active research area. If S is the set of all available sensors, $c(s)$ is the cost of placing sensor $s \in S$, and $f : 2^S \rightarrow \mathbb{R}$ is a set function that encapsulates the objective of sensor placement, such as detection or identification of specific failures, while assigning a weight to a sensor configuration $S \subseteq \mathcal{S}$, then a typical sensor placement problem can be stated as

$$\max_{S \subseteq \mathcal{S}} f \quad \text{s.t.} \quad \sum_{s \in S} c(s) \leq M. \tag{11.3}$$

Here M is the allowed budget for the cost of sensor placement. In the context of sensor placement, faults can refer to a wide variety of scenarios. For instance, in water networks, faults can be pipe bursts, leakages, malicious introduction of contaminants, etc.

Short- and long-term forecasting. Integration of demand, generation, and storage is key to efficient – accessible, reliable, and flexible – operation of modern infrastructure networks, such as smart grids [14, 15]. Precise short- and long-term forecasting of generation and demand variables through intelligent and adaptive elements is required to achieve this objective. Using continuous traces of usage patterns that are unique to each premise, along with the database of accumulated observations, thanks to the web of WSN, better design and control of demand response actions can be instantiated. In fact, in many applications, a more proactive approach, which relies on control decisions based on the possible future states of a system, is rather an essential requirement. Moreover, accurate forecasting of system parameters not only leverages planning for growth and changes, but also creates better awareness among customers of their own consumption patterns.

The forecasting problem of state variables in the short and long term is very much related to the state estimation discussed above. Using the system model as given in (11.1) and (11.2), if states are estimated for some time in the future, i.e., $\hat{x}_{k+t}$ for $t > 0$, then the problem becomes a forecasting problem, also referred to as the *prediction* problem.

Theft detection and anti-fraud in utility networks. As a result of the integration of WSN with advanced metering infrastructure technologies, high-resolution and fine-grained data is now available to utility service providers such as electricity distribution companies. Using data analytic approaches, utilities are aiming to regulate as well as implement theft detection and anti-fraud mechanisms, which could not be realized through traditional physical checks, to minimize non-technical losses [16, 17].

A simple formulation of theft detection of the quantity under consideration is as follows. Assume that there are two types of consumers: genuine and fraudulent. The goal is to detect fraudulent consumers based on the measurements collected from sensing or metering devices. Let y_t^i be the measurement of the quantity consumed by consumer i at time t, where $t \in \{1, \ldots, K\}$, and K is the time interval between billing cycles. Moreover, the probability density functions of y_t^i corresponding to genuine and fraudulent customers are p_g and p_f, respectively. The hypotheses that consumer i is genuine (H_g) and consumer i is fraudulent (H_f) can then be evaluated using the likelihood ratio test:

$$\frac{\prod_{t=1}^{K} p_g(y_t^i)}{\prod_{t=1}^{K} p_f(y_t^i)} \underset{H_g}{\overset{H_f}{\gtrless}} \gamma. \tag{11.4}$$

Here, γ reflects the service provider's trade-off between missing a fraudulent consumer detection and an incorrect detection.

11.2.2 Examples

In what follows, we discuss two example applications of WSN for CPS.

Smart grid. An electrical (or power) grid is a system for generating, transmitting, distributing, and controlling electricity [18]. Traditionally, these systems are used to carry power in one direction, from a few central generators to a large number of consumers. *Smart grids*, on the other hand, use two-way flows of both electricity and information in order to provide widely distributed and automated energy-delivery networks. Compared to traditional grids, smart grids can deliver power in more efficient ways by utilizing modern information technologies. Furthermore, smart grids can respond to a broad range of events that may occur anywhere in the grid; for example, they may change the power flow automatically when a medium-voltage transformer fails unexpectedly.

Reliable and online information, which can be obtained from sensors monitoring the grid, is crucial to the operation of a smart grid [19] Since the installation and maintenance costs of wired monitoring systems may be very high, wireless sensors may act as a key enabling technology. WSNs can provide a cost-effective sensing and communication platform, enabling the grid to respond to events and changing conditions in a timely and proactive manner.

Intelligent traffic control. Traditionally, traffic signals were controlled by disconnected devices with fixed schedules, which did not have the ability to adapt to the varying traffic situation. However, in recent years, these controllers have been connected both with each other and with sensor devices. This has led to the formation of intelligent traffic-control systems, which provide substantial benefits in terms of wasted time, environmental impact, and public safety [20]. Unfortunately, the installation and maintenance costs of wired sensing systems can be prohibitively high, which hinders large-scale deployment [21]. Since sensor nodes with integrated sensing, computing, and wireless communication capabilities offer much lower installation costs, WSN may revolutionize the field of traffic control.

The direct goal of traffic monitoring is to provide an accurate estimate of the current traffic situation. However, in order to be able to perform proactive, dynamic traffic control, we also need to be able to forecast traffic [22]. In the literature, a large number of models have been proposed for this prediction problem. For example, Xie *et al.* [23] recently proposed to use Gaussian processes and demonstrated their applicability using real-world data.

11.3   Attacks against sensors

In this section, we present an overview of various attacks that can be mounted against a WSN, which may cause serious damage if successful. The consequences of these attacks may range from deterioration or even complete disruption of services offered by the WSN to complete failure of the physical process whose control depends on the sensory data.

11.3.1   Types of attacks

In the literature on WSNs, a multitude of attacks have been studied. We divide these attacks into two main groups: denial-of-service attacks and data falsification attacks.

11.3.1.1 Denial-of-service (DoS) attacks

Denial-of-service attacks try to break the availability security property of the system. Keeping in view the fact that, in CPSs, the CIA security paradigm is often modified to AIC, DoS attacks in sensor networks, which directly impact the availability of services, become a prominent threat to CPS operations. These attacks can be both physical and cyber.

Typically, a classification of various DoS attacks is based on different layers in the protocol stack of the sensor nodes. Details of such a classification can be found in [24–26].

Physical layer attacks. *Wireless jamming* is a typical physical layer attack used by an adversary to put a certain number of nodes in the system out of service by disrupting valid communication between sensor nodes. Various *active* and *reactive* jamming strategies can be used by an adversary for this purpose [27]. Active jamming revolves around the theme of keeping the channel busy for longer periods irrespective of the traffic patterns on the channel, and includes constant jamming, deceptive jamming, and random jamming strategies. Active jamming is relatively easier to detect. Reactive jammers, on the other hand, sense traffic on the communication channel and transmit when the channel is active, and remain idle in the case when the channel is free. Consequently, they become hard to detect. *Physically destroying a sensor* is also a physical layer DoS attack that renders the node out of service. Similarly, *tampering* with the hardware or software configurations of nodes to put them out of service also results in the denial of service.

Link/medium access control layer attacks. An adversary might attack the link layer (or media access control (MAC) layer) by introducing malicious collisions, often referred to as *collision attacks*, resulting in repeated retransmissions of the frames associated with various MAC protocols. This not only decreases the network throughput, but also targets sensor nodes' power supplies through *denial-of-sleep* attacks, in which the batteries of nodes are exhausted much earlier than expected due to a large number of retransmissions.

Network and routing layer attacks. Some of the DoS attacks on the network layer involve *spoofing* or altering the routing information to cause routing disruptions, with detrimental consequences on the overall network performance, such as extending or shortening the source routes, increasing end to end latency, which is crucial in CPS keeping in view the significant requirement of making decisions in real time, decreasing network throughput, etc. *Black hole attacks* and *sinkhole attacks* are examples in which compromised nodes establish an important role by becoming a part of several routes and then dropping packets through them. Another attack with serious ramifications is a *wormhole attack* [28, 29], in which an attacker receives a message from one point and then replays it from another point in the network after passing it from the source to the new point through a low-latency link (wormhole link). In many protocols, nodes broadcast hello messages to determine one-hop neighbors for routing purposes. An attacker may compromise a node, and through a high-power transmitter may inform other nodes that it is their one-hop neighbor which can provide a superior route to the base station. As a result, many nodes attempt to route their traffic through the compromised node even though the compromised node does not lie within their radio range. This sort of attack is often referred to as a *hello flood attack*.

Transport layer attacks. At the transport layer, which is mainly responsible for maintaining reliable end to end connections between nodes, some protocols require nodes at either end of the connection to maintain state. An attacker may bombard a node implementing such a protocol with new connection requests without ever completing them, also known as a *flooding attack*, which might exhaust the connection buffer of the node. In another attack, called a *desynchronization attack*, an attacker disrupts the existing connection by repeatedly spoofing the message to desynchronize the endpoints and cause retransmissions

Application layer attacks. If sensors are not transmitting observations at fixed intervals, but are rather triggered by some physical activity such as responding to an event, then an adversary can create an attack by *overwhelming* a sensor by artificially generating events. This may result in battery exhaustion and bandwidth consumption. Moreover, on the application layer side, sensors typically utilize simple and not highly sophisticated operating systems to run various applications. As a result, software vulnerabilities of sensor nodes can be exploited by an attacker through *mal-packets* [30] or *sensor worm attacks* [31, 32], in which specially crafted data can exploit memory-related vulnerabilities and use application code in a sensor to further propagate in the network, resulting in malicious behavior and node failures. Propagation of such worms is typically done either by scanning vulnerable devices, or by spreading through topological neighbors [33].

11.3.1.2   Data falsification attacks

Data falsification attacks try to breach the integrity security property of the system through modifying sensor data or injecting false/fake data. These are typically cyber attacks. An example of such a scenario is a *byzantine failure*, in which different information and falsified data is passed to different nodes from an attacked sensor, which might result in an incorrect state estimation or subjugation of control information.

In many CPS, such as power grids, estimating the state of unknown system variables by analyzing measurements obtained through various metering devices is a crucial process in controlling the overall system operations. An attacker, in the case of access to these meters, may successfully inject malicious measurements, also known as *false data injection attacks* [34, 35], and exploit the configuration of a power system, thus misleading the state estimation process. The false information can include an incorrect observation, incorrect timing information, and an incorrect sender identity. An instance of an attack where an adversary illegitimately claims multiple identities, either by impersonating other nodes or by generating an arbitrary number of node identities, is a *sybil attack* [36] or *node replication attack* [37]. These attacks are particularly effective in thwarting the redundancy mechanisms in sensor networks.

In the context of control of CPS, integrity refers to the trustworthiness or authenticity of the sensors and controllers. By compromising the integrity, an adversary can cause a *deception attack* in which a controller or a node within the network relies on and believes the false information received from one or multiple sensing devices [38]. In a similar context, *stealthy deception attacks* have been studied for the water SCADA (supervisory control and data acquisition) systems in [39, 40]. Another data falsification attack is a *replay attack*, in which an attacker attempts to control the system by compromising a set of sensors, recording their observations for a certain time, and then replaying them while injecting exogenous control inputs [41, 42]. In sensor networks, data collected from various sensing nodes is eventually aggregated, in either a distributed or centralized manner. The control action of the controller is based on the aggregated data received. Thus, if attackers have a high-level knowledge of the aggregation schemes and their parameters, they can conduct a *collusion attack* [43], in which aggregated data is modified due to the false data injected through a set of compromised nodes. This may lead to a control decision that can damage the physical plant, causing financial or physical losses.

11.3.2 Examples of attacks and vulnerabilities in CPS

Stuxnet. Stuxnet is a computer worm targeting programmable logic controllers (PLCs), which was discovered in June 2010 [44]. The worm is widely regarded as a milestone in cyber security, since it was the first malware that was designed to cause physical damage. Although it has not been confirmed who developed the worm, it is believed that the developers had nation-state support and backing, owing to the highly sophisticated design of the worm, and its usage of an unprecedented four zero-day exploits and two digital certificates stolen from separate well-known companies [45]. To infect computers, the worm is able to propagate through both infected removable drives, such as USB flash drives, and local area networks. These propagation vectors allow it to infect sensitive control systems, which are traditionally supposed to be secured by the "air-gap" (i.e., by not connecting the sensitive systems to the public Internet).

Stuxnet reportedly targeted Iranian uranium-enrichment facilities at Natanz [46]. Once it has infected a computer controlling a PLC that meets specific configuration requirements, the worm performs a man-in-the-middle attack, which fakes industrial process control sensor signals. Next, the worm tries to damage uranium centrifuges by increasing and decreasing their rotor speeds well above and below their normal operating speed, probably in order to cause damaging vibrations. Even though the attack did not result in the total destruction of the centrifuges, it drastically decreased their lifetime: due to the attack, around one-fifth of Iran's nuclear centrifuges were reportedly destroyed.

Sensys traffic sensors. Sensys Networks is a company that supplies wireless traffic-detection and integrated traffic-data systems, and has deployed more than 50 000 devices in 45 US states and 10 countries. In 2014, a cyber security researcher, Cesar Cerrudo, discovered that the Sensys Networks VDS240 wireless vehicle-detection systems are vulnerable to multiple attacks [47, 48]. These systems comprise magnetic sensors embedded in roadways, which transmit traffic-flow data over a wireless channel to nearby access points or repeaters. These in turn pass the data to traffic signal controllers, allowing intelligent traffic control based on real-time traffic data.

However, the lack of proper security mechanisms allows attackers to tamper with traffic data or compromise the sensor devices. Firstly, the wireless traffic between sensors and access points is unencrypted, and an attacker can intercept and replay traffic data [49]. More importantly, the sensors accept software updates without checking the integrity of the supplied software code, which allows an attacker to compromise the devices. Such compromises enable the attacker to take complete control and send fake data to traffic-control systems. Furthermore, it might also be possible to develop worms that infect vulnerable sensors and then propagate to nearby devices on wireless channels, infecting all sensors in a large area starting from a single device [47].

Even though the vulnerabilities do not allow direct control of traffic signals, an attacker may cause traffic jams and problems at intersections. By supplying fake traffic data, it is possible to influence the timing of traffic lights, to have electronic signs to display incorrect instructions, and to cause ramp meters to allow cars on the freeways slower or faster. Since the discovery of these vulnerabilities, Sensys Networks has released software updates to remediate the vulnerabilities identified in their traffic sensors [49].

Oil-pipeline explosion. In August 2008, the Baku–Tbilisi–Ceyhan oil pipeline in eastern Turkey exploded, which the Turkish government publicly blamed on a mechanical failure [50]. Even though the pipeline was monitored by sensors and cameras, the blast did not trigger any distress signals and the cameras failed to capture the combustion. The complete failure of the alarm systems is very surprising, since critical indicators, such as pressure and flow, were transmitted via both a terrestrial wireless system and backup satellite links to a central control room.

However, according to experts familiar with the incident, the explosion was actually caused by a cyber attack [51]. The attackers allegedly used a vulnerability in the communication software of the cameras to gain entry to the system, which enabled them to access operational controls, without compromising the main control room. By gaining access to smaller industrial control systems at a few valve stations, the attackers were able to shut down the alarms, cut off communications, and super-pressurize the oil in the pipeline. Since no evidence of a physical bomb was found, it is very likely that the explosion was created by the high pressure alone.

Vulnerable traffic signals. Intelligent traffic-control systems not only provide benefits in terms of wasted time and resources, but also provide new opportunities for attackers. While traditional hardware systems were susceptible only to attacks based on direct physical access, systems now may be vulnerable to attacks through wireless interfaces, or even to remote attacks through the Internet. A recent case study by Ghena *et al.* [20] analyzed the security of traffic infrastructure in cooperation with a road agency located in Michigan. The agency operates around a hundred traffic signals, which are all part of the same wireless network, but the signals at every intersection operate independently of the other intersections. The study found three major weaknesses in the road agency's traffic infrastructure: lack of encryption for the network, lack of secure authentication due to the use of default usernames and passwords on the devices, and vulnerability to known exploits. Owing to the hardware-based failsafes, an attacker cannot put traffic lights into an unsafe configuration, but may be able to cause disastrous traffic congestion.

11.4 Notion of attack resilience

In CPS, as a result of the integration of the cyber and physical domains, physical processes are directly influenced by the integration of IT systems comprising sensing elements and communication networks. Consequently, the physical dynamics of the system can be manipulated through cyber means. This is advantageous, on the one hand, as it stipulates enhanced monitoring and efficient control. However, on the other hand, it raises new threat channels, against which the system needs to be secured. To account for modeling uncertainties and physical disturbances in the control of physical processes, several tools have been developed in classical control, including robust control and stochastic control. However, the added dimension of cyber domain vulnerabilities, which are capable of sabotaging the overall operations and control of CPS, pose new challenges and make these tools insufficient. Thus, there is an imminent need for a systematic and thorough analysis of security and resilience against faults, failures, and adversarial actions in CPS.

11.4.1 Security and resilience

Cyber security (or computer security) is traditionally concerned with preventing attacks from succeeding. *Security* can be attributed as the ability of a system to avert attacks and remain protected against malicious behavior. *Resilience*, on the other hand, relates to the behavior of a system once its security has been compromised. More specifically, resilience can be thought of as a system's ability to recover online in case of an attack and continue to operate with minimal and tolerable disruption, i.e., to gracefully degrade its operations. Note that security is a *pre-event* property, whereas resilience is a *post-event* attribute.

Since perfect security against all attacks is, in practice, either impossible or financially unattainable, there is always a possibility of a successful attack or malicious intrusion. Recent examples, such as the Stuxnet worm, have shown that determined and resourceful attackers can penetrate even highly secured and secluded systems. In CPS, a successful attack might cause serious physical damage to the system and disrupt services, such as energy supply and water distribution, thereby incurring execrable financial losses. Since control decisions in CPS are made in real time by controllers based on observations received from a WSN, resilience is a critical issue in CPS. Consequently, any compromise or disruption in WSN due to malicious attacks or failures of sensor nodes would severely impact the physical process as illustrated in Figure 11.3. Thus, *a resilient WSN is imperative for a resilient CPS.*

In light of the above discussion, merely securing WSN against attacks, such as the ones mentioned in Section 11.3, is not sufficient to ascertain the operational normalcy in CPS. In fact, a resilient design of WSN that maintains the task of

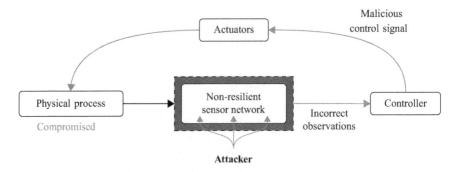

Figure 11.3   Consequences of attacks against non-resilient WSN in CPS

transmitting correct sensor data in a timely manner to the controller on the physical side, despite adversarial attacks, is a crucial requirement. In a broader perspective, the resilience property of networks in general, and for WSN in CPS in particular, can be attributed to the following three factors:

- functional correctness of the network (design)
- ability to sustain under reliability failures (faults), and
- ability to survive against security failures (attacks)

11.4.2   Random failures and intentional attacks

In this section, we formulate general problems for resilient WSN designs. As mentioned above, resilience in WSN can be against *random failures*, which correspond to sensor faults and device malfunctions, and *intentional attacks*, which correspond to sensors being attacked by an adversary.

A very important aspect of the resilient design problem is the *trade-off* between cost and resilience: in practice, making a system more resilient generally entails incurring additional costs. For example, if we achieve resilience through redundancy, we have to deploy additional devices and links, which all have deployment and maintenance costs. As another example, if we achieve resilience by changing the architecture of a system from the most efficient one to a more resilient one, then we incur efficiency loss, which we can model as a form of cost. Consequently, finding the optimal design in practice means striking the right balance between maximizing resilience and minimizing cost.

However, in many cases, we are given a fixed *budget M* for building a system, which means that we can choose only from those designs whose cost does not exceed the budget. Consequently, in these cases, we have to find the most resilient design from the set of designs that satisfy the budget constraint. In other words, we have to solve a *constrained optimization* problem instead of a tradeoff problem.[1] In

---

[1]We can also formulate the resilient design problem as the problem of minimizing cost while achieving a threshold level of resilience. In terms of the computational complexity, since this formulation is equivalent to having a budget constraint for most problems, we do not discuss it in detail.

this chapter, we will formulate the resilient design problem in this way, that is, as a constrained optimization problem based on an exogenous budget value. Note that, given a method for solving the constrained optimization problem, one can solve the tradeoff problem in practice by searching for the budget value that optimizes resilience and cost.

Random failures. First, we discuss failures that are not caused by adversarial action, which we call *random failures*. These include, for example, software crashes due to random faults in static random-access memory (SRAM) and wireless-link failures due to extreme weather conditions. Note that, since our focus is on attack resilience, we disregard systematic failures in this chapter, and present only a simplistic model of random failures. The common characteristic of these failures is that they occur independently of defensive and adversarial actions, i.e., the probability of each event is exogenous to the problem.

Now, we present a simple formulation of the failure-resilient design problem. We let $\mathcal{F}$ denote the set of possible failures in the system, and for each failure $f \in \mathcal{F}$, we let β_f denote the probability of the given failure occurring. Assuming that the failures are independent, we can express the probability of a set F of failures occurring as

$$\Pr[F] = \left(\prod_{f \in F} \beta_f \right) \left(\prod_{f \notin F} (1 - \beta_f) \right).$$

Next, we let $\mathcal{D}$ be the set of possible designs, we denote the cost of design d by $C(d)$, and we denote the loss sustained by a system based on design $d \in \mathcal{D}$ when failures in F occur by $Loss(d, F)$. Then, if our goal is to minimize the expected amount of losses sustained by the system, the failure-resilient design problem can be formulated as

$$\operatorname*{argmin}_{d \in \mathcal{D}: C(d) \leq M} \mathrm{E}[\mathrm{Loss}(d, \cdot)] \tag{11.5}$$

$$= \operatorname*{argmin}_{d \in \mathcal{D}: C(d) \leq M} \sum_{F \subseteq \mathcal{F}} \Pr[F] \mathrm{Loss}(d, F) \tag{11.6}$$

$$= \operatorname*{argmin}_{d \in \mathcal{D}: C(d) \leq M} \sum_{F \subseteq \mathcal{F}} \left(\prod_{f \in F} \beta_f \right) \left(\prod_{f \notin F} (1 - \beta_f) \right) \mathrm{Loss}(d, F). \tag{11.7}$$

Intentional attacks. Contrary to random failures, intentional attacks never follow an exogenous probability distribution. In fact, adversaries can take into consideration the design of a system and the deployed defensive countermeasures, and strike at the weakest point. Consequently, the occurrence of attacks is endogenous to the model, since they depend on the chosen design, which makes the attack resilience problem more complex.

Similarly to the design options, the adversary's actions are also costly. For example, to launch a physical attack, such as wireless jamming, the adversary may have to buy special equipment and spend effort to carry out the attack. As another example, to launch a cyber attack, the adversary has to discover a security vulnerability in the system, which again requires spending effort or hiring experts. However, in practice, adversaries have limited amounts of resources available for their attacks. Consequently, similarly to the defender's problem, we can model the adversary's decision as a constrained optimization based on an exogenous budget constraint B. Note that, as an alternative, we can also model the adversary's decision as a tradeoff problem.

Finally, we need to formulate the adversary's objective. Here, we will assume a worst-case adversary whose goal is to maximize the losses sustained by the system. Let $\mathcal{A}$ be the set of possible adversarial actions, let the cost of action $a \in \mathcal{A}$ be $c(a)$, and let the loss sustained by the system based on a design d when the adversary carries out the actions in A be denoted by $\text{Loss}(d, A)$. Then, for a given design d, the adversary's decision problem can be formulated as

$$\underset{A \subseteq \mathcal{A}: \sum_{a \in A} c(a) \leq B}{\text{argmax}} \quad \text{Loss}(d, A), \tag{11.8}$$

and the attack-resilient design problem is

$$\underset{d \in \mathcal{D}: C(d) \leq M}{\text{argmin}} \left(\underset{A \subseteq \mathcal{A} : \sum_{a \in A} c(a) \leq B}{\text{max}} \text{Loss}(d, A) \right). \tag{11.9}$$

11.4.3   Challenges

In this subsection, we discuss the main challenges in solving attack-resilient problems for WSN as formulated above.

Computational complexity. According to the widely accepted Cobham–Edmonds thesis [52], solving a computational problem is feasible only if we can solve the problem in polynomial time. However, in many attack-resilient CPS problems, the set of possible designs grows exponentially as the size of the problem increases. For instance, in a sensor placement problem, the number of feasible placements is an exponential function of the number of locations at which sensors can be placed. Consequently, the number of designs to choose from quickly becomes "astronomical" as the size of the problem increases. As an example, if we can place at most 100 sensors in 500 possible locations, then the number of feasible subsets is approximately 2×10^{107}, more than the number of atoms in the observable universe, which is only 10^{80}. For such problems, since an exhaustive search is computationally infeasible, we have to find a more intelligent, polynomial-time algorithm.

Unfortunately, many attack-resilient design problems are NP-hard, which means that they cannot be solved in polynomial time (given that P $\neq$ NP, which is a very widely accepted conjecture). To tackle such computationally challenging problems, in practice we have to use approximation and heuristic algorithms. For example, the optimal design of a WSN that is used for pipeline leakage detection in a water distribution network is NP-hard even without an attacker. Let the set of possible sensor locations be $\mathcal{S}$, let the cost of placing a sensor at location $s \in \mathcal{S}$ be $c(s)$, let L be the set of possible leakages (e.g., the set of pipeline sections that may be damaged), and let the set of leakages detected by a sensor at location s be $L_s \subseteq L$. Then, we can show that it is NP-hard to determine whether there exists a placement S that can detect all leakages and whose cost $\sum_{s \in S} c(s)$ does not exceed our budget M.

We prove NP-hardness by reducing a well-known NP-hard problem, the set cover problem, to the above decision problem. Given a base set U, a set ε of subsets of U, and a number k, the set cover problem is to determine whether there exists a set $\mathcal{C} \subset \varepsilon$ such that $\mathcal{C}$ covers all elements of U (i.e., for every $u \in U$, there is an $E \in \mathcal{C}$ such that $u \in E$) and $|\mathcal{C}| \leq k$. We can reduce an instance of the set cover problem to our decision problem as follows. Let the set of possible leakages L be U, let the set of possible sensor locations S be ε, let the set of leakages L_s detected by a sensor at location s be the corresponding subset from ε, let the cost of placement for all locations be $c(s) = 1$, and let the budget M be k. Then, it is easy to see that a feasible sensor placement detecting all leakages exists if and only if there exists a set cover.

Estimating model parameters. In order to apply theoretical results to a real-world system, we must be able to map the parameters of the model to real-world data, which can prove to be challenging. Firstly, we need to be able to quantify the potential losses that arise from the various security incidents that might happen. These values can be quantified either as financial damage due to physical losses (e.g., increase in operating costs due to suboptimal control or cost of replacing damaged devices) or as liability/penalty to be paid (which can be estimated from past settlements). Note that, if we formulate the design problem using a budget constraint, we only need to be able to compare the possible outcomes to each other; hence, we have to estimate only the relative losses for the various incidents.

Secondly, we need to be able to estimate the cost of the attacker's actions, and, if we model the attacker's decision using a budget constraint, the value of this budget constraint as well. The latter can be very challenging. If we underestimate the attacker's budget, our system will not be resilient to more powerful attacks and may suffer intolerable losses. On the other hand, if we overestimate the attacker's budget, the resilience of our system against the actual attacks will be suboptimal. We can elude this problem by modeling the attacker's decision as a tradeoff problem; however, in this case, we need to be able to compare the cost of the attacker's actions to the potential loss values.

Finally, we need to be able to estimate the budget and the cost of the various design choices and defensive countermeasures. This is a relative easy task, since the budget (if there is one) is given, and we can assume that a defender knows the various deployment and maintenance costs of its system.

Modeling attackers. In order to design optimal attack-resilient systems, we must be able to model all the possible actions that an attacker may take. For physical attacks, this task is generally tractable, as an attacker is always bound by the laws of physics. For cyber attacks, however, establishing bounds on what an attacker might do can be more challenging. Cyber attacks are usually possible due to vulnerabilities that were unintentionally introduced into the software, that is, cyber attacks usually happen because one or more of our assumptions on how a system may be used have already failed. Consequently, we have to be careful about what further assumptions we make. We can overcome this paradox by using more abstract, higher-level attack models. For example, the attacker's actions can correspond to compromising a node or the integrity of a link, abstracting away from the specific types of exploits that an attacker might use and the specific access rights that it might gain. However, such higher-level models increase the difficulty of estimating the cost and success probability of the attacker's actions.

So far, we have assumed rational attackers with complete information. On the one hand, this is a very robust model security-wise, since it is based on pessimistic assumptions, which is in accordance with the principles of security (e.g., Kerckhoffs' principle [53]). On the other hand, these assumptions may be too strong in practice, where attacks are planned by human adversaries, who have bounded rationality and limited information. Modeling human adversaries' irrationality, biased decisions and perceptions, limited observations and computational capabilities, etc. using mathematical tools can be very challenging. These "imperfections" may be modeled most naturally using the game theory nomenclature – see, for example, the work of Pita *et al.* [54] and Freund *et al.* [55].

11.5   Attack-resilient design

In Section 11.4, we discussed the notion of attack resilience in WSN for CPS, and presented various models and challenges in solving attack-resilient problems for WSN. In this section, we discuss different approaches toward attack-resilient design of WSN, and also present two example problems in detail.

11.5.1   *Approaches for attack resilience*

Attack resilience can be achieved in a WSN using multiple approaches, ranging from resilient sensor placement to resilient data-aggregation algorithms. Here, we discuss three main approaches, noting that one may find many others in the literature. For example, attack resilience can be achieved by using resilient routing algorithms [56–58].

Sensor placement. We can mitigate attacks that compromise or incapacitate individual nodes via attack-resilient sensor placement, i.e., by choosing the locations of the sensor nodes so that the resulting WSN is attack-resilient. Such attacks have multiple adversarial effects, which we must take into account when planning node placement. Firstly, sensors impaired by an attack will no longer supply correct observational data, which can affect the controller and, consequently, the physical process of the CPS. To moderate losses arising from missing observations, we have to place sensors in such a way that the remaining sensors can supply adequate information in the case of an attack, which we call *resilient coverage*. Secondly, as the impaired nodes may have been used for data forwarding, the connectivity of the sensor network can also decrease, which necessitates designing the network topology to be resilient. Here, we will focus on resilient coverage, and we will discuss attack-resilient network topologies later. For a general survey on node placement in WSNs, we refer the reader to the survey of Younis and Akkaya [59].

The most widely used objective for resilient coverage is k-coverage: we say that a sensor network has k-coverage if every location is within the sensing range of at least k sensors [60]. In other words, a sensor network that has k-coverage can withstand attacks that impair at most $k - 1$ sensor nodes. The k-coverage metric is widely used because it has the appeal of simplicity as an objective. For example, Wang and Tseng [61] study placement schemes for providing k-coverage for an area with the minimum number of nodes.

However, k-coverage may be too simplistic for practical applications. The main disadvantage of k-coverage is that it treats coverage as a binary property, i.e., as something that either is or is not provided. In practice, however, the amount of information gathered can decrease more gradually as more and more sensor nodes are impaired. For example, having multiple observations for the same location may provide more accurate measurements, which may lead to more efficient control. As another example, when we cannot observe a location, we may be able to infer missing values from observations taken at nearby locations, which allows us to retain control of the physical processes despite the attack. Consequently, for many applications, more fine-grained metrics are necessary. For instance, Dhillon and Chakrabarty [62] study coverage for surveillance applications using a probabilistic objective, which is based on the uncertainty associated with sensor detections.

For optimal resilient coverage for CPS, we must go one step further and incorporate *how sensor data is used* into our objective function. In other words, we have to place sensor nodes in such a way that, when some are impaired by an attack, the remaining nodes can provide data that will result in acceptable control decisions.

Network topology. Designing resilient network topologies for sensor networks, capable of maintaining a certain level of some global performance measure in the presence of adversarial attacks, is crucial to the overall resilient CPS design. In the case of node or edge removal attacks, which can be random or strategic, and may correspond to node destruction, exhaustion, or jamming, the objective is to design and control the network structure to preserve structural properties of the graph to a reasonable extent. The basic premise of resilient network topology design is as follows (e.g., [63–66]). A set of sensor network performance measures, such as energy consumption, connectivity, distance or communication delay between nodes, throughput, is considered. These performance goals are then translated into network topology-based parameters, such as node or vertex connectivity, persistence, centrality-based measures, distance-based measures, Kirchhoff index, etc. Using tools such as combinatorial optimization, network topology is then optimized with respect to the network-based measures under node or edge removal conditions. An important consideration in attack-resilient sensor network design for CPS, especially in selecting the sensor network performance-based measure and its translation into the network topology-based metric, is to incorporate the consequences of the sensor network being in a closed loop, as shown in Figure 11.2.

Data aggregation. In general, sensors are spatially distributed in a WSN, collecting data from many vantage points, which is then aggregated, such as average, median, maximum, minimum, rather in a distributed manner. In a CPS, in-network computations are performed and control decisions are taken in real time based on the aggregated data observed from the sensor network. Several aggregation algorithms for sensor networks have been reported in the literature. However, since the majority of these schemes were not originally designed to be resilient against malicious attacks, such as data falsification attacks, data aggregation results can be easily manipulated even if a small subset of nodes are compromised. For instance, in a simple linear consensus algorithm designed to ensure that all sensors converge to an average of their initial observations, a single misbehaving node can result either in convergence to a value far from the average, which might be a significant issue in safety-critical processes, or in no convergence at all. In this direction, designing *resilient* data aggregation algorithms that can withstand the compromise of a subset of nodes and maintain a certain notion of correctness in computations is a way to go [67].

Some of the approaches to achieve resilient data aggregation include *preprocessing* sensory data before applying aggregation schemes, such as selecting a subset of received observations for the algorithm rather than using all of the available data (e.g., [68, 69]). Another approach is to *detect* if received data is from a valid sensor or a compromised node before applying an aggregation scheme (e.g., [70–72]). In yet another approach, the idea is to relate the resilience of the data aggregation algorithm against a certain number of malicious attacks to the underlying network topology (e.g., [68, 69]). For instance, if some connectivity conditions, say C, are satisfied in the underlying graph structure, then the data aggregation algorithm will give correct results even if a certain number of nodes, say A, are compromised, where A depends on C.

11.5.2   Examples

Resilient sensor placement for prediction. Now, we discuss the resilient sensor placement problem for prediction and forecasting applications in more detail.

We let S denote the set of locations at which sensors can be placed, and let Y denote the target variable that we have to predict. For example, in a traffic-control CPS, the set S can model the road segments on which traffic-flow sensors can be placed, and variable Y can represent the future traffic situation. The design choice in this problem is to select a subset $S \subseteq S$ of locations at which sensors are placed, such that $|S| \leq M$, and the goal is to predict Y as accurately as possible. We formalize this goal as minimizing the posterior variance $\sigma^2_{Y|S}$ of the target variable Y given observations received from the sensors placed at S. Quantifying how inaccurate our predictions are using variance is reasonable, since minimizing variance minimizes the mean-squared error of predictions, and variance is also logarithmically proportional to the uncertainty of the target variable measured in entropy. Then, without an attacker, the sensor placement problem can be expressed as

$$\underset{S \subseteq \mathcal{S}: |S| \leq M}{\text{argmin}} \ \sigma^2_{Y|S}. \tag{11.10}$$

Next, we introduce an adversary who can launch denial-of-service type attacks that impair sensor nodes. We assume that the adversary has a limited budget B, which we express as a constraint on the number of impaired sensors. Formally, the adversary can attack any subset $A \subseteq S$ such that $|A| \leq B$. Finally, we assume that the adversary is a worst-case attacker, i.e., its goal is to maximize loss. Then, the resilient placement problem can be expressed as

$$\underset{S \subseteq \mathcal{S}: |S| \leq M}{\operatorname{argmin}} \left(\underset{A \subseteq S: |A| \leq B}{\max} \sigma^2_{Y|(S \backslash A)} \right). \tag{11.11}$$

So far, we have not discussed what model is used for predicting variable Y from the observations taken at locations S. Now, we assume that a Gaussian-process-based regression model [73] is used (such models have previously been applied successfully to, for example, traffic forecasting [23]). In this model, each possible sensor location is modeled as a random variable, which represents the (*a priori* unknown) observations that would be taken at the location. Next, let $\sum$ be the (prior) covariance matrix of the target variable and the random variables representing the locations S. Then, the posterior variance of Y given observations at S is

$$\sigma^2_{Y|S} = \sigma^2_Y - \mathbf{S}_{YS} \mathbf{S}_{SS}^{-1} \mathbf{S}_{SY}, \tag{11.12}$$

where σ^2_Y is the prior (i.e., without any observations) variance of Y, $\sum_{YS}$ (or $\sum_{SY}$) denotes the submatrix of $\sum$ formed by row Y (or column Y) and the columns (or rows) in set S, and $\sum_{SS}$ denotes the principal submatrix formed by rows and columns in S.

Unfortunately, this sensor placement problem is computationally hard, even without an adversary. Formally, determining whether there exists a subset of locations $S \subseteq \mathcal{S}$ such that the variance $\sigma^2_{Y|S}$ is less than or equal to a given threshold is an NP-hard problem in general [74]. Consequently, to tackle this problem, either we have to focus on certain special cases, which can be solved efficiently, or we have to use approximation algorithms or heuristics. Das and Kempe [74] showed that the non-resilient variant of the problem (i.e., without an attacker) can be solved in polynomial time for certain covariance matrices. For example, if we represent the covariance matrix as a graph, in which nodes correspond to variables and edges correspond to pairs of variables with non-zero covariance, and this graph is a tree, then the optimal solution can be found using dynamic programming. By generalizing this result, we can show that the optimal solution of the resilient problem can also be found using dynamic programming if the graph corresponding to the covariance matrix is a tree.

However, many practical instances of the problem do not fall into these special cases, which necessitates using some heuristic or approximation algorithm. One of the most commonly used heuristics is the greedy algorithm, since it acts as an approximation algorithm for problems where the objective function is a submodular set function. Unfortunately, the above objective function is not submodular. However, for the non-resilient variant of the problem, Das and Kempe [74] provided a provable bound on the quality of the solution yielded by the greedy algorithm, based on quantifying how close the objective function is to being submodular. Similarly to the special case results, we can generalize this result as well, and we can provide a bound on the greedy algorithm for the resilient problem.

Resilient consensus algorithm. As an example of data aggregation in sensor networks in the presence of adversaries, we consider a resilient consensus problem.

A sensor network in which each sensor is connected to a subset of other sensors can be modeled by an undirected graph $G(V, E)$, where V represents the set of sensor nodes, and E represents the set of communication links between nodes through which they exchange information with their neighbors. Each sensor node i has a state value at a given time k, denoted by $x_i(k)$, which can be a sensor measurement, position variable, opinion, etc. Based on the state values of neighbor nodes, each sensor updates its state with an objective that all nodes eventually converge to a common state value. However, some nodes are compromised and act maliciously by not updating their states as per the defined update law. Consequently, their objective is to prevent the normal nodes from reaching consensus.

Thus, our objective is to design an update rule for the normal nodes such that they all converge to a common value even in the presence of a number $\mathcal{A}$ of adversarial nodes. More precisely, a resilient consensus protocol needs to be designed that can achieve the following two objectives:

- As $k \to \infty$, then $x_i(k) = x_j(k) = x$ for all the normal nodes i, j (*agreement condition*).
- Let $x_{\min}(0)$ and $x_{\max}(0)$ respectively be the minimum and maximum of the initial values of the normal nodes, then $x_{\min}(0) \leq x_i(k) \leq x_{\max}(0)$, for all k and for any normal node i (*safety condition*).

Assume that $\mathcal{N}(i)$ represents the neighbors – nodes that are directly connected to node i in G – of node i. In a typical approach to solve linear consensus in a distributed way assuming that no misbehaving node exists, each node updates its state by taking some weighted average of the states of its neighbors. In the case of an undirected graph G, this approach guarantees consensus as long as the graph is connected. However, in the case of misbehaving nodes, if there is even one such node, consensus might not be achieved.

There are approaches for resilient consensus, such as the weighted mean-subsequence-reduced (W-MSR) algorithm [68], in which each node takes a weighted average of the states of a subset of its neighbors, selected in some clever way. If the graph satisfies certain connectivity conditions, then the proposed algorithm guarantees resilient consensus in the presence of a number $\mathcal{A}$ of adversaries, where $\mathcal{A}$ is related to the connectivity of the graph. Though this approach solves the resilient consensus problem, at the same time it requires the graph to be very highly connected, even for a very small number of adversarial attacks. Thus, in the cases of a higher number of misbehaving nodes and sparse networks, the application of this approach is limited.

In another approach [75], the notion of *trusted nodes* is introduced and a resilient consensus protocol in the presence of trusted nodes is proposed to address the limitations of the above scheme. The basic idea is to distribute a set of trusted nodes, which are nodes with a higher level of security against attacks and their state

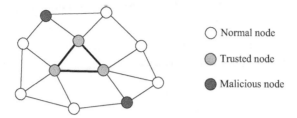

Figure 11.4   Each normal node is connected to a trusted node, and the set of trusted nodes induces a connected subgraph

values can be trusted, in such a way that consensus is achieved in the presence of *any* number of adversaries. As previously, a node updates its state by taking a weighted average of its own state and the states of its selected neighbors. Under this approach, to achieve consensus in the presence of any number of adversarial nodes, it is shown in [75] that the set of trusted nodes should form a *connected dominating set*. In other words, resilient consensus can be achieved for arbitrary A whenever each node is connected to at least one trusted node, and the set of trusted nodes induce a connected subgraph, as illustrated in Figure 11.4. However, this approach relies on the fact that trusted nodes are completely protected against attacks. Thus, both approaches have their own merits and limitations.

11.6   Conclusions and future directions

Resilience in WSNs is a critical issue, as these networks may act as a key enabling technology in CPS, but at the same time they pose substantial security challenges. The importance of this issue is elevated by the fact that WSN in CPS are part of a closed control loop, and thus control decisions may be manipulated by attackers that have gained unauthorized access to the WSN. In this chapter, we provided an overview of key topics related to the resilience of WSN in CPS. First, we discussed potential applications of WSN in CPS, followed by an overview of possible attacks against WSN. Then, we formalized the notion of attack resilience in WSN, and discussed the main challenges to attack-resilient design. Moreover, we presented various approaches to achieve attack resilience, along with detailed examples. Finally, in the remainder of this chapter, we outline directions for future research in the area of resilient WSN for CPS.

Intrusion detection systems. Intrusion detection systems (IDSs) are useful as they provide a way to detect an attack before it can cause damage. This early detection mechanism provides a way to take preventive measures before the system sustains any substantial losses, thereby elevating the attack resilience of the system. IDSs have been successfully deployed in various industrial control systems; however, their utility has not been widely explored in the domain of WSN for CPS.

One of the key obstacles in this regard is that the devices in the sensor network are resource-constrained. For instance, to maximize the lifetime of the network, battery power must be conserved. However, IDS can be computationally expensive, which limits their deployment in WSN. As a result of this limitation, we may not be able to deploy IDS at every single node and may not be able to run it at all times. Therefore, we need placement and scheduling of IDS in WSN that maximize the probability of attack detection while satisfying the resource constraints of the devices and the network. In the context of CPS, a key challenge here is that this optimization – placement and scheduling – should also take into account the characteristics of the physical process.

Moving target defense. We can make a WSN more resilient to attacks by minimizing the attackers' ability to learn and exploit the system parameters. One of the ways to achieve this objective is by continuously changing the defending strategy, i.e., system configurations and defensive countermeasures. For instance, in the IDS scheduling problem discussed above, if we use a predictable schedule, an attacker might be able to time its attack to avoid detection by learning the schedule on each node. On the other hand, if we use an unpredictable schedule, an attacker will not be able to time its attack, and we will be able to detect it with high probability. Nonetheless, to the best of our knowledge, moving target defense approaches have not been applied to WSN for CPS.

Comprehensive design. A defining feature of CPS is the tight integration of physical and cyber elements, which form a closed control loop. The attack resilience of each element – physical and cyber – can be optimized individually. In this approach, the optimization of each element depends on the design and parameters of the other element, and considers them to be exogenous to the problem. Consequently, in this approach, even though each element is individually optimal, the system as a whole is not necessarily optimal. Therefore, instead of optimizing each element of the closed control loop individually, we have to devise an approach for optimizing the system as a whole.

Acknowledgments

This work was supported in part by the National Science Foundation (CNS-1238959), the Air Force Research Laboratory (FA8750-14-2-0180), and the National Institute of Standards and Technology (70NANB13H169).

References

[1] Akyildiz I.F., Su W., Sankarasubramaniam Y., Cayirci E. Wireless sensor networks: a survey. *Computer Networks*, 38(4): 393–422, 2002.

[2] Lee E.A. Cyber physical systems: design challenges. In *Proceedings of the 11th IEEE International Symposium on Object Oriented Real-Time Distributed Computing (ISORC)*, pp. 363–369. IEEE, 2008.

[3] Mainwaring A., Culler D., Polastre J., Szewczyk R., Anderson J. Wireless sensor networks for habitat monitoring. In *Proceedings of the 1st ACM International Workshop on Wireless Sensor Networks and Applications*, pp. 88–97. ACM, 2002.

[4]   Son B., Her Y., Kim J.G. A design and implementation of forest-fires surveillance system based on wireless sensor networks for South Korea mountains. *International Journal of Computer Science and Network Security*, 6(9): 124–130, 2006.

[5]   Wu F.J., Kao Y.F., Tseng Y.C. From wireless sensor networks towards cyber physical systems. *Pervasive and Mobile Computing*, 7(4): 397–413, 2011.

[6]   Cardenas A.A., Amin S., Sastry S. Secure control: Towards survivable cyber-physical systems. In *Proceedings of the 28th International Conference on Distributed Computing Systems Workshops*, pp. 495–500. IEEE, 2008.

[7]   Sridhar S., Hahn A., Govindarasu M. Cyber-physical system security for the electric power grid. *Proceedings of the IEEE*, 100(1): 210–224, 2012.

[8]   McDaniel P., McLaughlin S. Security and privacy challenges in the smart grid. *IEEE Security Privacy*, 7(3/May): 75–77, 2009.

[9]   Rial A., Danezis G. Privacy-preserving smart metering. In *Proceedings of the 10th Annual ACM Workshop on Privacy in the Electronic Society (WPES'11)*, pp. 49–60. ACM, 2011.

[10]   Sankar L., Rajagopalan S.R., Mohajer S., Poor H.V. Smart meter privacy: A theoretical framework. *IEEE Transactions on Smart Grid*, 4(2): 837–846, 2013.

[11]   Alemdar H., Ersoy C. Wireless sensor networks for healthcare: A survey. *Computer Networks*, 54(15): 2688–2710, 2010.

[12]   Haykin S. (ed.) *Kalman Filtering and Neural Networks*. John Wiley, 2001.

[13]   Olfati-Saber R. Distributed Kalman filtering for sensor networks. In *Proceedings of the 46th IEEE Conference on Decision and Control*, pp. 5492–5498. IEEE, 2007.

[14]   Hernandez L., Baladrón C., Aguiar J., *et al.* A survey on electric power demand forecasting: Future trends in smart grids, microgrids and smart buildings. *IEEE Communications Surveys & Tutorials*, 16(3): 1460–1495, 2014.

[15]   Mirowski P., Chen S., Ho T.K., Yu C.N. Demand forecasting in smart grids. *Bell Labs Technical Journal*, 18(4): 135–158, 2014.

[16]   Amin S., Schwartz G., Cardenas A., Sastry S. Game-theoretic models of electricity theft detection in smart utility networks: Providing new capabilities with advanced metering infrastructure. *IEEE Control Systems*, 35 (1): 66–81, 2015.

[17]   McLaughlin S., Holbert B., Fawaz A., Berthier R., Zonouz S. A multi-sensor energy theft detection framework for advanced metering infrastructures. *IEEE Journal on Selected Areas in Communications*, 31(7): 1319–1330, 2013.

[18]   Fang X., Misra S., Xue G., Yang D. Smart grid – the new and improved power grid: A survey. *IEEE Communications Surveys & Tutorials*, 14(4): 944–980, 2012.

[19]   Gungor V.C., Lu B., Hancke G.P. Opportunities and challenges of wireless sensor networks in smart grid. *IEEE Transactions on Industrial Electronics*, 57(10): 3557–3564, 2010.

[20]   Ghena B., Beyer W., Hillaker A., Pevarnek J., Halderman J.A. Green lights forever: Analyzing the security of traffic infrastructure. In *Proceedings of the 8th USENIX Workshop on Offensive Technologies (WOOT'14)*, pp. 1–10. USENIX Association, 2014.

[21]   Tubaishat M., Zhuang P., Qi Q., Shang Y. Wireless sensor networks in intelligent transportation systems. *Wireless Communications and Mobile Computing*, 9(3): 287–302, 2009.

[22]   Smith B.L., Demetsky M.J. Traffic flow forecasting: Comparison of modeling approaches. *Journal of Transportation Engineering*, 123(4): 261–266, 1997.

[23]   Xie Y., Zhao K., Sun Y., Chen D. Gaussian processes for short-term traffic volume forecasting. *Journal of the Transportation Research Board*, 2165(1): 69–78, 2010.

[24]   Raymond D.R., Midkiff S.F. Denial-of-service in wireless sensor networks: Attacks and defenses. *IEEE Pervasive Computing*, 7(1): 74–81, 2008.

[25]   Wang Y., Attebury G., Ramamurthy B. A survey of security issues in wireless sensor networks. *IEEE Communications Surveys & Tutorials*, 8(8): 2–23, 2006.

[26]   Wood A., Stankovic J.A. Denial of service in sensor networks. *Computer*, 35 (10): 54–62, 2002.

[27]   Xu W., Ma K., Trappe W., Zhang Y. Jamming sensor networks: Attack and defense strategies. *Network, IEEE*, 20(3): 41–47, 2006.

[28]   Maheshwari R., Gao J., Das S.R. Detecting wormhole attacks in wireless networks using connectivity information. In *Proceedings of the 26th IEEE International Conference on Computer Communications (INFOCOM'07)*, pp. 107–115. IEEE, 2007.

[29]   Poovendran R., Lazos L. A graph theoretic framework for preventing the wormhole attack in wireless ad hoc networks. *Wireless Networks*, 13(1): 27–59, 2007.

[30]   Gu Q., Ferguson C., Noorani R. A study of self-propagating mal-packets in sensor networks: Attacks and defenses. *Computers & Security*, 30(1): 13–27, 2011.

[31]   Francillon A., Castelluccia C. Code injection attacks on Harvard-architecture devices. In *Proceedings of the 15th ACM Conference on Computer and Communications Security*, pp. 15–26. ACM, 2008.

[32]   Yang Y., Zhu S., Cao G. Improving sensor network immunity under worm attacks: a software diversity approach. In *Proceedings of the 9th ACM International Symposium on Mobile Ad Hoc Networking and Computing*, pp. 149–158. ACM, 2008.

[33]   Wang Y., Wen S., Xiang Y., Zhou W. Modeling the propagation of worms in networks: A survey. *IEEE Communications Surveys & Tutorials*, 16(2): 942–960, 2014.

[34]   Liu Y., Ning P., Reiter M.K. False data injection attacks against state estimation in electric power grids. *ACM Transactions on Information and System Security*, 14(1): 13, 2011.

[35] Xie L., Mo Y., Sinopoli B. False data injection attacks in electricity markets. In *Proceedings of the 1st International Conference on Smart Grid Communications*, pp. 226–231. IEEE, 2010.

[36] Newsome J., Shi E., Song D., Perrig A. The sybil attack in sensor networks: Analysis and defenses. In *Proceedings of the 3rd International Symposium on Information Processing in Sensor Networks*, pp. 259–268. ACM, 2004.

[37] Parno B., Perrig A., Gligor V. Distributed detection of node replication attacks in sensor networks. In *IEEE Symposium on Security and Privacy*, pp. 49–63. IEEE, 2005.

[38] Amin S., Cárdenas A.A., Sastry S.S. Safe and secure networked control systems under denial-of-service attacks. In *Hybrid Systems: Computation and Control*, pp. 31–45. Springer, 2009.

[39] Amin S., Litrico X., Sastry S., Bayen A.M. Stealthy deception attacks on water SCADA systems. In *Proceedings of the 13th ACM International Conference on Hybrid Systems: Computation and Control*, pp. 161–170. ACM, 2010.

[40] Amin S., Litrico X., Sastry S., Bayen A.M. Cyber security of water SCADA systems – Part I: Analysis and experimentation of stealthy deception attacks. *IEEE Transactions on Control Systems Technology*, 21(5): 1963–1970, 2013.

[41] Mo Y., Chabukswar R., Sinopoli B. Detecting integrity attacks on scada systems. *IEEE Transactions on Control Systems Technology*, 22(4): 1396–1407, 2014.

[42] Mo Y., Sinopoli B. Secure control against replay attacks. In *Proceedings of the 47th Annual Allerton Conference on Communication, Control, and Computing*, pp. 911–918. IEEE, 2009.

[43] Rezvani M., Ignatovic A., Bertino E., Jha S. Secure data aggregation technique for wireless sensor networks in the presence of collusion attacks. *IEEE Transactions on Dependable and Secure Computing*, 12(1): 98–110, 2015.

[44] Kushner D. The real story of Stuxnet. *IEEE Spectrum*, 50(3): 48–53, 2013.

[45] Kaspersky Lab. Kaspersky Lab provides its insights on Stuxnet worm, 24 September 2010. See http://www.kaspersky.com/about/news/virus/2010/ Kaspersky_Lab_provides_its_insights_on_Stuxnet_worm. Accessed: 1 May 2015.

[46] Kelley M.B. The Stuxnet attack on Iran's nuclear plant was "far more dangerous" than previously thought. *Business Insider*, 20 November 2013. See http://www.businessinsider.com/stuxnet-was-far-more-dangerous-than-previous-thought-2013-11. Accessed: 27 April 2015.

[47] Cerrudo C. Hacking US (and UK, Australia, France, etc.) traffic control systems. *IOActive Labs Blog*, 30 April 2014. See http://blog.ioactive.com/2014/ 04/hacking-us-and-uk-australia-france-etc.html. Accessed: 3 May 2015.

[48] Zetter K. Hackers can mess with traffic lights to jam roads and reroute cars. *Wired*, 30 April 2014. See http://www.wired.com/2014/04/traffic-lights-hacking/. Accessed: 3 May 2015.

[49]   ICS-CERT. *Sensys Networks Traffic Sensor Vulnerabilities*, 28 October 2014. Advisory (ICSA-14-247-01A). See http://ics-cert.us-cert.gov/advisories/ICSA-14-247-01A. Accessed: 3 May 2015.

[50]   Mouawad J. Conflict narrows oil options for West. *The New York Times*, 14 August 2008. See http://www.nytimes.com/2008/08/14/world/europe/14oil.html. Accessed: 3 May 2015.

[51]   Robertson J., Riley M.A. Mysterious '08 Turkey pipeline blast opened new cyberwar. *Bloomberg*, 10 December 2014. See http://www.bloomberg.com/news/articles/2014–12–10/mysterious-08-turkey-pipeline-blast-opened-new-cyberwar. Accessed: 27 April 2015.

[52]   Cobham A. The intrinsic computational difficulty of functions. In *Proceedings of the 1964 Congress for Logic, Methodology, and the Philosophy of Science*, pp. 24–30. North-Holland, 1965.

[53]   Kerckhoffs A. La cryptographie militaire. *Journal des Sciences Militaires*, IX(January): 5–38, 1883.

[54]   Pita J., Jain M., Tambe M., Ordóñez F., Kraus S. Robust solutions to Stackelberg games: Addressing bounded rationality and limited observations in human cognition. *Artificial Intelligence*, 174(15): 1142–1171, 2010.

[55]   Freund Y., Kearns M., Mansour Y., Ron D., Rubinfeld R., Schapire R.E. Efficient algorithms for learning to play repeated games against computationally bounded adversaries. In *Proceedings of the 36th Annual Symposium on Foundations of Computer Science (FOCS)*, pp. 332–341. IEEE, 1995.

[56]   Deng J., Han R., Mishra S. A performance evaluation of intrusion-tolerant routing in wireless sensor networks. In *Proceedings of the 2nd International Workshop on Information Processing in Sensor Networks (IPSN)*, pp. 349–364. Springer, 2003.

[57]   Al-Karaki J.N., Kamal A.E. Routing techniques in wireless sensor networks: A survey. *Wireless Communications*, 11(6): 6–28, 2004.

[58]   Akkaya K., Younis M. A survey on routing protocols for wireless sensor networks. *Ad Hoc Networks*, 3(3): 325–349, 2005.

[59]   Younis M., Akkaya K. Strategies and techniques for node placement in wireless sensor networks: A survey. *Ad Hoc Networks*, 6(4): 621–655, 2008.

[60]   Cardei M., Wu J. Coverage in wireless sensor networks. In *Handbook of Sensor Networks: Compact Wireless and Wired Sensing Systems*, eds Ilyas M., Mahgoub I., pp. 422–433. CRC Press, 2004.

[61]   Wang Y.C., Tseng Y.C. Distributed deployment schemes for mobile wireless sensor networks to ensure multilevel coverage. *IEEE Transactions on Parallel and Distributed Systems*, 19(9): 1280–1294, 2008.

[62]   Dhillon S.S., Chakrabarty K. Sensor placement for effective coverage and surveillance in distributed sensor networks. In *Proceedings of the 2003 IEEE Wireless Communications and Networking Conference (WCNC)*, pp. 1609–1614. IEEE, 2003.

[63]   Abbas W., Egerstedt M. Robust graph topologies for networked systems. In *Proceedings of the 3rd IFAC Workshop on Distributed Estimation and Control in Networked Systems*, pp. 85–90. Elsevier, 2012.

[64] Dekker A.H., Colbert B.D. Network robustness and graph topology. In *Proceedings of the 27th Australasian Conference on Computer Science*, vol. 26, pp. 359–368. Australian Computer Society, 2004.

[65] Laszka A., Buttyán L., Szeszlér D. Designing robust network topologies for wireless sensor networks in adversarial environments. *Pervasive and Mobile Computing*, 9(4): 546–563, 2013.

[66] Santi P. Topology control in wireless ad hoc and sensor networks. *ACM Computing Surveys*, 37(2): 164–194, 2005.

[67] Wagner D. Resilient aggregation in sensor networks. In *Proceedings of the 2nd ACM Workshop on Security of Ad Hoc and Sensor Networks,* pp. 78–87. ACM, 2004.

[68] LeBlanc H.J., Zhang H., Koutsoukos X., Sundaram S. Resilient asymptotic consensus in robust networks. *IEEE Journal on Selected Areas in Communications*, 31(4): 766–781, 2013

[69] Sundaram S., Hadjicostis C.N. Distributed function calculation via linear iterative strategies in the presence of malicious agents. *IEEE Transactions on Automatic Control*, 56(7): 1495–1508, 2011.

[70] Marano S., Matta V., Tong L. Distributed detection in the presence of byzantine attacks. *IEEE Transactions on Signal Processing*, 57(1): 16–29, 2009.

[71] Olfati-Saber R., Franco E., Frazzoli E., Shamma J. Belief consensus and distributed hypothesis testing in sensor networks. In *Networked Embedded Sensing and Control,* eds Antsaklis P.J., Tabuada P., Lecture Notes in Control and Information Science, pp. 169–182. Springer, 2006.

[72] Vempaty A., Tong L., Varshney P. Distributed inference with byzantine data: state-of-the-art review on data falsification attacks. *IEEE Signal Processing Magazine*, 30(5): 65–75, 2013.

[73] Rasmussen C.E., Williams C.K.I. *Gaussian Processes for Machine Learning.* MIT Press, 2006.

[74] Das A., Kempe D. Algorithms for subset selection in linear regression. In *Proceedings of the 40th Annual ACM Symposium on Theory of Computing*, pp. 45–54. ACM, 2008.

[75] Abbas W., Vorobeychik Y., Koutsoukos X. Resilient consensus protocol in the presence of trusted nodes. In *Proceedings of the 7th International Symposium on Resilient Control Systems*, pp. 1–7. IEEE, 2014.

Chapter 12

Case studies of WSN-CPS applications

Fang-Jing Wu[1], Tie Luo[2] and Hwee Pink Tan[3]

Abstract

The most representative form of cyber-physical systems (CPSs) involves wireless sensor networks (WSNs) as the main means to interact with physical entities. This chapter reviews a number of such WSN-CPS applications and reveals how these applications bridge the gap between sensing information in the cyber world and diverse entities in the physical world. We divide these applications into five categories: smart space systems, healthcare systems, emergency response systems, human activity inference, and smart city systems. Smart space systems monitor energy usage, temperature, and various other attributes of appliances in an indoor space. Healthcare systems assist people to improve physical and emotional well-being through automatic sensing and sense-making technologies. Emergency response systems search for and rescue people as soon as possible in emergency situations such as fire outbreaks. Human activity inference systems interpret human intention behind sensing information to facilitate human daily activities related to social events, road safety, mood detection, interactive games, etc. Smart city systems concentrate on city dynamics such as urban environmental monitoring, human mobility, and transport information. Our discussion in this chapter is steered from simple to complex systems in terms of networking technologies, service ranges, system integration, and human engagement. We conclude by discussing important technical components, future trends, and open issues in WSN-CPS applications in order to provide readers with technical pointers for designing next-generation WSN-CPS applications.

[1]Institute for Infocomm Research (I²R), Agency for Science, Technology and Research (A*STAR), Singapore, e-mail: fang-jing.wu@neclab.eu
[2]Institute for Infocomm Research (I²R), Agency for Science, Technology and Research (A*STAR), Singapore, e-mail: luot@i2r.a-star.edu.sg
[3]Institute for Infocomm Research (I²R), Agency for Science, Technology and Research (A*STAR), Singapore, e-mail: hptan@smu.edu.sg

12.1   Introduction

Cyber-physical systems (CPSs) which incorporate wireless communications, micro-electromechanical systems (MEMS), intelligent decision making, ubiquitous computing, and integration control among diverse entities have been boosting many promising applications and open up more opportunities to enrich the inter-action between the cyber world and the physical world. This chapter will system-atically review wireless sensor network (WSN)-CPS applications from static toward dynamic networks, from small-scale to large-scale service coverage, and from simpler toward more complex system interaction, as shown in Figure 12.1, including smart space systems, healthcare systems, emergency response systems, human activity inference, and smart city systems. Figure 12.2 gives an overview of coarse-grained classification based on the three different criteria: network flexibility, service coverage, and human engagement. Generally, for network flexibility and service coverage, a more dynamic system can operate in a more large-scale area. Moreover, based on the degree of human engagement, the former two categories are much simpler and emphasize how sensors interact with hardware devices as well as humans, while the latter three categories are more complex and focus on how humans interact with a whole wireless sensor network, people, and a whole city. The main goal of this chapter is to bring important factors and com-parison from system-level, service-level, and the level of human engagement

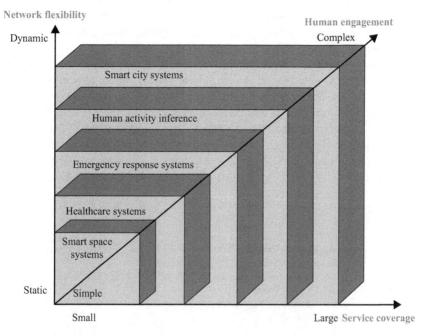

Figure 12.1   General analysis of WSN-CPS applications

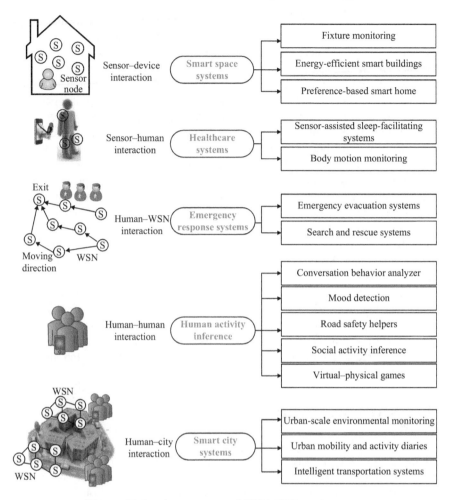

Figure 12.2   An overview of WSN-CPS applications

perspectives to audiences' attention through reviewing some promising WSN-CPS applications. We further discuss some future trends and open challenges in future WSN-CPS applications.

This chapter is organized as follows. Section 12.2 will review three types of smart space systems: fixture monitoring systems, energy-efficient smart buildings, and preference-based smart home. These systems will control the usage of household appliances and utilities in a certain intelligent way. The control principles in the first two types are to avoid waste of resources, while control principles in the last type will depend on personal preference. In Section 12.3, two types of healthcare systems will be discussed: sensor-assisted sleep-facilitating systems and body motion monitoring. The goal of the first type is to improve sleep quality of humans, while the second type concentrates on identifying body motion patterns.

Section 12.4 studies the integration of real-time monitoring and intelligent decision making in emergency response systems that provide people with adequate instructions when a dangerous event happens (e.g. a fire). Based on the response diversity and actuation capabilities, the existing emergency response systems can be classified into two types: emergency evacuation systems and search and rescue systems. The former exploits pre-deployed sensor networks to guide people to exits, while the latter integrates mobile platforms (e.g. robots), opportunistic communications, and parallel computing to support diverse actions. Furthermore, as off-the-shelf smartphones equipped with various sensors are able to bridge data generated in the cyber world and human activities in the physical world, in Section 12.5 we comprehensively study how to infer everyday human activities automatically. We will review five types of applications including conversation behavior analyzer, mood detection, road safety helpers, social activity inference, and virtual–physical games. The first three types aim at human behavior inference, while the last two types focus on inferring social intention to facilitate human-to-human interaction. Section 12.6 reviews interesting applications in a smart city including urban-scale monitoring, urban mobility and activity diaries, and intelligent transportation systems. The former two types are intended to derive deep knowledge behind sensing data for better understanding of city dynamics, while the last type is to provide convenient transportation-related information for improving daily commutes. Finally, Section 12.7 discusses fine-grained classification based on some technical features and requirements of WSN-CPS systems and highlights some important challenges for future systems.

12.2   Smart space systems

This section will review three types of smart space systems based on different requirements in a smart environment. The first type is to monitor the statuses of appliances, the second type adaptively adjusts the temperature for the energy-saving purpose, and the third type is more flexible to fit preferences of multiple users.

12.2.1   Fixture monitoring systems

Monitoring of electrical and water fixtures in smart space is necessary for conserving energy or water cost. Such systems typically comprise three technical stages: (1) fixture discovery, (2) fixture recognition, and (3) fixture disaggregation. The fixture discovery is to infer the existence of electrical and water fixtures in a house automatically. For example, [1] deploys multi-modal sensors including motion sensors, light sensors, water meters, and power meters in a house so that each fixture will have a distinctive usage profile, called 'fixture profile', that is a combination of multi-modal sensing data instead of single-modal data from a single smart meter or an ambient sensor. Since a fixture usually creates a pair of 'ON' and 'OFF' events, these multi-modal sensors and smart meters will collaborate to discover the number of fixtures in a house and their fixture profiles through data fusion and matching algorithms.

The fixture recognition is to identify when a particular fixture is turned on or off. For example, in [2] and [3] a sensor is attached to a wall socket and a hose to monitor high-frequency signals in the voltage and water pressure, respectively. In the training phase, a user will manually turn on/off each fixture so that the system can learn the fixture profiles based on the usage of fixtures. Afterwards, the system will be able to recognize those fixtures automatically when they are used. Finally, fixture disaggregation is to identify how much energy or water is used by each individual fixture. For example, in [4] and [5] a sensor is attached to each electrical (or water) fixture to recognize when they are used and also a smart meter is used on the electrical (or water) mains to monitor aggregated energy (or water) usage in the entire house. Since the total energy (or water) usage in a house is equal to the sum of energy (or water) usage of each individual fixture, the system can compute the quantity of energy (or water) used by each individual fixture.

12.2.2   Energy-efficient smart buildings

In the United States, 40–50% of the energy consumption in buildings is used for heating, ventilation, and air-conditioning (HVAC) systems. Therefore, optimizing the energy usage of HVAC systems in buildings is critical from both cost saving and sustainability perspectives. Two types of intelligent HVAC systems are designed from two different perspectives, namely occupancy-based HVAC systems and comfort-based HVAC systems. The former utilizes WSNs and ambient Wi-Fi infrastructure to facilitate HVAC control and actuation based on the occupancy estimation, while the latter takes the feedback from occupants (i.e. the comfort level) into consideration for HVAC actuation.

12.2.2.1   Occupancy-based HVAC systems

Since current HVAC systems operate based on a static schedule regardless of whether the room is occupied or empty, the 'occupancy level' (i.e. the number of people inside a building) is considered by some intelligent HVAC systems to facilitate HVAC control in a building [6, 7]. Technically, occupancy estimation can be accomplished by either passive infrared (PIR) sensor networks, camera sensor networks, or existing Wi-Fi infrastructure. However, some challenges arise in a WSN-based occupancy estimation system. The PIR sensors can only provide binary occupancy detection in the sense that an occupied room is assumed to be fully occupied and it is hard to know how many people inside the room, while the camera sensors can only be deployed along public hallways due to privacy issues. Thus, in [6], wireless camera sensors and PIR sensors are combined to estimate the number of people in a building so as to control the HVAC system optimally. Figure 12.3(a) shows the architecture of the occupancy-based HVAC system, where the PIR sensors are deployed on the ceiling to detect if a room is occupied and the camera sensors serve as optical turnstiles to measure the number of people transiting from one area to other areas. Figure 12.3(b) shows the workflow of the system, where a fusion algorithm will estimate the current occupancy level based on the sensed occupancy from the combined PIR camera sensor network. To avoid

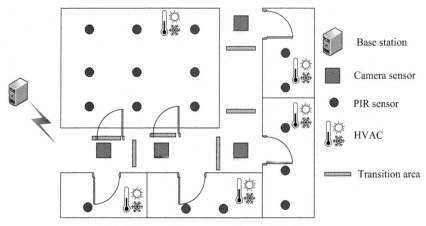

(a) Architecture of the occupancy-based HVAC system.

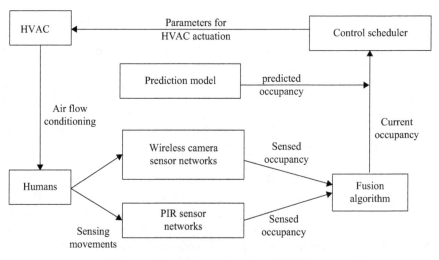

(b) The workflow of the occupancy-based HVAC system.

Figure 12.3   An overview of the occupancy-based HVAC system

the control delay due to the thermal ramp-up or down in a room, a prediction model is designed to predict the occupancy level in the near future that will be combined with the current occupancy estimated by the fusion algorithm. The final estimated occupancy level will be the input to the control scheduler to adjust the parameters for HVAC actuation. However, WSN-based solutions rely on costly sensor deployment and maintenance. In [7], a system utilizes the existing Wi-Fi networks in a building and the smartphones carried by occupants to infer the occupancy level for HVAC actuation. In that system, an offline phase will carefully mark the boundaries of each Wi-Fi access point while a smartphone may move and handover between different Wi-Fi access points. Each room is associated with a Wi-Fi access point that can detect user appearance in the room. A user is assumed to be in his/her room whenever he/she is detected by the Wi-Fi access point of this room. Then, the HVAC actuation server controls the ventilation of a room only when its occupancy changes.

12.2.2.2 Comfort-based HVAC systems

Some HVAC systems follow a comfort-based industry standard, American Society of Heating, Refrigerating and Air-Conditioning Engineers (ASHRAE) Standard 55 [8], to evaluate the comfort index, where multiple parameters such as humidity, temperature, and air flow are considered to estimate how warm or cold occupants feel on a discrete scale from -3 to $+3$. Positive values indicate that occupants are warm, while negative values indicate that occupants are cold. A zero value indicates that occupants are comfortable. However, instead of interaction with sensors or devices, such a system considers the concept of human-as-sensors to adjust temperatures adaptively for improving occupant comfort [9, 10]. To collect feedback from occupants, a mobile application runs on a user's smartphone that allows the user to give a vote at each feedback period, where votes are valued from -3 to $+3$ representing seven different levels of comfort from 'hot', 'warm', 'slightly warm', 'neutral', 'slightly cool', 'cool', and 'cold', respectively. With the collected user feedback, the system will learn the correction offsets of temperatures for different moments of the day to adjust the HVAC system in a building adaptively. For example, a room will need to adjust temperature if the user feedback indicates the room is hot. Therefore, the system can adjust temperature adaptively throughout the day according to these correction offsets.

12.2.3 Preference-based smart home

Since a one-size-fits-all control system in a smart home environment is infeasible, [11, 12] bring the concept of 'user preference' into a smart home environment. In [11], based on the historical hot water usage of each individual household, a just-in-time hot water supply system is designed to determine when the hot water recirculation pump should operate. Generally, there is a short period of waiting time before hot water comes in a water recirculation system when people want to use hot water in their houses. The waiting time may cause a waste of water and can be an annoyance to people. The system thus leverages the fact that every household has unique patterns of hot water usage at predictable time intervals (e.g. mornings and evenings) to design the hot water supply system. In this system, the hot water recirculation pump is connected to an electric motor that will generate current when the hot water is used. A sensor is responsible for monitoring the current generated by the electric motor. A micro-controller is responsible for learning and predicting the timing of hot water usage. A naive Bayes learning algorithm is used to construct the prediction model of the hot water usage for each household, where the following five features are considered to predict whether hot water will be used in the near future: (a) time of day, (b) day of the week and the total amount of time in which hot water was used in the past, (c) 15 minutes, (d) 60 minutes, and (e) 120 minutes.

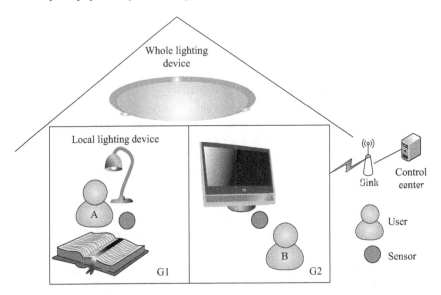

Figure 12.4   An example of preference-based light control systems

Alternatively, some systems consider an intelligent lighting system. Since the illumination requirements for family members are different according to their activities, a personalized light control system in a house is designed to meet different user preferences [12]. Figure 12.4 shows an example, where user A is reading in G_1 and user B is watching television in G2. In this scenario, both of them require sufficient background illumination from the whole lighting device, and user A requires concentrated illumination from the local lighting device for reading. In this system, a home is modeled as multiple grids, and each grid deploys a light sensor to monitor the light intensity which is provided by the background and concentrated lighting. The user requirement of illumination is modeled as a combination of an interval of illumination and a coverage range of illumination. If the provided light intensity is in the specified interval for all of the grids in the specified coverage range, the system considers that the user is satisfied. However, it may not be possible to satisfy all users simultaneously. In this case, the proposed algorithms will gradually relax illumination intervals of users until all users are satisfied.

12.3   Healthcare systems

Next, we will review how healthcare systems assist static human sleep and track dynamic body motions.

12.3.1   Sensor-assisted sleep-facilitating systems

People with sleep disorders usually suffer from various symptoms, ranging from impaired concentration, memory lapses, loss of energy, fatigue, lethargy, to emotional instability. These can lead to even more serious consequences such as social problems and traffic accidents. Recently, many research efforts have been invested to improve sleep quality, which is one of the important issues in our daily life. Two types of sensor-assisted systems are considered to improve sleep quality of individuals: sleep environment monitoring [13] and sleep disorder detection and treatment [14]. The former exploits sensors to better understand the sleep environment, while the latter detects sleep disorder events using electrocardiograph (ECG) sensors or pulse oximeters and adjusts the sleep posture of the user in a non-invasive way. A traditional diagnosis of sleep disorders, polysomnography, is a multi-parametric sleep study that is usually conducted in a sleep center to evaluate the sleep quality of individuals. However, such an evaluation cannot determine actual environmental factors at individuals' homes. Thus, in [13] a system is designed to help people identify when and why their sleep was interrupted at home. To track environmental factors associated with sleep quality, including light, sound, temperate, air quality, and disruptions by others in the household over time, a sensor suite is deployed on the user's night stand to collect sleep environmental factors. The sensor suite consists of several types of sensors including an infrared (IR) camera, two passive infrared (PIR) motion detectors, two upward-facing light sensors, a microphone, and a temperature sensor.

The system provides expert doctors with patients' sleep habits and detailed environmental factors for further treatment. Patients can also track their data through a sleep-monitoring user interface. However, rather than external environmental factors, disordered sleep might be resulted from physiological factors such as 'sleep apnea' (a disturbance in breathing during sleep). Thus, in [14] an auto-adjustable smart pillow system is designed which changes the height and shape of a user's pillow to relieve sleep apnea, where a pulse oximeter is used to detect blood oxygen saturation (also called SpO_2) level for sleep apnea detection in real time. Figure 12.5 shows the system architecture of the smart pillow system which is composed of a pulse oximeter, a smartphone, and an adjustable pillow. The pulse oximeter is attached to the user's fingertip to monitor the user's SpO_2 continuously while the user is sleeping. The SpO_2 and heart rate are collected at a sampling rate of 60 Hz. The data will be transmitted to the user's smartphone through Bluetooth communications. The sleep apnea detection algorithm will detect sleep apnea events based on predefined thresholds of SpO_2. If continuous sleep apnea events are detected for a long period of time, a pillow adjustment decision will be made. The smartphone will send out pillow adjustment commands to the adjustable pillow through Bluetooth communications. Otherwise, the system will not adjust the pillow since the patient can recover from the sporadic events automatically. The adjustable pillow consists of five bladders. Through extensive experiments on the adjustable pillow, bladder 2 and bladder 5 contribute most to the apnea alleviation. Thus the system will adjust the shapes of bladder 2 and bladder 5 according to a sequence of combinations of their shapes.

12.3.2 Body motion monitoring

Some healthcare systems focus on identifying injury patterns caused by body motions and muscle usage through body sensor networks. The potential applications may help athletes reduce their risk of injury and facilitate home rehabilitation

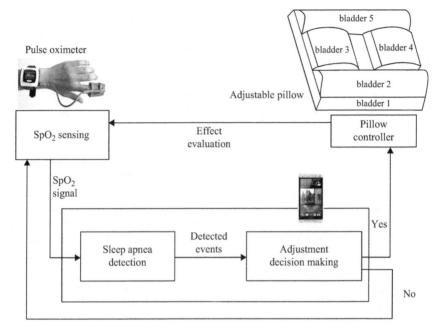

Figure 12.5   Architecture of the smart pillow system

remotely [15, 16]. In [15], wearable sensors and a sink are attached on a user's muscles for muscular activity recognition and motion tracking, as shown in Figure 12.6. Each sensor node consists of a three-axis accelerometer, gyroscope, and magnetometer. The sensing data is sent to the sink and then to the back-end server. The sink is responsible for performing a time division multiple access (TDMA) protocol to schedule the communications between sensors and the sink. The back-end server will conduct muscle activity recognition and motion tracking. The accelerometer data alone is used to perform muscle activity recognition as it provides significant features; accelerometer, gyroscope, and magnetometer data are all considered together for motion tracking. To recognize muscle activities, the system extracts time-domain and frequency-domain features to build a decision tree which will classify the type of muscle activities for newly arrived sensing data. The selected time-domain features contain root mean square of the accelerometer data and cosine correlation between the accelerometer axes, while the selected frequency-domain features are frequency domain entropy and power spectral density. To visualize and render the body motions, the accelerometer, gyroscope, and magnetometer collaborate to compute accurate orientations of these sensors through sensor data fusion algorithms. A similar system is designed in [16] which helps a patient conduct his/her rehabilitation program at home. Through an interactive program, the system will estimate how well a patient can achieve a certain level of body rehabilitation. This way, patients will no longer need to stay in a hospital as traditional rehabilitation requires.

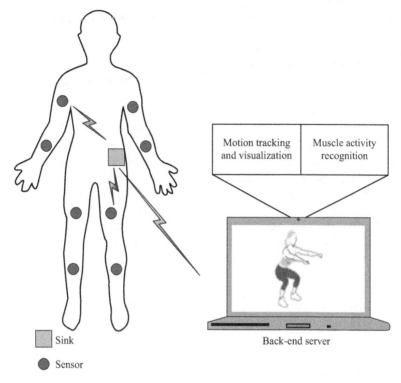

Sink

Sensor

Back-end server

Figure 12.6 The overview of the musculoskeletal monitoring system

12.4 Emergency response systems

In emergencies, the interactions among people and the environment become much more diverse and the complexity of the emergency responses also becomes much greater. Thus, we review two types of emergency response systems: WSN-aided evacuation systems and mobility-supported search and rescue systems. The former type relies on a static WSN to guide people to exits, while the latter type introduces mobile entities to conduct search and rescue tasks.

12.4.1 Emergency evacuation systems

This type of system exploits WSNs to find a safe path to exits in emergencies. Considering a fire, in [17] a distributed protocol is designed to coordinate sensors for computing the evacuation paths. The evacuation principle in [17] is to provide a user with the safest path bypassing hazardous regions instead of the shortest path which may be very close to the sources of hazards. To achieve this goal, each sensor node will maintain a potential value, which is a level of danger, to guide people to the neighboring sensor node with the lowest potential value. Initially, each sensor is assigned a potential value according to its distance to the nearest exit.

In case of emergencies, sensors within a certain distance from the emergency locations will form hazardous regions by raising their potential values so that sensors near the exits will have smaller potential values and sensors near the emergency locations will have higher potential values. The distributed protocol will identify the evacuation paths to exits along sensors with higher potential values to those with lower potential values. Moreover, [18] extends [17] to a 3D environment, where sensors are categorized into three classes: normal sensors, exit sensors and stair sensors. A sensor is considered to be in a hazardous region if either (1) it is within D hops away from hazards or (2) it is a stair sensor and its downstairs sensors are in a hazardous region. The evacuation principle is to guide people to the rooftops if there are no safe paths to the downstairs. However, such an emergency evacuation system may suffer from a congestion problem and an oscillation problem. To solve the two problems, the objective of the former is to evacuate people as soon as possible in a load-balancing way, while the objective of the latter is to avoid guiding people to move back and forth. To solve the congestion problem, [19] proposes a distributed protocol to balance the number of evacuees between multiple paths to different exits. Each sensor knows its location and is able to detect the number of people within its sensing coverage using image-processing technologies. Similarly, each sensor maintains a potential value to find an evacuation direction toward its neighbors based on the number of people around itself. A sensor with a larger potential value implies that there are more people within its neighborhood. Therefore, each sensor will select the neighboring sensor with the lowest potential value to be its evacuation direction. Here, the potential value of each sensor is computed based on its current potential value, the number of people detected by the sensor, and the total number of people detected by its neighboring sensors. Since the evacuee density may affect the walking speed during evacuation, [20] extends [19] to reduce the congestion level by incorporating walking velocity into the potential value of a sensor, where the walking velocity is determined by a mapping function from the evacuee density to human walking speeds. Moreover, [21] and [22] focus on estimating evacuation time accurately. The system in [21] proposes a distributed protocol to estimate the evacuation time based on pre-stored corridor lengths and the moving velocity derived from the current evacuee density, while [22] analyzes the evacuation time based on a guiding tree of sensors rooted at the exit sensor, the corridor capacity and lengths, exit capacity, and evacuation distribution. On the other hand, a user may move back and find an alternate route since the hazard is spreading. To solve this oscillation problem, the system in [23] predicts the dangerous spreading to compute a path to the exit with the minimal number of oscillations.

12.4.2   Search and rescue systems

By integrating mobility entities and parallel computing, emergency response systems will be able to support more dynamic search and rescue tasks during an emergency. Since the dynamics of hazard spreading may force people to move, and the injured may need to communicate with the external world when the communication infrastructure fails, identifying the number of people and where they are in an emergency is usually the first step before rescue. In [24] a robot-sensor network system is designed to track people autonomously without a prior localization infrastructure. In this system, people generate detectable signals such as heat, CO_2, or sounds; the sensors are responsible for detecting if some people are around them, and the robots will move around to find these people through sensor navigation. Some prototyping platforms provide firefighters with safety navigation while they are expediting rescue missions [25]. Two major components, namely ultrasonic beacons and ultrasonic trackers, are adopted to guarantee safe movements of firefighters. Each firefighter wears an ultrasonic tracker to receive signals from ultrasonic beacons. Three types of ultrasonic beacons are designed in the system for different purposes: firefighter beacons, exit beacons and auxiliary beacons. Each firefighter has a firefighter beacon so that injured firefighters can be found by other firefighters. Exit beacons are used to mark exits, while auxiliary beacons are used to mark way-points inside a building or injured/trapped people along a return path. However, since the pre-deployed sensor infrastructure provides limited information and reduced reliability in case of structural collapse, [26] designs a controllable flying sensing platform in support of search and rescue missions in an indoor emergency. On the other hand, while most existing systems consider an indoor fire emergency, the system in [27] implements a system, termed CenWits (Connectionless Sensor-Based Tracking System Using Witnesses), to search for lost or injured hikers in a large wilderness area. Instead of a well-connected network, all hikers form an opportunistic network to exchange their witnesses which indicate the encounter information with each other so as to locate missing hikers. This system consists of a number of sensors, access points (APs), location points (LPs), and an external processing center. Each hiker carries a sensor with a GPS receiver and an RF transmitter for communicating with other sensors, APs and LPs. A set of APs are deployed at predefined locations (e.g. intersections of footpaths or resting areas), and each AP is connected to the external processing center. A few LPs are deployed at particular locations to update sensors' locations in case GPS cannot work. The external processing center is responsible for collecting the witnesses from all APs. When two sensors are within communication range, they will record the each other's presence and also exchange their earlier witnesses, where each record in the witnesses including the encountered node ID, the current time, the encountered location, and the number of transferred hops. Once a sensor meets an AP, all of its witnesses will be uploaded to the AP. Based on the witnesses, the system can estimate the possible locations of a missing hiker to perform rescue missions. Moreover, search and rescue systems may need to provide real-time and in-situ information for remote rescuers and commanders. However, processing and providing global detailed information is a computation-intensive task. Thus, [28] considers grid computing technology to support parallel computing in an emergency response system. This system is composed of four major components: data acquisition and storage, simulation component, agent-based command-control component, and grid middleware. The data acquisition and storage component collects raw sensing data from multiple types of sensors (e.g. smoke, temperature, and gas sensors), transforms the raw sensing data into an appropriate form (e.g. transforming thermocouple voltage readings into temperatures), and ensures database accuracy and reliability. The simulation component supports the prediction of fire spreading in a parallel and distributed manner. The agent-based command-control component provides remote rescuers and commanders with a user interface in support of query-and-response operations between them and the system. The grid middleware component provides a unified interface for communications between the simulation component and other components.

12.5   Human activity inference

Signals behind human activities provide emerging hints for modern CPS that will incorporate richer human input to design-promising applications. This section will review five different types of systems, each of which considers different signals of human activities to design CPS. The first type aims at conversation patterns, the second considers voice signals, the third pays attention to travel experience, the fourth targets social behavior, and the fifth looks at interactive gaming activities.

12.5.1   Conversation behavior analyzer

As conversation is an important part of daily human activities, many application systems focus on monitoring human conversation for social purposes or verbal behavior therapy that helps children to speak better. Conversation behavior is studied from three perspectives: conversation group identification [29], speaker identification [30], and conversation pattern analysis and consultation [31, 32]. The first one is to figure out the number of conversation groups nearby, the second one is to recognize the person who is speaking in a group, and the third one extracts conversation features of users and reminds users to slow down conversation or listen to conversation more than speaking.

Conversation group identification: In [29], multiple smartphones collaborate with each other to find out the conversation groups nearby. Figure 12.7 shows the workflow of the system. Initially, all smartphones discover each other based on a threshold of Bluetooth signal strength. When a smartphone perceives sounds, it will conduct a local classification algorithm to determine if the sounds are voices from the phone owner based on historical information, e.g. average level of loudness. Here, this system assumes that the voices of the phone owner are usually louder than the voices recorded from other users. Once a smartphone detects a voice segment from the phone owner, it will request other smartphones to verify its voice detection through a collaborative voting mechanism. If the smartphone receives positive votes from all other smartphones, it will generate a voice vector with a start time T_i and an end time T_j and share the voice vector with all smartphones for conversation clustering. Finally, the system will cluster those users who do not speak at the same time and are not mostly silent at the same time into a single conversation group, since voice segments in the same conversation group are well synchronized and happen one after the other. One of the potential applications for such a system is to provide a communication topology analysis in the real world

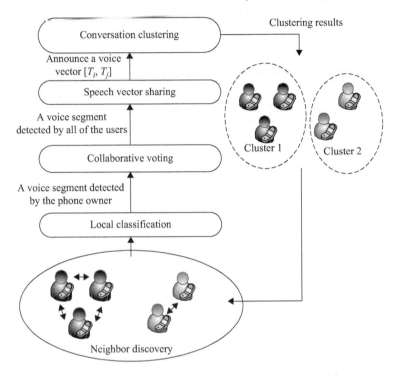

Figure 12.7   The workflow of conversation group identification

since conversations imply real-world social connections and are more reliable than online social networks in the cyber world. For example, the system can further find out the social center with the most connections to others.

Speaker identification: The system in [30] exploits continuous audio sensing to identify the person you are talking with in order to avoid the awkward situation of forgetting his/her name. However, continuous sensing and data processing will quickly drain the smartphone battery since both are computation-intensive tasks conducted by the main processor of a smartphone. Thus, multiprocessor hardware architecture is considered to reduce the energy consumption of continuous sensing in the background, where lightweight sensing and data pre-processing is offloaded to a low-power processor. This system operates on a sequence of two stages from lower power requirements to higher power requirements. The low-power processor is attached with an external microphone and is responsible for sound and speech detection. Once human speech is detected, the low-power processor will wake up the main processor to conduct computation-intensive tasks including identification of high-quality speech frames, feature extraction from the speech frames, and speaker classification, where the speaker classification models are learned from daily phone calls and face-to-face conversations.

Conversation pattern analysis and consultation: In a single conversation group, [31] exploits multiple smartphones to monitor conversational turns for better

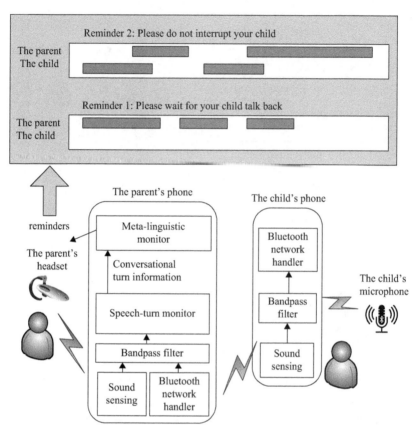

Figure 12.8   The workflow of the speech-language-therapy system

understanding personal social conversation behavior, so as to remind a user to listen to conversation from a particular group member. Here, a 'conversational turn' is a continuous segment of human speech with a start time and an end time. When a conversation happens among a group, group members' smartphones will perceive different patterns of voice strength from each other depending on their positions. Based on the observation, the system can learn the signatures of each group member's conversation to recognize the speaker and the duration of the conversation. This system analyzes personal conversation patterns including the number of people you are talking with, the number of conversational turns in a group, and the number of conversational turns between each pair of participants. This kind of technology is considered to treat children's language delay through meta-linguistic analysis of parent–child conversation in a real-time manner [32]. Figure 12.8 shows the system architecture of the speech-language-therapy system. The parent has a smartphone paired with a Bluetooth headset, while the child has a smartphone associated with a Bluetooth microphone. Each smartphone collects sounds continuously and extracts the human voice using a bandpass filter. The smartphone on the child will send the filtered voice to the parent's phone in real time for further data analysis. Meanwhile, the speech-turn monitor will transform the collected voice data into conversational turn information including speaker, start time, duration and speech rate based on predefined thresholds and then send the information to the meta-linguistic monitor. Based on the previous turn histories, the meta-linguistic monitor will trigger the reminders based on some reaction rules.

12.5.2 Mood detection

This type of system exploits the physiological signals (e.g. voices) collected by smartphones to infer human psychological states (i.e. human mood).

As psychological and affective states (such as stress and mood) are significant element in driving social behavior and influencing physical and emotional well-being, [33] and [34] focus on sensing the psychological states of humans. Conventionally, the detection of symptoms of stress relies on biological sensors in an intrusive way, such as chemical analysis, skin conductance readings, and electro-cardiograms. However, the use of such intrusive technologies may incur additional stress. Generally, when people feel stressed, their voice changes and provides significant patterns for detecting symptoms of stress. Based on the observations, [33] exploits smartphones to recognize stress from human voice unobtrusively. A two-phase approach, including an offline training phase and an online estimation phase, is designed to achieve this goal. In the offline training phase, a voice-based stress classifier is built based on eight voice features including standard deviation of pitch, difference between the maximal and the minimal pitch, perturbation in pitch, fundamental frequency of the voice spectrum, ratio of frequency above 500 Hz, rate of speech, the power spectrum of a short-term voice, and the level of regularity. In the online estimation phase, the voice-based stress classifier determines if a person feels stressed according to the real-time sensed voices. However, since a smartphone is able to capture richer information in daily life, such as when and where we have been, whom we have been talking with, what applications we have been using, and even more, [34] infers moods of a user by analyzing communication history and usage patterns of applications in addition to assistance of built-in sensors. To build a mood estimation model, comprehensive data collection is launched to gather a participant's feature patterns including everyday mood scores and smartphone usage. In the course of data collection, a user can input his/her current mood state with two five-level scores representing the two mood dimensions in [35]: the pleasure dimension and the activeness dimension. For the pleasure dimension, scores of 1~5 indicate 'very displeased', 'displeased', 'neutral', 'pleased', and 'very pleased', respectively. For the activeness dimension, scores of 1~5 indicate 'very inactive', 'inactive', 'neutral', 'active', and 'very active', respectively. A user's smartphone also captures records of social interaction including SMS information, email information, call information, application usage, web browsing, and user locations to build a multi-linear regression model based on the statistical usage for estimating the mood of a user.

12.5.3   Road safety helpers

Many research projects paid attention to improving personal travel experience through crowdsourced data and enhancing the safety of pedestrians and drivers using smartphones. In [36], users can share their journeys with people who have mobility patterns and everyday activities similar to their own. To enrich the information behind collected data and enhance data usage, the system provides users with an interactive interface for getting feedback along their journeys (e.g. traffic accidents or congestion) and exchanging instant messages between users. A publish/subscribe framework is designed to allow a user to access trips contributed to by a community. While travel safety is a critical issue, [37] and [38] focus on detecting unsafe conditions for pedestrians and drivers. A person engaged in a phone call while crossing the road is generally more at risk than others because the phone blocks the view of the user. To improve the safety of people who walk and talk, [37] uses the back camera of the user smartphone to detect vehicles approaching the user. The vehicle detection is based on image recognition technologies with two separate phases, namely an offline training phase and an online detection phase. In the offline training phase, positive image samples and negative image samples are collected. A positive image sample contains the rear or frontal view of a car, while a negative image sample shows side views of cars or random urban environments. Both positive and negative image samples are considered to build a classification model through a machine learning algorithm. Then, the online car detection algorithm running on the smartphone is composed of four steps: (1) image capture, (2) image preprocessing, (3) car detection, and (4) user alert. The system triggers image capture only during an active phone call for energy-saving purposes. The image preprocessing step uses the accelerometer data to estimate the orientation of the smartphone and performs the image alignment according to the direction of gravity. The car detection determines whether an image represents a vehicle based on the classification model built by the offline training phase. If the smartphone detects a vehicle, an alert will be issued from the smartphone to remind the user of the car approaching. On the other hand, an approach using dual-camera smartphones to track dangerous driving behavior is an efficient way to reduce risk of traffic injury. In [38], the front and back cameras on a smartphone are properly scheduled to monitor dangerous driving conditions inside and outside a car. The front camera estimates the head direction and blinking rate of the driver by tracking the head poses and eyes to infer whether the driver is tired and distracted. The back camera monitors the distance between cars to detect whether the car is too close to the car ahead. When either situation is detected by the smartphone, it will change the color bar of driving status on the screen and announce an alert to remind the driver.

12.5.4   Social activity inference

Some systems exploit smartphones to detect social activities and infer social intention. Two types of system are designed for this purpose. The first one is to find out social intention behind group activities, while the second one focuses on how to infer intention for device pairing for automatic data exchanges between two users.

12.5.4.1 Group social intention

In [39], a group-based navigation system is designed to help users find a particular person in a social venue. Generally, since most people stand and walk around together in a social event (e.g. a conference or a party), this system assumes that the moving traces of users in the same group have high similarity. Based on the similarity of moving traces, the system can show the relative positions of users in a social event for finding a person. Figure 12.9 shows the system overview of the group-based navigation system. Each user carries a smartphone which will continuously collect samples of accelerometer, digital compass, and Bluetooth received signal strength (RSS) from the neighboring clients. The accelerometer and digital compass data are considered together to estimate step vectors of a user based on step counts, personal stride length, and direction information. Then, the estimated step vectors and Bluetooth RSS will be reported to the back-end server for further activity inference. The back-end server will analyze the collected sensing data to create a grouping graph based on proximity and trace similarity. Two smartphones are in proximity if both of them receive Bluetooth RSS from each other greater than a predefined threshold. To evaluate the similarity between two user traces, a distance function is defined based on the number of insert, delete, and replace operations needed to convert traces of a user into the traces of another user. For a given pair of proximity and trace similarity, the back-end server will compute the group likelihood between a pair of user clients. If the group likelihood between

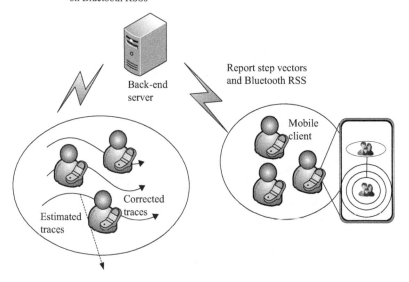

Figure 12.9   An overview of the group-based navigation system

them is greater than a predefined threshold, there is an edge between them in the grouping graph. Meanwhile, the trace similarity is adopted to correct the estimated initial traces because of the property of group moving together. Once the grouping graph is created, the back-end server will estimate the relative positions of these user clients in the social event and show the group information on each user's smartphone.

In addition to walking together, a group of people may take photos together in a social activity. The system in [40] identifies people who appear in the same photo and tags more detailed activity information in the photo automatically. A promising application of this system is to automatically share and tag human activities in online social networks (e.g. Facebook) through the analysis of sensing data. The main difference between this system and other automatic knowledge extraction systems (e.g. video surveillance systems) is richer human social interaction and human input, while other automatic knowledge extraction systems are focusing more on specific even detection. Thus, the system will tag a photo with a format of photo-taking time, photo-taking place, photo-taking participants, and activities in the photo to enrich the information behind the photo. Figure 12.10 shows a scenario of the automatic image tagging system, where user D is ready to take a photo for user A and user B, and user C is in the proximity. When user D activates the smartphone's camera to take a photo, user D's smartphone will broadcast to request all of the users' smartphones in the proximity to collect sensing data from the microphone, GPS, compass, light sensor and accelerometer. When user D clicks the camera, these smartphones will record all sensing data for a short period of time for further image tagging. The image tags will be generated by the four modules: location detection, time detection, participant recognition, and activity recognition. The former two modules tag the location and time of the photo based on existing

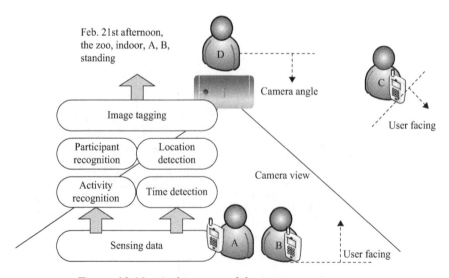

Figure 12.10   Architecture of the image-tagging system

localization technologies and the system timestamp of smartphones, where the light intensity is considered to determine if the photo is taken indoors. The participant recognition infers people in the photo based on the motion signatures of users captured by the accelerometer, user facing extracted from the compass data, and motion vectors of moving objects extracted from consecutive snapshots. For example, the motion signatures of users A and B will be very different from the motion signatures of user C at the photo-taking moment because user C may move around at that moment, the photo-taking participants' facing angles are usually opposite to the photographer's, and the motion vectors of moving objects from these consecutive snapshots will have high correlation between these participants' accelerometer data when they are playing a sport (e.g. table tennis). The activity recognition conducts activity classification to classify the user activity based on the accelerometer data and acoustic data, where limited types of activities including standing, sitting, walking, jumping, biking, playing, talking, music, and silence are considered.

12.5.4.2 Device pairing intention

Some systems infer social intention behind body motions and gestures to facilitate information sharing automatically. In [41], a user can pair his/her mobile device with another nearby user's mobile device by pointing his/her smartphone toward the intended person. Figure 12.11 shows how to detect user intention of device pairing before data exchanges, where user A points the smartphone toward user B with two consecutive beep signals emitted at P and P′ positions. Each nearby smartphone will compute the elapsed time of arrival (ETOA) between the two beep signals. Finally, user A's smartphone will select the one with the maximal ETOA

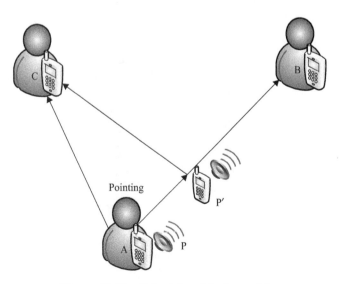

Figure 12.11 Principle of device pairing

difference between the two beep signals for device pairing based on the triangular inequality $d_{PP} (= d_{PB} - d_{P'B}) > d_{PC} - d_{P'C}$.

Another system in [42] automatically infers human handshake behavior in a social event to enable natural information exchange after detecting handshaking behavior between two persons. As shown in Figure 12.12, in the physical world, the handshake behavior between two people implies that a social link will be authenticated between them before they exchange personal information (e.g. exchange of business cards). On the other hand, in the cyber world, a handshake procedure is adopted by two nodes to authenticate each other before they exchange data. The system follows the concept of 'handshake' to design an authentication mechanism to facilitate automatic data exchanges between two users following the handshake behavior. In this system, the similarity of the accelerometer data between two user smartphones is considered to determine if they have handshake behavior. Each user carries a smartphone and wears a watch-like sensor node with an accelerometer on his/her wrist. Each sensor node is associated with the user smartphone through Bluetooth, while the communications between sensors are through IEEE 802.15.4. Each sensor node is responsible for detecting handshaking events and reporting accelerometer data to its smartphone. Upon receiving accelerometer data from

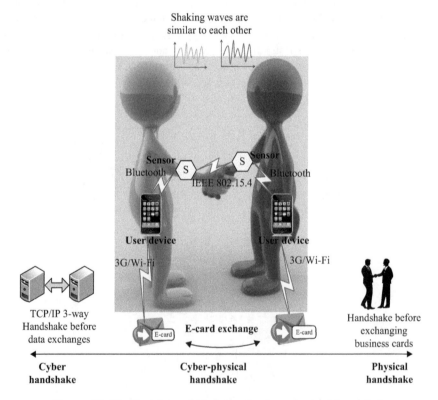

Figure 12.12 System architecture of cyber-physical handshake

sensors, each smartphone will compute the similarity between the two users' accelerometer data. If the similarity is greater than a predefined threshold, the smartphone will exchange the user's personal contact information with the other user automatically.

12.5.5   Virtual-physical games

Recently, the confluence of sensing capabilities of mobile devices and wireless networking technologies has made social gaming systems more user-friendly, where people carry portable gaming devices with built-in sensors to interact with remote users anytime, anywhere. Exploiting diverse devices to enhance game interfaces opens up many opportunities to interweave body motions in the physical world with the fabric of social games in the virtual world. Thus, this kind of application focuses on designing virtual-physical gaming systems to achieve such sophisticated interactions.

In [43], a social exergame supporting multiple exercise devices is designed for playing repetitive exercises among several users, where each user can choose a preferred device to play the game such as treadmill running, stationary cycling, hula hooping, and rope jumping. Figure 12.13 shows the system architecture of the

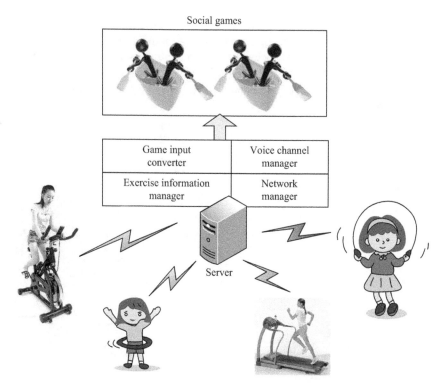

Figure 12.13   System architecture of a social game with multiple heterogeneous controllers

gaming system, where a user's exercise intensity can be measured using standard metrics, e.g. rotations per minute for hula hoops, rope jumps, and stationary bikes, or speed (km/h) for treadmills. There are four key components in this system: the game input converter, the voice channel manager, the network manager, and the exercise information manager. The game input converter will map the intensity of body exercises from devices to input values of the game. The voice channel manager provides voice communications among users to facilitate social interaction during game play. The network manager supports communication fairness among users due to network delay variation. The exercise information manager summarizes exercise statistics, e.g. duration and total calories burned.

Instead of repetitive exercises, [44] exploits body sensor networks (BSNs) to build a virtual-physical social network platform which can facilitate group Tai-Chi exercises, a popular sport in Chinese communities with continuous and diverse body motions. As shown in Figure 12.14, through this system, users can share with each other remotely their Tai-Chi motions on conventional social networks (e.g. Facebook). In this system, there are three major components: (1) the BSNs, (2) the social network, and (3) clients. In a BSN, each user wears nine sensors and a sink node. Each sensor has a three-axes accelerometer and a digital compass, and the sink node runs a polling protocol to collect sensory data from these sensors to the social network. The social network contains two components, the Tai-Chi engine and the community engine. The Tai-Chi engine is responsible for computing and rendering users' motions, and the community engine provides a web service embedded in the conventional social network to facilitate social interaction among users (i.e. sharing users' Tai-Chi motions). To enhance user experiences of gaming systems, [45] exploits BSNs incorporated with multiple game screens to broaden

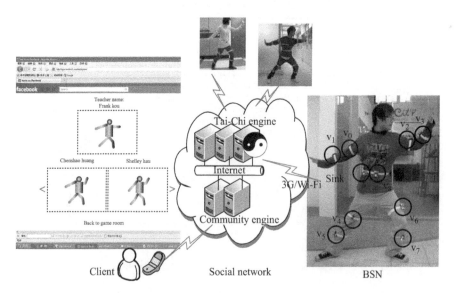

*Figure 12.14   System architecture of virtual-physical Tai-Chi exercise.
Adapted from [44]*

Figure 12.15   A multi-screen video game. Adapted from [45]

players' views and provide more realistic interaction with the fabric of games. As shown in Figure 12.15, the player wears four inertial sensor nodes, one on the broomstick, one on the forearm, one on the upper arm, and one on the club to play the Quidditch sport from the *Harry Potter* movie. In addition to the BSN, the game engine is responsible for computing the orientations of sensors to represent angles of four cameras (east, west, north, and south) for providing a 360-degree panorama of the game.

12.6   Smart city systems

Finally, we review three types of urbah-scale systems that will involve large-scale environmental data, long-term human activities, and urban transportation behavior.

12.6.1   Urban-scale monitoring

Many research efforts exploit crowdsourcing and participatory sensing (e.g. using sensor-equipped mobile devices) to collect urban-scale sensing data such as noise levels, air quality and network connectivity. In general, smartphones can collect dynamic sensing data at an incredible rate to generate a huge amount of data, contributing to the so-called Big Data. While data quality is a significant concern, which is addressed in [46], sensing capability has spurred the development of promising applications that can extract knowledge from Big Data to reflect city dynamics.

12.6.1.1   Noise monitoring

Noise pollution is one of the important problems in urban environments which will affect human mobility, well-being and health. Conventionally, a noise-monitoring system deploys a few sound level meters at certain locations to measure noise levels and creates a noise map by extrapolating city-wide noise levels from local measurements. However, the typical approach is error-prone, costly, and only available for outdoor places. Emerging sensing methodologies consider the better support of participation and engagement of citizens to collect fine-grained and city-wide sound data using smartphones. The system described in [47] uses smartphones as noise sensors and involves citizens who carry them to measure, locate, and collect qualitative sound data intermittently. Data collection is conducted by a mobile application running on a smartphone to collect sound data from a micro-phone, location data from the GPS sensor, timestamp, and user inputs at given intervals. The collected data is then sent to the back-end server for further data analysis. The collected sound data will be visualized using three colors which indicate the health risk of the current exposure level based on predefined thresh-olds, where green is for 'no risk', yellow means 'be careful', and red is for 'risky'. In addition to the measured sound data, a user is allowed to input free text (e.g. car, home, or offices) to provide richer information in the collected data.

12.6.1.2   Air quality monitoring

The monitoring of air quality in an urban area has also attracted many researchers' attention recently. The system in [48] concerns the real-time and fine-grained information of air quality in a city area based on spatial and temporal features extracted from existing monitoring stations and diverse data sources observed in the city which have a strong correlation with air qualities such as meteorology, traffic flow, human mobility, structure of road networks, and point of interests (POIs). Figure 12.16 shows the framework of the air quality inference system that consists of two major stages: offline learning and online inference. This system models the city area as grids, where some of the grids contain air quality stations to collect air quality index (AQI) records. A mapping algorithm is incorporated to represent the meteorological data, POIs, and vehicle trajectories on the road

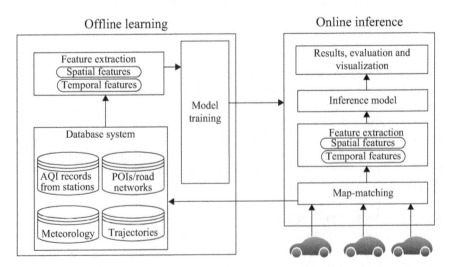

Figure 12.16   The framework of the urban air quality inference system. Adapted from [48]

network for further feature extraction. The temporally related and spatially related features will be extracted from the collected data. The temporally related features (i.e. features that vary with time) are extracted from meteorological data and the spatial trajectories including temperature, humidity, and average speed of vehicles, while the spatially related features (i.e. features that vary with location) are extracted from POIs and road network databases including the density of POIs and the length of roads in a region. After the inference model is created in the offline stage, the online stage will infer the air quality for those grids without air quality stations and visualize the city-wide air quality information for further use cases.

12.6.1.3   Network quality assessment

Network quality (in terms of signal strengths, link speeds, connectivity stability, etc.) is an important aspect of smart cities. As operators tend to overclaim network qualities such as connection speed, and actual measurements on devices are error-prone, assessing network quality in terms of real user experience may yield a more 'useful' reference. Therefore, some systems exploit sensing capabilities of smartphones to crowd-source for such user experience of Wi-Fi [49] and cellular networks [50]. In [49], a Wi-Fi advisory system, called WiFi-Scout, has been developed on Android to fulfill this purpose. It crowd-sources from smartphone users their ratings ('fast', 'medium', 'slow') on Wi-Fi hotspots, as well as (implicitly) obtains various other useful data including locations, SSIDs (service set identifiers), signal strengths, link speeds, uploading and downloading speeds of Wi-Fi access points.

WiFi-Scout provides three advisory modes: (1) offline search, (2) online review, and (3) gamification-based Wi-Fi map. The first mode allows users to search for available Wi-Fi access points in the proximity of specified queried regions even when users do not have Internet connection. The second node allows users who already connect to Wi-Fi access points to report their experience on using these Wi-Fi access points through their smartphones. The last mode displays the crowd-sourced locations of Wi-Fi access points on a city map, but unlike other similar applications, each access point is represented by a user who has contributed the most useful information to it. The contributions of users are quantified using a social-economic scheme [51], which provides incentives for users to report and improves the trust level of user reports. Another system [50] assesses quality for cellular networks, also using crowd-sourcing techniques. Other than the locations of cellular towers, it also provides assessments of signal quality and coverage for nearby cellular towers.

12.6.2   Urban mobility and activity diaries

Many research efforts pay attention to human mobility data collected by smartphone which is represented as an activity diary for better understanding of city dynamics and facilitating urban planning. We study two types of activity diaries: transportation activity diaries and everyday life diaries. The former type focuses on everyday commute patterns, while the latter finds out more about various patterns in our daily life.

12.6.2.1   Transportation diaries

This type of system exploits smartphones to figure out the transportation behavior of individuals. In [52], a mobile sensing system is designed to collect personal cycling experience and share cycling-related data among cycling communities through the developed web service. This system consists of three tiers: the mobile sensor tier, the sensor access point tier, and the back-end server tier. For the mobile sensor tier, each bicycle is equipped with several types of sensors including a GPS, a CO_2 meter, an accelerometer, a microphone, a magnetic sensor, a pedal speed meter, and a Bluetooth/802.15 4 gateway, while the cyclist carries a mobile phone. These sensors form a bicycle-area network (BAN). The intra-BAN communications between sensors are through IEEE 802.15.4, while the sensor data is sent from the DAN to the mobile phone via a Bluetooth/IEEE 802.15.4 gateway. The sensor access point tier consists of a number of mobile phones and Wi-Fi access points for providing communications via either GSM/GPRS networks or IEEE 802.11 networks. This tier provides reliable network access to convey sensing data from the mobile sensor tier to the back-end server tier. The back-end server tier implements data-mining and data-visualization algorithms incorporated with a query-response handler to display detailed information about cyclist experience, such as cyclist routes on the map, current speed, average speed, distance traveled, calories burned, and CO_2 levels along the cycling routes.

In addition to cycling experience, [53], [54], and [55] figure out everyday commute behavior. The system in [53] uses smartphones to automatically carry out transportation activity survey which investigates when, where and how people travel in an urban area. This system is composed of two major components, namely a front-end sensing system and a back-end data analysis system. To optimize energy usage of a smartphone, the front-end sensing avoids using the GPS sensor in the user's long-stay places. To achieve the objective, a place-learning algorithm is implemented on each smartphone to collect the Wi-Fi signatures of a place if the user stays in the place for a long period of time. When the user enters a place, the user smartphone will conduct place matching based on the learned Wi-Fi signatures and avoid using GPS if the current place has the same Wi-Fi signatures as one of the learned places. In the back-end data analysis system, clustering algorithms are implemented to detect if a user stops at certain locations. In addition, a decision-tree-based classification algorithm is considered to detect the transportation modes of users, where the maximal speed, between-stop average speed, accelerometer force variance, and distance to the closest bus and mass rapid transport (MRT) stops are extracted to construct the decision tree. However, GPS-based detection of transportation mode has some essential limitations on energy consumption, availability in indoor/underground environments, and detection accuracy. Thus, [54] considers accelerometer-only approaches to detect transportation modes, where the gravity is estimated based on the accelerometer measurements. This system designs a hierarchical classification algorithm incorporating three classifiers from coarse-grained toward fine-grained to detect the transportation mode of a user. The first classifier detects if a user is walking. If not, the second classifier will detect if the user is stationary. If not, the last classifier will perform fine-grained detection to classify the current transportation behavior into one of five transportation modes: bus, train, metro, tram, or car. In [55], a route-sharing and recommendation system is constructed, where users can contribute and search fine-grained elevation and distance information along their routes to know if a route is suitable for a certain mode of transportation (e.g. hiking or cycling).

12.6.2.2   Everyday life diaries

In addition to transportation activities, this type of system considers more diverse mobility and activity patterns [56, 57]. The system in [56] constructs a text-searchable diary which transforms collected GPS data points into textual descriptions of semantic locations and activity categories so that users can search their historical activities using text inputs (e.g. 'where did I have dinner?'). There are four phases to extract meaningful information from continuous and massive GPS raw data: (1) segmentation of moving patterns, (2) trajectory clustering, (3) creation of semantic places, and (4) activity matching. Since the collected continuous GPS signals contain a lot of redundant information, the first phase represents these continuous GPS signals as a sequence of linear routes with non-uniform representative GPS points. The second phase links these segments into a small number of trajectories based on the spatial correlation between these segments, where each trajectory is represented by a pair of 'begin point' and 'end point' segments. In the third phase, to transform these GPS locations into semantic places, the system conducts reverse geo-coding which maps the GPS coordinates into textual descriptions (e.g. Starbucks). Finally, the last phase infers possible user activities in a certain location by matching the location categories provided by Yelp (which is a location-based service) and collects user reviews and recommendations of restaurants, shopping, nightlife and entertainment. On the other hand, [57] focuses on identifying live points of interest (LPOIs) which are real-time activity hotspots in a city. This system uses smartphones to collect audio clips and location information through GPS, Wi-Fi, and cellular networks in those places where people spend a significant amount of their time. The audio data is used to infer the gender of a participant. Once the location data is sent to the back-end servers, a density-based clustering algorithm is adopted to find out the activity hotspots and the detailed information of participants (e.g. 20% males and 80% females).

12.6.3   Intelligent transportation systems

As intelligent transportation systems are important elements in a smart city, we will review some systems from four perspectives of services including finding a taxi or passengers, carpooling services, traffic monitoring, and finding parking lots.

12.6.3.1   Taxi/passenger finder

There are two essential requirements of taxi services in a city area: (1) finding the best locations where a taxi driver can find passengers easily and (2) finding the best locations or road segments where people get a taxi easily. To meet the two requirements, [58] proposes a recommendation system for finding passengers and vacant taxis based on historical GPS trajectories of taxis. This system consists of an offline data-mining component and an online recommendation component. The offline data-mining component is responsible for collecting and detecting parking places of taxis to learn statistical results of taxis' pick-up/drop-off behavior (e.g. parking places, time interval between two consecutive vacant taxis, and queue length of passengers) and passenger mobility patterns (e.g. when and where passengers get on/off a taxi). The online recommendation component incorporates a taxi recommender and a passenger recommender to provide recommendation of taxi services. The taxi recommender provides taxi drivers with the better locations and routes to these locations so that taxi drivers can maximize the profit of the next trip, while the passenger recommender provides a user with the nearby parking places of taxis with the minimal waiting time.

12.6.3.2 Carpooling services

Finding the best route schedule for taxi carpooling is an efficient means to reduce transportation cost and air pollution. To achieve the goal, an urban-scale taxi carpooling service is considered in [59]. Figure 12.17 shows the framework of the taxi carpooling service which considers a dynamic scheduling problem of carpooling in a city. Each taxi will update its status including its ID, the current time, the geographical location, the number of on-board passengers, and its current schedule if the taxi driver is willing to join the carpooling service. Each mobile user can submit a user query anytime, anywhere, where the user query will be associated with the submission time, the pick-up point, the drop-off point, and the early and late bounds of pick-up and drop-off times. When a user query is submitted by a mobile user, the carpooling search and scheduling components will search the candidate

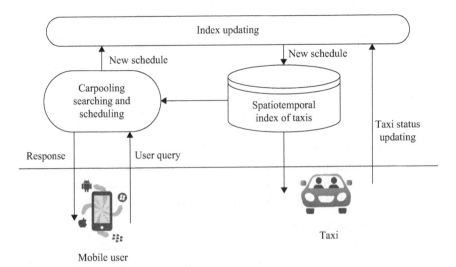

Figure 12.17   The framework of the dynamic taxi carpooling service.
Adapted from [59]

taxi that satisfies the user query and has the minimum additional incurred travel distance. This system incorporates a spatiotemporal index of taxis which will speed up the searching process based on a grid-based road model. If the user query is satisfied, the system will update the spatiotemporal index of taxis and inform the corresponding taxi of the new schedule. However, as many systems have focused on how to exploit the mobility patterns of taxis/passengers to schedule carpooling routes, [60] integrates software and hardware design as well as a win-win fare model which is an incentive mechanism [61, 62] to encourage both taxi drivers and passengers to join the carpooling service. In this system, there are three components: passenger clients, Cloud server, and onboard TaxiBox. The passenger client will provide delivery requests to the Cloud server for taxi dispatch. Based on the delivery requests provided by passengers, the Cloud server will return a carpooling option with a reduced fare for passengers' approval, along with a non-carpooling option with a regular fare for comparison. When a passenger approves taxi carpooling, the Cloud server will find a suitable taxi for carpooling and send out the route schedule to the taxi's onboard TaxiBox. The onboard TaxiBox is equipped with several types of sensors including alcohol/smoke sensors, a three-axis accelerometer, a camera, a microphone, a GPS sensor, and a communication module. The onboard TaxiBox is responsible for reporting the taxi's physical status (e.g. locations and speeds) and the delivery status (e.g. delivery distance, the number of passengers, fare, working duration, start time, end time, pick-up locations and drop-off locations) to the Cloud server. A unique feature of this system is the win-win fare model which shares the benefit of taxi carpooling among the taxi driver and all of the passengers proportionally. For example, three passengers request a taxi with 17, 3, and 45 non-carpooling fare, respectively. If they join carpooling, the total carpooling fare is 52. Thus, the total benefit of carpooling is $(17 + 32 + 45) - 52 = 42$. Let α denote the sharing percentage of all passengers. Thus, $1 - \alpha$ is the sharing percentage of the driver. Accordingly, each passenger will get the benefit proportionally based on his/her travel distance.

12.6.3.3 Traffic monitoring and navigation

Providing real-time and in-situ traffic information for remote users is an essential requirement in a smart city. A participatory CPS prototype system, called 'ContriSense: Bus', is presented in [63] for public transportation. It provides bus commuters with information such as estimated time to arrival and bus speed in order to ease travel planning and improve travel experience for bus commuters. It employs RESTful API and designs algorithms for near real-time sensing and mapping of GPS readings to correct sequences of bus stops. One key feature of the system is that the traffic information is crowdsourced from the public mass, i.e. bus commuters. This makes incentives critical to such crowdsourcing or participatory sensing systems, which are addressed by [64] using a game-theoretical approach and by [62] using an auction-based approach, respectively.

The system in [64] considers a vehicle sensor network for traffic monitoring. Each vehicle is equipped with a set of sensors including a GPS sensor, a Wi-Fi communication module, and a camera. These vehicles collect traffic-related sensing data and report to the back-end server through opportunistic communications in the sense that the sensing data is allowed to be exchanged among vehicles and between a vehicle and a wireless access point if they are within communication range of each other. A remote user can acquire traffic information through a visualization interface. Since the bandwidth and connection are not always available, this system allows users to specify how to prioritize data (e.g. preferring to deliver a summary before detailed values). In [65], GPS data of vehicles are exploited to estimate real-time speed information of roads, where the data uploaded by parked cars and cars waiting for traffic lights are given lower weights in determining the road speed. The Waze system [66] is a community-based traffic navigation system, where crowdsourced traffic information is shared among users to improve driving experience in their daily life.

12.6.3.4   Parking finder

Finding available parking lots sometimes wastes time and fuel consumption in daily life. To find an available parking lot efficiently, [67] considers WSNs to monitor availability of parking lots and provide drivers with real-time parking information. Figure 12.18 illustrates the system architecture. Each car park is deployed with a WSN with a sink, where each sensor node is composed of a heat sensor, a light sensor, a variation sensor, and an RF communication module. Each sensor is powered by a solar system. These sensors will cooperate to detect if a parking lot is occupied and send the collected information to the sink in a multi-hop

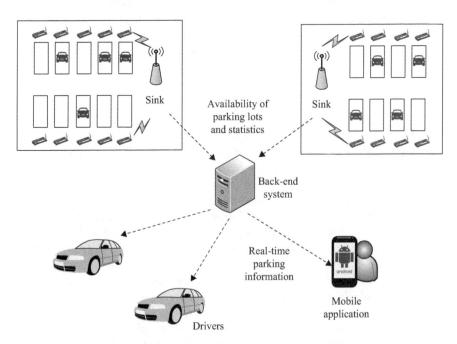

Figure 12.18   System architecture of a parking finder application

way. Then, the information of parking lots will be collected to the back-end system for further analysis. A driver can view the number of available parking lots and their locations through a mobile application.

12.7 Conclusion: discussion, comparison, and future challenges

Finally, we discuss fine-grained classification based on some technical features and requirements of these systems and point out opportunities and challenges in future systems.

12.7.1 Discussion and comparison

Technically, we classify these systems from the following six perspectives: network configuration, communication patterns, sensing techniques, information diversity, decision-making techniques, and service ranges, as shown in Table 12.1. For network configuration, smart space systems, healthcare systems, and emergency response systems are usually deployed in a particular place with fixed network deployment; emergency response systems, human activity inference, and smart city systems contain some mobile entities in the network (e.g. robots or mobile phones); nodes and mobile entities in human activity inference and smart city systems can join/depart "into/from" the network dynamically. For communication patterns, the traffic patterns in the former three types of systems are collecting sensing data periodically; the communications in emergency response systems and human activity inference sometimes are on-demand only when some specific events are detected (e.g. hazards and human conversations); emergency response systems and smart city systems may have cross-network information flows among heterogeneous networks. For sensing techniques, the former three types of systems rely on WSN-based sensing techniques, human activity inference exploit built-in sensors on smartphones, and smart city systems incorporate WSN-based and mobile sensing techniques together. For information diversity, the former four types of systems usually perceive data in a small-scale area; the latter three have multi-modal data sources; and dynamic human data input (e.g. conversation voices and human mobility) is an important factor in the last two types of systems. For decision-making techniques, the former three can be solved using some deterministic algorithms or in-network decision-making; the latter three sometimes need non-deterministic algorithms, data mining, and machine learning in support of decision-making in an uncertain environment and situation; the last one even relies on Big Data analytics to extract knowledge behind sensing information. For service ranges, the former two are usually in a home area; the emergency response systems and human activity inference serve people in multi-stair buildings or small-scale road segments; the last one works in a city-scale area.

12.7.2 Opportunities and challenges beyond

Finally, we look ahead to some potential research challenges and open issues for next-generation WSN-CPS applications and emerging Internet of Things (IoT) applications.

Table 12.1 Features of WSN-CPS applications

Aspects for system classification	Features	Smart spaces systems	Healthcare systems	Emergency response systems	Human activity inference	Smart city systems
Network configuration	Place-centric deployment	✓	✓	✓	✓	✓
	Network with mobile entities		✓	✓	✓	✓
	Dynamic network formation					
Communication patterns	Periodic communications	✓	✓	✓	✓	
	On-demand communications			✓		✓
	Cross-network data flows			✓		
Sensing techniques	WSN-based sensing	✓	✓			
	Mobile sensing				✓	✓
	WSN + mobile sensing			✓		
Information diversity	Small-scale data monitoring	✓	✓	✓	✓	✓
	Multi-modal data sources			✓	✓	✓
	Human data input				✓	✓
Decision-making techniques	Deterministic algorithms and in-network decision-making	✓	✓	✓		✓
	Non-deterministic algorithms, data mining, and machine learning			✓	✓	✓
	Big Data analytics					
Service ranges	Home areas	✓	✓			
	Multi-stair buildings or small-scale road segments			✓	✓	✓
	City-scale areas					✓

12.7.2.1   Cross-domain intelligence for urban Internet of Things

As the European Smart Cities Project [68] has been considering different service sectors such as Smart Governance, Smart Mobility, Smart Utilities, Smart Buildings and Smart Environment to assess the level of smartness of European cities, cross-domain technique integration will be an essential requirement in the future of WSN-CPS applications, where data sensing, knowledge extraction, and data visualization techniques are important technical elements in future systems. The data sensing techniques will aim at how to collect multidimensional and high-quality data effectively, how to collect data without compromising personal privacy, and how to design incentive sensing models which may incorporate participatory sensing, crowdsourcing, cooperative and opportunistic sensing technologies for novel applications. The knowledge extraction will focus on how to figure out deep information behind data through data mining, machine learning, and knowledge discovery methods, how to infer human intentions and activities, and how to design distributed, parallel, and scalable algorithms to handle large, multi-modal, heterogeneous and distributed streams of data. The data visualization will emphasize how to visualize heterogeneous streaming data in a real-time way, how to represent data in a more intuitive way, and how to abstract the relationship between data.

12.7.2.2   Software-defined networking for future Internet

With the Internet Engineering Task Force (IETF) and the Telecommunication Standardization Sector of the International Telecommunication Union (ITU-T) standardization efforts for software-defined networking (SDN) [69, 70], SDN architecture provides more flexible networking operation for the future Internet, where the control plane is decoupled from the data plane. To accomplish such an SDN architecture, OpenFlow [71] is one mechanism to facilitate the communications between the control plane and the data plane. The SDN represents potential trends to support future IoT applications for conveying information between different entities.

12.7.2.3   Machine-to-machine communication issues

As some machine-to-machine (M2M) standard groups are making efforts on a common M2M service layer [72], a lightweight publish/subscribe messaging transport protocol [73], and M2M architecture and interworking technologies [74], automated M2M communications and intelligence enables a wide variety of applications incorporating wired or wireless communications, sensors, devices, computers, robots, mobile equipment, and the Cloud to communicate and exchange information efficiently. Future trends are to enable WSN-CPS with standard-compliant M2M technologies to develop more scalable, flexible, secure, and cost-efficient WSN-CPS applications.

Coupling cyber security and physical privacy: Cross-architecture data flow is a basic concept of future IoT systems that will considerably increase the difficulty of protecting system security and personal privacy. Since all of the physical entities are interconnected in the cyber world, an abstraction layer in-between will be an essential requirement to convey information between physical entities and cyber systems so that information flow security in the cyber world and personal privacy in the physical world are guaranteed simultaneously. For example, when an entity A commits an actuation to an entity B in a condensed and privacy-preservation way, the abstraction layer will be able to authenticate the interdependence of behavior; meanwhile, entities can keep pre-fetched content (e.g. an aggregated map with location-enhanced information) [75] via a pull-based information flow instead of pushing the content of itself.

Flexible human–computer interaction: As the promising M2M applications attract a lot of research and ad industry attention, human input becomes a critical commodity. Therefore, cross-platform human–computer interfaces will be important elements in next-generation systems to bridge human intention in the physical world and actuation in the cyber world. Moreover, with portable, wearable, and mobile human–computer interfaces becoming popular, innovative human–computer interfaces will be able to facilitate the interaction between humans and systems naturally. For example, a magnet could be a kind of human–computer interface to cooperate with a magnetic sensor grid that can recognize distribution of the applied magnetic field and further infer human intention [76].

References

[1] V. Srinivasan, J. Stankovic and K. Whitehouse. FixtureFinder: Discovering the existence of electrical and water fixtures. In *Proceedings of the IEEE International Symposium on Information Processing in Sensor Networks*, pp. 115–128, Philadelphia, PA, USA, 8–11 April 2013.

[2] S. Gupta, M. S. Reynolds and S. N. Patel. ElectriSense: Single-point sensing using EMI for electrical event detection and classification in the home. In *Proceedings of the ACM International Conference on Ubiquitous Computing*, pp. 139–148, Copenhagen, Denmark, 26–29 September 2010.

[3] J. Froehlich, E. Larson, T. Campbell, C. Haggerty, J. Fogarty and S. N. Patel. HydroSense: Infrastructure-mediated single-point sensing of whole-home water activity. In *Proceedings of the ACM International Conference on Ubiquitous Computing*, pp. 235–244, Orlando, FL, USA, 2009.

[4] Y. Kim, T. Schmid, Z. M. Charbiwala, J. Friedman and M. B. Srivastava. Nawms: Nonintrusive autonomous water monitoring system. In *Proceedings of the ACM International Conference on Embedded Networked Sensor Systems*, pp. 309–322, Rayleigh, North Carolina, USA, 5–7 November 2008.

[5] Y. Kim, T. Schmid, Z. M. Charbiwala and M. B. Srivastava. ViridiScope: Design and implementation of a fine grained power monitoring system for homes. In *Proceedings of the ACM International Conference on Ubiquitous Computing*, pp. 245–254, Orlando, FL, USA, 2009.

[6] V. L. Erickson, S. Achleitner and A. E. Cerpa. POEM: Power-efficient occupancy-based energy management system. In *Proceedings of the IEEE*

International Symposium on Information Processing in Sensor Networks, pp. 203–216, Philadelphia, PA, USA, 8–11 April 2013.

[7]   B. Balaji, J. Xu, A. Nwokafor, R. Gupta and Y. Agarwal. Sentinel: Occupancy based HVAC actuation using existing WiFi infrastructure within commercial buildings. In *Proceedings of the ACM International Conference on Embedded Networked Sensor Systems*, pp. 17:1–17:14, Rome, Italy, 11–14 November 2013.

[8]   American Society of Heating, Refrigerating and Air Conditioning Engineers (ASHRAE) Standard 55: Thermal environmental conditions for human occupancy. ASHRAE, 2004.

[9]   F. Jazizadeh and B. Becerik-Gerber. Toward adaptive comfort management in office buildings using participatory sensing for end user driven control. In *Proceedings of the ACM Workshop on Embedded Systems for Energy-efficient Buildings*, pp. 1–8, Toronto, Canada, 6–9 November 2012.

[10]   V. L. Erickson and A. E. Cerpa. Thermovote: Participatory sensing for efficient building HVAC conditioning. In *Proceedings of the ACM Workshop on Embedded Systems for Energy-efficient Buildings*, pp. 9–16, Toronto, Canada, 6 November 2012.

[11]   A. Frye, M. Goraczko, J. Liu, A. Prodhan and K. Whitehouse. Circulo: Saving energy with just-in-time hot water recirculation. In *Proceedings of the ACM Workshop on Embedded Systems for Energy-efficient Buildings*, pp. 16:1–16:8, Rome, Italy, 13–14 November 2013.

[12]   M.-S. Pan, L.-W. Yeh, Y.-A. Chen, Y.-H. Lin and Y.-C. Tseng. A WSN-based intelligent light control system considering user activities and profiles. *IEEE Sensors Journal*, 8(10): 1710–1721, 2008.

[13]   M. Kay, E. K. Choe, J. Shepherd, B. Greenstein, N. Watson, S. Consolvo and J. A. Kientz. Lullaby: A capture & access system for understanding the sleep environment. In *Proceedings of the ACM International Conference on Ubiquitous Computing*, pp. 226–234, Pittsburgh, PA, USA, 5–9 September 2012.

[14]   J. Zhang, Q. Zhang, Y.Wang and C. Qiu. A real-time auto-adjustable smart pillow system for sleep apnea detection and treatment. In *Proceedings of the IEEE International Symposium on Information Processing in Sensor Networks*, pp. 179–190, Philadelphia, PA, USA, 8–11 April 2013.

[15]   F. Mokayay, B. Nguyen, C. Kuo, Q. Jacobson, A. Rowey and P. Zhangy. MARS: A muscle activity recognition system enabling self-configuring musculoskeletal sensor networks. In *Proceedings of the IEEE International Conference on Information Processing in Sensor Networks*, pp. 191–202, Philadelphia, PA, USA, 8–11 April 2013.

[16]   Y.-C. Tseng, C.-H. Wu, F.-J. Wu, C.-F. Huang, C.-T. King, C.-Y. Lin, J.-P. Sheu, C.-Y. Chen, C.-Y. Lo, C.-W. Yang and C.-W. Deng. A wireless human motion capturing system for home rehabilitation. In *Proceedings of the IEEE International Conference on Mobile Data Management*, pp. 359–360, Taipei, Taiwan, 18–21 May 2009.

[17]   Y.-C. Tseng, M.-S. Pan and Y.-Y. Tsai. Wireless sensor networks for emergency navigation. *IEEE Computer*, 39(7): 55–62, 2006.

[18]   M.-S. Pan, C.-H. Tsai and Y.-C. Tseng. Emergency guiding and monitoring applications in indoor 3D environments by wireless sensor networks. *International Journal of Sensor Networks*, 1(1/2): 2–10, 2006.

[19]   W.-T. Chen, P.-Y. Chen, C.-H. Wu and C.-F. Huang. A load-balanced guiding navigation protocol in wireless sensor networks. In *Proceedings of the IEEE Global Telecommunications Conference*, pp. 1–6, New Orleans, LA, USA, 30 November–4 December 2008.

[20]   P.-Y. Chen, Z.-F. Kao, W,-T. Chen and C.-H. Lin. A distributed flow-based guiding navigation protocol in wireless sensor networks. In *Proceedings of the International Conference on Parallel Processing*, pp. 105–114, Taipei, Taiwan, 13–16 September 2011.

[21]   Y. Chen, L. Sun, F. Wang and X. Zhou. Congestion-aware indoor emergency navigation algorithm for wireless sensor networks. In *Proceedings of the IEEE Global Telecommunications Conference*, pp. 1–5, Houston, TX, USA, 5–9 November 2011.

[22]   L.-W. Chen, J.-H. Cheng, Y.-C. Tseng, L.-C. Kuo, J.-C. Chiang and W.-J. Lin. LEGS: A load-balancing emergency guiding system based on wireless sensor networks. In *Proceedings of the IEEE International Conference on Pervasive Computing and Communications*, pp. 486–488, Lugano, Switzerland, 19–23 March 2012.

[23]   L. Wang, Y. He, Y. Liu, W. Liu, J. Wang and N. Jing. It is not just a matter of time: Oscillation-free emergency navigation with sensor networks. In *Proceedings of the IEEE International Symposium on Real-Time Systems*, pp. 339–348, San Juan, 4–7 December 2012.

[24]   J. Reich and E. Sklar. Robot-sensor networks for search and rescue. In *Proceedings of the IEEE International Workshop on Safety, Security and Rescue Robotics*, Gaithersburg, MD, USA, 22–25 August 2006.

[25]   Summit safety. See http://www.summitsafetyinc.com/.

[26]   A. Purohit, Z. Sun, F. Mokaya and P. Zhang. SensorFly: Controlled-mobile sensing platform for indoor emergency response applications. In *Proceedings of the IEEE International Symposium on Information Processing in Sensor Networks*, pp. 223–234, Chicago, IL, USA, 12–14 April 2011.

[27]   J.-H. Huang, S. Amjad and S. Mishra. CenWits: A sensor-based loosely coupled search and rescue system using witnesses. In *Proceedings of the ACM International Conference on Embedded Networked Sensor Systems*, pp. 180–191, San Diego, USA, 2–4 November 2005.

[28]   L. Han, S. Potter, G. Beckett, G. Pringle, S. Welch, S.-H. Koo, G. Wickler, A. Usmani, J. L. Torero and A. Tate. FireGrid: An e-infrastructure for nextgeneration emergency response support. *Journal of Parallel and Distributed Computing*, 70(11): 1128–1141, 2010.

[29]   C. Luo and M. C. Chan. SocialWeaver: Collaborative inference of human conversation networks using smartphones. In *Proceedings of the ACM*

International Conference on Embedded Networked Sensor Systems, pp. 20:1–20:14, Rome, Italy, 11–14 November 2013.

[30] H. Lu, A. J. B. Brush, B. Priyantha, A. K. Karlson and J. Liu. SpeakerSense: Energy efficient unobtrusive speaker identification on mobile phones. In *Proceedings of the International Conference on Pervasive Computing*, pp. 188–205, San Francisco, CA, USA, 12–15 June 2011.

[31] Y. Lee, C. Min, C. Hwang, J. Lee, I. Hwang, Y. Ju, C. Yoo, M. Moon, U. Lee and J. Song. SocioPhone: Everyday face-to-face interaction monitoring platform using multi-phone sensor fusion. In *Proceedings of the ACM International Conference on Mobile Systems, Applications, and Services*, pp. 375–388, Taipei, Taiwan, 25–28 June 2013.

[32] I. Hwang, C. Yoo, C. Hwang, D. Yim, Y. Lee, C. Min, J. Kim and J. Song. TalkBetter: Family-driven mobile intervention care for children with language delay. In *Proceedings of the ACM International Conference on Computer Supported Cooperative Work and Social Computing*, pp. 1283–1296, Baltimore, MD, USA, 15–19 February 2014.

[33] H. Lu, D. Frauendorfer, M. Rabbi, M. S. Mast, G. T. Chittaranjan, A. T. Campbell, D. Gatica-Perez and T. Choudhury. StressSense: Detecting stress in unconstrained acoustic environments using smartphones. In *Proceedings of the ACM International Conference on Ubiquitous Computing*, pp. 351–360, Pittsburgh, PA, USA, 5–9 September 2012.

[34] R. LiKamWa, Y. Liu, N. D. Lane and L. Zhong. MoodScope: Building a mood sensor from smartphone usage patterns. In *Proceedings of the ACM International Conference on Mobile Systems, Applications, and Services*, pp. 389–402, Taipei, Taiwan, 25–28 June 2013.

[35] J. A. Russell. A circumplex model of affect. *Journal of Personality and Social Psychology*, 39(6): 1161–1178, 1980.

[36] M. Harding, J. Finney, N. Davies, M. Rouncefield and J. Hannon. Experiences with a social travel information system. In *Proceedings of the ACM International Conference on Ubiquitous Computing*, pp. 173–182, Zurich, Switzerland, 8–12 September 2013.

[37] T. Wang, G. Cardone, A. Corradi, L. Torresani and A. T. Campbell. WalkSafe: A pedestrian safety app for mobile phone users who walk and talk while crossing roads. In *Proceedings of the ACM Workshop on Mobile Computing Systems and Applications*, pp. 5:1–5:6, San Diego, CA, USA, 12–13 February 2012.

[38] C.-W. You, M. M. de Oca, T. J. Bao, N. D. Lane, H. Lu, G. Cardone, L. Torresani and A. T. Campbell. CarSafe: A driver safety app that detects dangerous driving behavior using dual-cameras on smartphones. In *Case studies of WSN-CPS applications Proceedings of the ACM International Conference on Ubiquitous Computing*, pp. 671–672, Pittsburgh, PA, USA, 5–9 September 2012.

[39] T. Higuchi, H. Yamaguchi and T. Higashino. Clearing a crowd: Context-supported neighbor positioning for people-centric navigation. In *Proceedings*

of the International Conference on Pervasive Computing, pp. 325–342, Newcastle, UK, 18–22 June 2012.

[40]   C. Qin, X. Bao, R. R. Choudhury and S. Nelakuditi. TagSense: A smartphone-based approach to automatic image tagging. In *Proceedings of the ACM International Conference on Mobile Systems, Applications, and Services*, pp. 1–14, Washington, DC, USA, 28 June–1 July 2011.

[41]   C. Peng, G. Shen, Y. Zhang and S. Lu. Point&Connect: Intention-based device pairing for mobile phone users. In *Proceedings of the ACM International Conference on Mobile Systems, Applications, and Services*, pp. 137–150, Krakow, Poland, 22–25 June 2009.

[42]   F.-J. Wu, F.-I. Chu and Y.-C. Tseng. Cyber-physical handshake. In *Proceedings of the ACM Special Interest Group on Data Communication (SIGCOMM) Conference*, pp. 472–473, Toronto, Canada, 15–19 August 2011.

[43]   T. Park, I. Hwang, U. Lee, S. I. Lee, C. Yoo, Y. Lee, H. Jang, S. P. Choe, S. Park and J. Son. ExerLink – enabling pervasive social exergames with heterogeneous exercise devices. In *Proceedings of the ACM International Conference on Mobile Systems, Applications, and Services*, pp. 15–28, Windermere, Cumbria, UK, 2012.

[44]   F.-J. Wu, C.-S. Huang and Y.-C. Tseng. My Tai-Chi book: A virtual-physical social network platform. In *Proceedings of the IEEE International Symposium on Information Processing in Sensor Networks*, pp. 428–429, Stockholm, Sweden, 12–16 April 2010.

[45]   C.-H. Wu, Y.-T. Chang and Y.-C. Tseng. Multi-screen cyber-physical video game: An integration with body-area inertial sensor networks. In *Proceedings of the IEEE International Conference on Pervasive Computing and Communications Workshops*, pp. 832–834, Mannheim, Germany, 29 March–2 April 2010.

[46]   C.-K. Tham and T. Luo. Quality of contributed service and market equilibrium for participatory sensing. In *Proceedings of the IEEE International Conference on Distributed Computing in Sensor Systems (DCOSS)*, pp. 133–140, Cambridge, MA, USA, 21–23 May 2013.

[47]   N. Maisonneuve, M. Stevens, M. E. Niessen and L. Steels. NoiseTube: Measuring and mapping noise pollution with mobile phones. In *Proceedings of the International Symposium on Information Technologies in Environmental Engineering*, pp. 215–228, Thessaloniki, Greece, 28–29 May 2009.

[48]   Y. Zheng, F. Liu and H.-P. Hsieh. U-Air: When urban air quality inference meets big data. In *Proceedings of ACM SIGKDD International Conference on Knowledge Discovery and Data Mining*, pp. 1436–1444, Chicago, IL, USA, 11–14 August 2013.

[49]   F.-J. Wu and T. Luo. WiFiScout: A crowdsensing WiFi advisory system with gamification-based incentive. In *Proceedings of the IEEE International Conference on Mobile Ad Hoc and Sensor Systems (MASS)* (Demo paper), Philadelphia, PA, USA, 27–30 October 2014.

[50]   OpenSignal. See http://opensignal.com/

[51]   T. Luo, S. S, Kanhere and H.-P. Tan. SEW-ing a simple endorsement web to incentivize trustworthy participatory sensing. In *Proceedings of the IEEE International Conference on Sensing, Communication, and Networking (SECON)*, 30 June–3 July, Singapore, 2014.

[52]   S. B. Eisenman, E. Miluzzo, N. D. Lane, R. A. Peterson, G.-S. Ahn and A. T. Campbell. Bikenet: A mobile sensing system for cyclist experience mapping. *ACM Transactions on Sensor Networks*, 6(1): 6:1–6:39, 2009.

[53]   F.-J. Wu, H. B. Lim, F. Pereira, C. Zegras and M. E. Ben-Akiva. A user-centric mobility sensing system for transportation activity surveys. In *Proceedings of the ACM International Conference on Embedded Networked Sensor Systems*, pp. 74:1–74:2, Rome, Italy, 11–14 November 2013.

[54]   S. Hemminki, P. Nurmi and S. Tarkoma. Accelerometer-based transportation mode detection on smartphones. In *Proceedings of the ACM International Conference on Embedded Networked Sensor Systems*, pp. 13:1–13:14, Rome, Italy, 11–14 November 2013.

[55]   RouteYou. See http://www.routeyou.com/.

[56]   D. Feldman, A. Sugaya, C. Sung and D. Rus. iDiary: From GPS signals to a text-searchable diary. In *Proceedings of the ACM International Conference on Embedded Networked Sensor Systems*, pp. 6:1–6:12, Rome, Italy, 11–14 November 2013.

[57]   E. Miluzzoy, M. Papandreax, N. D. Lanez, A. M. Sarroffy, S. Giordanox and A. T. Campbell. Tapping into the vibe of the city using VibN, a continuous *Case studies of WSN-CPS applications* sensing application for smartphones. In *Proceedings of the International Symposium on Digital Footprints to Social and Community Intelligence*, pp. 13–18, Beijing, China, 17–21 September 2011.

[58]   N. J. Yuan, Y. Zheng, L. Zhang and X. Xie. T-Finder: A recommender system for finding passengers and vacant taxis. *IEEE Trans. Knowledge and Data Engineering*, 25(10): 2390–2403, 2013.

[59]   S. Ma, Y. Zheng and O. Wolfson. T-Share: A large-scale dynamic taxi ridesharing service. In *Proceedings of the IEEE International Conference on Data Engineering*, pp. 410–421, Brisbane, Australia, 8–11 April 2013.

[60]   D. Zhang, Y. Li, F. Zhang, M. Lu, Y. Liu and T. He. coRide: Carpool service with a win-win fare model for large-scale taxicab networks. In *Proceedings of the ACM International Conference on Embedded Networked Sensor Systems*, pp. 9:1–9:14, Rome, Italy, 11–14 November 2013.

[61]   T. Luo and C.-K. Tham. Fairness and social welfare in incentivizing participatory sensing. In *Proceedings of IEEE SECON*, pp. 425–433, 18–21 June 2012.

[62]   T. Luo, H.-P. Tan and L. Xia. Profit-maximizing incentive for participatory sensing. In *Proceedings of IEEE INFOCOM*, Toronto, Canada, 27 April–2 May 2014.

[63]   J. K.-S. Lau, C.-K. Tham and T. Luo. Participatory cyber physical system in public transport application. In *Proceedings of CCSA, IEEE/ACM*

International Conference on Utility and Cloud Computing, pp. 355–360, Melbourne, Australia, 5–7 December 2011.

[64] B. Hull, V. Bychkovsky, Y. Zhang, K. Chen, M. Goraczko, A. Miu, E. Shih, H. Balakrishnan and S. Madden. Cartel: A distributed mobile sensor computing system. In *Proceedings of the ACM International Conference on Embedded Networked Sensor Systems*, pp. 125–138, Boulder, CA, USA, 31 October–3 November 2006.

[65] C.-H. Lo, W.-C. Peng, C.-W. Chen, T.-Y. Lin and C.-S. Lin. CarWeb: A traffic data collection platform. In *Proceedings of the IEEE International Conference on Mobile Data Management*, pp. 221–222, Beijing, China, 27–30 April 2008.

[66] Waze. See https://www.waze.com/.

[67] P. Lee, H.-P. Tan and H. Mingding. A solar-powered wireless parking guidance system for outdoor car parks. In *Proceedings of the ACM International Conference on Embedded Networked Sensor Systems*, pp. 423–424, Seattle, Washington, USA, 1–4 November 2011.

[68] European smart cities. See http://www.smart-cities.eu/. Accessed: 8 July 2015.

[69] IETF, forwarding and control element separation (forces). See https://data tracker.ietf.org/wg/forces/documents/. Accessed: 12 August 2015.

[70] ITU-T, Software-defined Networking (SDN). See http://www.itu.int/en/ ITU-T/sdn/Pages/default.aspx. Accessed: 3 January 2016.

[71] OpenFlow. See https://www.opennetworking.org/sdn-resources/openflow. Accessed: 12 November 2015.

[72] ITU-T Focus Group M2M. See http://www.itu.int/en/ITU-T/focusgroups/ m2m/Pages/default.aspx. Accessed: 5 December 2015.

[73] OASIS Standard-MQTT Version 3.1.1. See https://www.oasis-open.org/ news/announcements/mqtt-version-3-1-1-becomes-an-oasis-standard. Accessed: 15 January 2016.

[74] ETSI-M2M. See http://www.etsi.org/technologies-clusters/technologies/m2m. Accessed: 13 June 2015.

[75] S. Amini, J. Lindqvist, J. Hong, J. Lin, E. Toch and N. Sadeh. Cache: Caching location-enhanced content to improve user privacy. In *Proceedings of the ACM International Conference on Mobile Systems, Applications, and Services*, pp. 197–210, Washington, DC, USA, 28 June–1 July 2011.

[76] C.-H. S. C.-T.W. B.-Y. C. D.-N. Y. Rong-Hao Liang and Kai-Yin Cheng. Gausssense: Attachable stylus sensing using magnetic sensor grid. In *Proceedings of the ACM User Interface Software and Technology Symposium*, pp. 319–326, Cambridge, MA, USA, 7–10 October 2012.

Chapter 13

Medical cyber-physical systems

Daniel Sonntag[*]

Abstract

In this chapter, we explain the background of medical cyber-physical systems (MCPSs) and discuss some recent promising directions. This is then followed by an overview of some of the issues investigated recently, and the kind of solutions and architectures that have been proposed, including related work with some technical discussions. Finally, there is a section on challenges and opportunities that describes some of the future issues that need to be addressed. Humans-in-the-loop MCPS are of particular interest: active input modes include digital pens, smartphones and automatic handwriting recognition for a direct digitisation of patient data. Passive input modes include sensors of the clinical environment and/or mobile smartphones. This combination of knowledge acquisition input sources (while using machine learning techniques) has not yet been explored in clinical environments and is of specific interest because it combines previously unconnected information sources for individualised treatments. Model-based developments and user-centred design have major roles to play in achieving this goal.

13.1 Introduction

The next generation of medical care devices and procedures will almost certainly include several methods developed in the field of medical cyber-physical systems (MCPS). These can be described as innovative, real-time, networked medical device systems to improve safety and efficiency in healthcare. The specific advantage of the concepts of cyber-physical systems (CPS) hereby involves the use of both real-time sensor devices (e.g., monitoring devices such as bedside monitors) and real-time actuation devices (such as infusion pumps). In this way, MCPS collect information from the monitoring sensors and actuate by, for example, adjusting the setting of actuation devices, or fire an alarm, or provide decision support to caregivers. In recent years, physiological closed-loop devices have been proposed

[*]German Research Center for Artificial Intelligence (DFKI), Stuhlsatzenhausweg 3, 66123 Saarbruecken, Germany, e-mail: sonntag@dfki.de

which feature important standard safety analysis methods in CPS. Patient safety is the primary concern in closed-loop MCPS. Closed-loop MCPS can be slower than open-loop systems, and the constant feedback process of the system to adjust its parameters must be supervised in some cases. We will review some of those systems. In the coming years, computational approaches are expected to mature to the point of being able to automatically control such networks of sensors and actuators, not only to monitor physiological states (and provide alarm systems), but also to automatically control actuators, such as blood glucose regulation, within the operation region, thereby partially or fully closing the loop and becoming faster than the open-loop solution.

From the caregiver's perspective, we are not only interested in automatic model-based closed-loop MCPS frameworks, as many interventions require expert feedback and supervisory control by a human. Therefore, we are particularly interested in MCPS with humans-in-the-loop, where open-loop safety can be guaranteed and active expert input can be processed at the same time. Humans-in-the-loop systems can have two different forms. (i) Either a humans-in-the-loop MCPS infers the user's intent by measuring and inferring something about human cognitive activity through, for example, body and brain sensors, which are mostly passive user input. Then, the embedded system translates the intent into control signals to interact with the physical environment [1]. Passive input modes also include sensors of the clinical environment and mobile smartphones. (ii) Alternatively, users provide active user input in the form of a supervisory control signal by a human, thereby closing the loop.

Figure 13.1 shows the combined picture: the patient, the caregiver, monitoring and delivery devices, and the network. The semantic patient model includes physiological models and patient records, which are expressed in a structured ontological representation. It also stores a collection of data about the sensed information to which machine learning can be applied. The input–output (I/O) behaviour of monitoring and delivery devices is specified by the medical application domain. The controller is a technical module for decision making that changes an output in the closed-loop control system that affects the controlled actuators (outer ring in Figure 13.1). In open-loop systems, the last step ('control commands') is missing, and the MCPS starts again with the monitoring device ('physiological, emotional and cognitive signals'). Open-loop controllers give no feedback to the system, i.e., no measurement of the system output that could go to the patient model, the monitoring devices or the controller is necessary. Smart alarm systems are implementations of open-loop systems, as they stop by alerting the caregiver ('alerts') – caregivers can then focus on making the clinical decisions, reducing the chances of missing critical events. In future MCPS, however, we are particularly interested in feedback control strategies for automatic closed-loop systems with humans-in-the-loop: (1) applications where humans directly control the system ('active input'); (2) medical applications where the system passively monitors humans ('passive input') and takes appropriate actions (outer cycle); or (3) hybrid systems that passively monitor human behaviour, take appropriate actions and also take occasional human inputs from the patient or the caregiver ('active input') to optimise

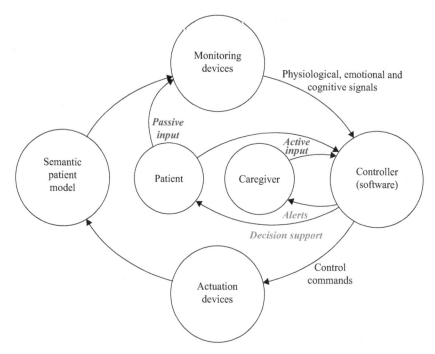

Figure 13.1   Conceptual architecture: networked closed-loop medical cyber-physical system with humans-in-the-loop extension

the controller software. Humans-in-the-loop feedback control strategies allow us to analyse properties of the whole MCPS using passive and active input together with new feedback control strategies.

Active input modes for the caregiver or the patient include digital pens, smartphones and automatic handwriting recognition for direct digitisation of patient data in real time [2]. This combination for knowledge acquisition and decision support (while using machine learning techniques) has not yet been explored in clinical environments and is of specific interest because it combines previously unconnected information sources. The innovative aspect is a holistic view on individual patients, whereby individual active and passive patient data obtained from sensors can be taken into account for clinical decision support [3, 4].

MCPS should address the knowledge acquisition bottleneck to perform better over time: future clinical care relies on digitised patient information. This information can be collected manually (active input) by a doctor or patient, or automatically (passive input) by using suitable physiological sensors, e.g., on a portable device the patient is carrying. Portable devices for patient monitoring include electrocardiogram, pulse oximetry, transcutaneous oxygen, pulmonary testing and spirometry.[1] Therefore, an MCPS must include a combination of those

---

[1]Commercial products for the consumer market or assisted-living environments, such as those found at http://www.spire.io/ and http://www.hexoskin.com/, are just emerging.

information sources for a combined decision-support model in an open- or closed-loop fashion. For example, a patient's recovery from a back injury can be monitored by their movement patterns while climbing steps and combined with the digitised patient disease record at the hospital. Patients will get direct treatment according to a direct data acquisition and interpretation workflow.

In the end, the doctor's decision support will be provided according to the data the MCPS collects from individual patients over time (to be stored and further processed in the semantic patient model). This approach will be scalable, will extend patient monitoring to data collections at home by using portable sensors that provide information about a patient's recovery status, and will influence the healthcare of the future.

Although having humans-in-the-loop has its advantages, modelling human behaviours is extremely challenging, owing to the complex physiological, psychological and behavioural aspects of human beings [5, 6]. This applies to both active and passive user input, and we will report on our recent developments.

This chapter is organised as follows. Section 13.2 provides the background and related work, including model-based developments, medical guidelines and user-centred design. Section 13.3 contains the technical components, including functional active and passive data acquisition methods, as well as functional sensor interpretation methods. Section 13.4 is about recent promising directions towards cognitive prostheses. Then we discuss challenges and opportunities in Section 13.5, including real-time clinical decision support, and conclude in Section 13.6.

13.2   Background and related works

Sensor network technologies, as described in the other chapters of this book, define the backbone of MCPS. The topics of their specialisation in the medical domain range from medical device software modelling to medical device synthesis for safety, assurance, security and control. Recent conferences and workshops[2] highlight the following technological background subfields. In order to complement the technical chapters of this book on cyber-physical systems by medical domain-specific details, we will draw particular attention to the subtopics highlighted in *italic* below.

1. High-confidence medical device software development and assurance: design and implementation, verification, validation and testing; robustness and fault tolerance; system integration and interoperability of heterogeneous systems; *open- and closed-loop control*; systems of systems; *clinical data management* and data security.
2. Modelling and simulation of operational scenarios: *modelling of failures (i.e., anomalies) in medical devices*; *modelling caregivers' and patients' behaviour*; high-fidelity patient models for design and testing; *machine learning models*.

---

[2]See, for example, http://iccps.acm.org/, http://workshop.medcps.org/ and http://www.fda.gov/MedicalDevices/NewsEvents/.

3. Embedded, real-time, networked systems: architecture, platform and middleware; resource management and quality of service (QoS), including medical device plug-and-play (PnP) operation; *distributed control and functional programming.*
4. Enabling technologies for future medical devices: *telemedicine and biosensor technologies*; implantable devices; *energy harvesting* and remote powering devices; medical ultrasound systems; robotic surgery.
5. Medical practice: *medical and clinical guidelines*; *user-centric design*; use and misuse of medical devices; risk understanding and *management of failures*; certification, *pre-clinical testing* and clinical evaluation; approval of non-deterministic and *self-adaptive medical devices*.

An excellent review of the challenges in MCPS, covering most of the above-mentioned topics, including high assurance in system software, interoperability, context-aware intelligence, autonomy, security and privacy, and device certifiability, in particular, was published in 2012 [7]. We will address some of those challenges (the highlighted challenge topics) by proposing an implementation of a technical architecture to provide a first reference architecture for long-term viability of MCPS. These developments require additions for new design, composition, verification and validation techniques.

According to the stated goal of developing techniques for building safe and effective MCPS for individualised treatments, model-based design techniques should be complemented by humans-in-the-loop techniques. Essentially, devices and communication in MCPS must take into account input from patients and caregivers, which we believe is of paramount importance for individualised treatments.

The second main goal for MCPS is continuous monitoring and care. Owing to the high cost associated with in-hospital care, there has been increasing interest in alternatives, such as home care, assisted living, telemedicine and sport–activity monitoring. Mobile monitoring and home monitoring of vital signs and physical activities allow caregivers to assess health conditions remotely at all times. Also, there is a growing popularity of sophisticated technologies, such as body sensor networks, to measure training effectiveness and athletic performance based on physiological data. However, most of the current systems operate in store-and-forward mode, with no real-time diagnostic capability. Physiological closed-loop technology, which employs passive input by monitoring devices, should allow diagnostic evaluation of vital signs in real time and make continuous care possible.

Traditional clinical scenarios can be viewed as closed-loop systems where caregivers are the controllers. New medical devices can act as sensors and actuators. But how to aid the caregivers in controlling the actuation part? There are two fundamental directions: (i) MCPS actuation (only) supports decision making; and (ii) MCPS actuation directly changes the patient's treatment (which means usually changing the patient's physiological state). The first direction emphasises the strong connection between MCPS and medical decision-support systems – traditionally implemented by a kind of alarm system that generates alarms for medical emergencies according to continuous data gathering and management. These smart alarm systems bring the caregiver into the loop around the patient. The second, more interesting, direction, however, is that the computational intelligence of the decision-support system can directly propose a new actuation step to the doctor. This second direction can be extended even further: the MCPS is enabled to actuate therapies automatically or according to the doctor's expert feedback signal. In this regard, building MCPS for safe closed-loop control for patient care delivery is not the only concern. Enabling actuation for therapies, to be done safely and effectively, often involves human supervision and judgement. We propose a real-time dialogue server for this humans-in-the-loop MCPS task.

13.2.1 Model-based developments

Clinical workflows should be given precise operational semantics. The analysis of such precise descriptions of scenarios should allow us to ensure that the instructions of caregivers are unambiguous and cover all possible situations, and ensure that devices can interact with each other as desired. We are also interested in exploring the effects of faults and user errors. Because of the strong coupling between device and patient, model-based frameworks that explicitly model devices have been proposed. In addition, interaction with the environment and with the patient would lead to safer, higher-confidence devices.

Model-based developments have been proposed in several case studies: safety-assured development of the generic patient-controlled analgesic (GPCA) infusion pump (model-driven implementation of infusion pump controller software) [8]; false alarm treatment [9]; and influence of human factors and ergonomics on patient safety and care quality [10]. When implementing open-loop systems, a seizure smart alarm [11, 12], smart alarms at home [13] and wireless sensor networks for assisted living and residential monitoring (Alarmnet) [14] have been proposed. Instances of continuous real-time telemonitoring of patients with chronic heart failure [15] and camera networks for healthcare [16] also belong to model-based developments.

The goal of model-based design in general is to address problems associated with designing complex control and signal processing in complex communication systems such as MCPS. It provides a common design environment to facilitate MCPS verification [17]. More precisely, after the process-based model is verified by a simulation, the MCPS has to perform a validation step of the resulting physical implementation with respect to the system requirements. In addition to testing functional behaviour, time constraints have to be checked (whether or not they are within the tolerances specified in the requirements). A very promising direction for software implementation is worst-case execution time (WCET) prediction [18]. WCET, typically used in reliable real-time systems where understanding the worst-case timing behaviour of software is important for reliability or correct functional behaviour, is used for infusion pumps or regulators in MCPS, for example. Modern functional programming platforms, e.g., Scala,[3] which bring together modern

---

[3]See http://www.scala-lang.org/.

object-oriented programming with functional programming principles, can be used for this purpose. Concerning WCET, if there is a model of the medical process which is in accord with the functional requirements of the medical device, then WCET can be predicted properly. If the model cannot be specified beforehand, then a model has to be learned by employing machine learning techniques. There are methods for testing that overestimate the necessary resources under certain conditions [19]. In all cases, certain risks need to be considered. Then state-of-the-art methods can make complex probability estimations that can be updated as time passes [20]. In order to quantify the uncertainties in human actions, one recent idea is to cast them as uncertainties in terms of model parameters [21].

A related concept is that of the compositionality of MCPS modules. New advances suggest that, if combinators in functional programming satisfy the strong (traditional) side-effect-free conditions for input–output systems and if the input–output modules themselves satisfy the side-effect-free condition in the weak sense, then the whole system remains side effect free in the weak sense [20, 22]. Here "weak sense" means that the input–output unit has a robust controller, which is capable of keeping it side effect free with epsilon precision.

13.2.2 Medical guidelines

From a technical viewpoint, the clinical state of the art in medical decision making is represented in medical guidelines. Guidelines reflect common best practice, are based on the results of well-understood clinical trials, and are at the core of the implementation of evidence-based medicine. There have been some efforts to formalise medical guidelines and to make them computer-accessible. Computerised clinical guidelines have contributed to the formalisation and automation of clinical data and knowledge. Of particular interest are guidelines to generate medical logic modules [23] because we interpret medical data as data which can feed personalised decision-support systems by reasoning mechanisms. A first step in this direction will be to better understand how current recommendations are written and to test the adequacy of guideline models [24] towards clinically relevant reasoning and decision support. Modelling is used increasingly in healthcare to increase shared knowledge, to improve the processes and to document the requirements of the solutions related to health information systems (HIS). There are numerous modelling approaches which aim to support these objectives, but a careful assessment of their strengths, weaknesses and deficiencies is needed. We mention three model-centric approaches in the context of HIS development: process modelling with the model-driven architecture BPMN (Business Process Modelling Notation) [25]; multiple models for treatment recommendations [26]; and the management of exceptional flows in medical processes [27].

MoKi[4] is a tool that supports the creation of articulated enterprise models through structured Wiki pages. Moki enables heterogeneous teams of (medical) experts with different knowledge engineering skills to actively collaborate by

---

[4]See http://moki.fbk.eu/.

inserting knowledge, transforming knowledge and revising knowledge at different degrees of formality. Active collaboration is guaranteed by an automatic translation between formal and informal specifications of the different (medical) experts, and by facilitating an integrated construction of the different parts of the integrated model. The main features that we use are modelling clinical guidelines in the context of the management of exceptional flows in medical processes [27]:

1. support for the construction of integrated domain and process models,
2. easy editing of a Wiki page by means of forms accessible to medical experts,
3. automatic import and export in Web Ontology Language (OWL) and BPMN,
4. easy import of lists of elements organised according to predefined semantic structures (taxonomy or partonomy),
5. term extraction functionalities,
6. graphical browsing/editing of the domain and process models and
7. fully integrated model evaluation functionalities (model checklist and quality indicators).

We modelled the American College of Radiology (ACR) guidelines as a first integration example of digital medical guidelines.[5] From the MCPS perspective, for routine mammography, the following turned out to be the most valuable for clinicians: textual input pages with additional clinical guideline annotations (conditions/actions), textual input pages with formal semantics (semi-formal fragments) and Wiki template pages.

13.2.3   User-centred design

User-centred design, ergonomics and ease-of-use issues in human–device interfaces are important factors that should be considered throughout the design process of MCPS. This type of interaction design in MCPS can be referred to as shared control or humans-in-the-loop, since both controllers (human and computer) act on the same dynamic system. Investigations include new user-centred control interfaces and shared control methods that can effectively delegate tasks and blend the control between the automatic controller and the human operator.

The design pillars of non-functional requirements are explained in [28]. In recent works [3, 29, 30], we have already studied the usability of novel real-time interaction design strategies in the medical field, such as speech interaction on a mobile device for medical image annotation, or classical tools, i.e., writing on mammography forms with a digital pen. Essentially, the various approaches all share the same underlying goal to make expert knowledge acquisition fast and easy, and to retrieve important case information in real time. Examples include the following: a head-mounted multimodal augmented reality system for learning and recalling faces [31]; integrating digital pens in breast imaging for instant

---

[5]See http://www.acr.org/Quality-Safety/Standards-Guidelines/Practice-Guidelines-by-Modality/Breast-Imaging/.

knowledge acquisition [30, 32]; as well as related work describing the application of multimodal user input modes to medical cyber-physical environments [4].

13.3 Technical components

Functional active and passive data acquisition methods have been developed, as well as functional sensor interpretation methods. Experts predict the change of the medical healthcare sector in Europe (in accordance with US developments) in three main stages: (1) digitisation of patient data, (2) usage of environmental (medical) sensor technology and (3) distributed data access and real-time clinical decision support. We have developed a first MCPS reference architecture and testbed system, which combines patient data from different sources that are acquired both manually and automatically, and includes data intelligence methods for clinical decision support. The main technical contributions include the real-time dialogue server and technical components for aggregating digitised patient information, combining manual data acquisition and sensory data; and the integration of data acquisition technology into the clinical testbed environment. Figure 13.2 shows the technical architecture of the MCPS, including the real-time information sources.

13.3.1 Real-time dialogue server

The real-time dialogue server is a software platform for medical device interoperability and user interaction. The CPS expert assembles the treatment system together with the clinician by connecting appropriate sensor and actuator devices.

Our main contribution to MCPS is the integration of distinct input channels consisting of the following active modalities: (i) speech-based interface, (ii) see-through head-mounted display (HMD) interface, (iii) pen-based interface, (iv)

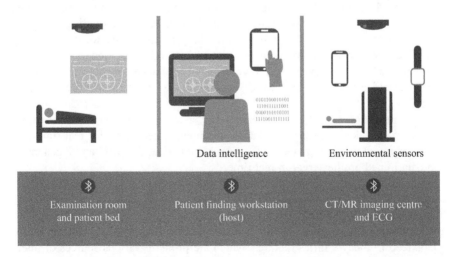

Data intelligence   Environmental sensors

Examination room   Patient finding workstation   CT/MR imaging centre
and patient bed   (host)   and ECG

Figure 13.2   Technical architecture of the MCPS; see http://www.dfki.de/ MedicalCPS/

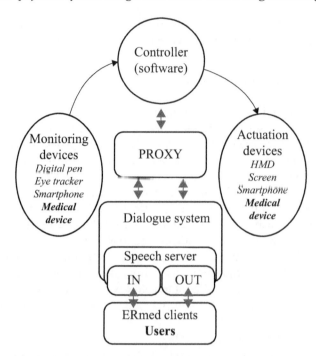

*Figure 13.3   Input–output real-time dialogue server connected to the controller;
based on http://www.dfki.de/RadSpeech/ERmed/*

gesture-based interface (Kinect) and (v) touch-based interfaces (for iOS- and Android-based handheld devices such as iPad, iPhone and Samsung Galaxy Note).

Figure 13.3 visualises the interplay of the active user input components, monitoring sensors and actuation devices in the context of the multidevice infrastructure. In the centre of the humans-in-the-loop infrastructure we have a proxy that is responsible for routing and forwarding method invocations to any target recipient, i.e., an I/O device. The invocation of the actual method on the target user end device is passed on through the network to the controller by means of the proxy.

The proxy architecture includes a new version of the mobile sensor architecture (mobile devices, sensor packages and mobile clients), so that real-time data can be captured and interpreted on the server. Individual datasets can be generated and data mined on the server in real time.

A major problem, however, concerns the run-time allocation of processing resources in a real hospital communication network. In addition to general problems of connecting disparate information systems in a hospital's picture archiving and communication system (PACS), the real-time processing aspect cannot be guaranteed in today's infrastructures, which are prone to various sources of interference from general communication traffic. Current challenges therefore include the streaming from multiple devices into central, integrated repositories, in real time. Note that this challenge is directly related to the use of clinical decision-support systems: often time-critical decision support must be implemented.

13.3.2 Mobile sensor architecture for continuous monitoring and care

One of the most important needs of modern medicine is to develop medical devices capable of providing continuous care (i.e., monitoring, decision support and delivery of therapy). Our proposed sensor system for real-time tasks distributed on multiprocessors [33] consists of three parts. (1) The activity sensor measures acceleration data in three dimensions and transmits them via a Bluetooth Low Energy (BLE) link to the activity sensor server. Connection and disconnection happen automatically as soon as the sensor and activity sensor server are in range. In the case of the loss of a connection, reconnection procedures are initiated. (2) The activity sensor server receives the streaming data. It converts the raw acceleration signals from the sensors into physical units. The signal is then processed, transformed to the frequency domain and activity detection performed. (3) From the detected activities, activity detection events are generated. These events contain timestamps together with the detected activities and are forwarded to the system server. The activity sensor server runs on a Linux PC and can be installed on the same physical machine as the system server or on a different computer connected by a standard network (local-area network, Internet). Optionally, raw sensor data can be transmitted from the activity sensor server to the system server to allow further processing in the sense of real-time data mining and distributed activity recognition. The sensor signals to be captured include preprocessing modules that classify four major vital signs commonly monitored in an Intensive Care Unit (heart rate, blood pressure, blood oxygen saturation and respiratory rate), which can be grouped into categories based on ranges (low, normal, high, very high, etc.) for "smart alarms" [13].

The motivation for additional environmental sensors lies in tracking the behaviour of patients or care-home residents and detecting abnormal living patterns. In addition, the MCPS perspective includes the monitoring of clinical care of bedridden patients. Our experimental tests show that compact algorithms based on nearest-neighbourhood classifiers and filter banks with infinite impulse response (IIR) filters or Haar wavelets can identify the state of the bedridden user in the form of postures and activity levels [34].

13.3.3 Data mining suite

Data mining can be efficient for a range of MCPS events, from simple to sophisticated ones, depending on a number of factors: (i) the information it receives (can be a very broad range), (ii) the expert knowledge provided in the form of medical ontologies (can be very little), (iii) the variety of the cases, (iv) the size of the database available and (v) the expected predictions. Our MCPS scenario will, eventually, require sophisticated spatio-temporal data mining tools. The use of sensory information (active and passive) and events, such as being at places (rooms, corridors, toilet), moving between places and medical events listed by experts, define a special spatio-temporal data mining environment. Predictions, however, can be made on various levels. This is a relevant aspect for CPS in general, because communication bottlenecks and the resulting delays can corrupt synchronisation within and between levels.

The requirements for a successful data mining application are as follows: data mining needs expert information about analogies, such as the set of sensors that involve the notification of, for example, the decision support when the sensor runs *in the alarm region*. The following methods have been proposed. Our data mining uses entropy estimation to estimate predictability. We started from entropy estimation of places (or spatial processes according to medical guidelines). Transition probabilities and the probabilities of event series have been estimated and utilised through prediction by a partial matching method, an adaptive statistical data compression technique based on context modelling and prediction [35, 36]. Prediction by partial matching (PPM) models use a set of previous symbols in the uncompressed symbol stream to predict the next symbol in the stream. We extended the set of information mined by using the InfoGain algorithm together with greedy selection, since it is known to be close to optimal for our conditions [37].

13.3.4   Semantic model for clinical information and patients

In order to enable technical access to clinical data, we need to provide means that allow us (a) to store the comprehensive dataset in a semantic consistent manner as well as (b) to capture the meta-information describing the relevant content from unstructured clinical data to enable their subsequent automatic processing for a seamless integration within clinical applications.

To address these requirements, as well as to overcome the given constraints regarding the sharing of patient data, we developed a generic clinical data access strategy within this MCPS activity that allows us to seamlessly and flexibly align patient data from various data sources, including sensor data. Our approach relies on two building blocks:

- First, an integrated clinical information model that establishes the foundation to integrate and structure clinical data by providing concepts covering meta-information and interpretations of clinical patient data (the semantic patient model).
- Second, a natural language processing (NLP) information extraction pipeline that provides the basic algorithms needed to extract the relevant information from unstructured clinical textual documents. In the project THESEUS MEDICO, we already addressed the challenge of extracting meta-information for medical images from textual ontology descriptions [38]. In MCPS we mainly focus on the extraction of information from clinical textual documents for their combination with sensory data towards real-time clinical decision support.

Through the semantic representation of clinical and administrative data, we established the basis for flexible aggregation and integration of clinical data with sensor data. Currently, we are in the process of aligning the semantically described clinical data with the patient sensor data.

The proposed architecture tries to make data accessible at the data acquisition stage and to provide new chances for direct return of investments in terms of direct interaction (towards interactive decision support), knowledge discovery from textual documents, and intelligent information presentation. Increased data availability should make individualised treatments possible. The problem we face is that these data are not semantically integrated. As a result, most of the available data, such as sensor data, are simply not used at their full strength in clinical decisions in MCPS.

What we need is an integrated and standardised representation of clinical patient data reflecting health status, since this is the basis for various clinical MCPS applications like outcome analysis or other decision-support systems. A standardised representation requires the use of established ontologies, vocabularies or coding systems like, for example, the ICD10, LOINC1 or SNOMED CT2 (see footnote 6).

We identified the following requirements for a model attempting to represent clinical information and sensor data using existing ontologies according to the first clinical data acquisition tests of the MCPS. *Integration*: data from various sources and of different format are integrated and linked. *Standards*: data should be expressed using established coding systems and terminologies. *Interpretation*: the semantics of clinical data must be defined in a consistent way. *Coverage*: it should be possible to represent all clinical data using the model in combination with other ontologies and all clinically relevant high-level concepts.

Starting from these requirements, a semantic model for clinical information (MCI) based on the ontology of general medical science (OGMS) has been proposed [39]. MCI has the purpose of integrating and structuring clinical data by providing concepts that cover meta-information and interpretations of clinical patient data. In this way, all the basic concepts that are needed to describe clinical information objects on the meta-level, like diagnosis, findings, reports, healthcare provider, procedures, identifiers (IDs), are covered by MCI. The resulting comprehensive semantic patient model for MCPS includes real-time application data from active user input or passive sensors and is represented using MCI in combination with large domain ontologies and coding systems (Figure 13.4).[6] Recent developments take open linked data[7] for symptom-disease relationships into account [40].

---

[6]SNOMED CT, Systematized Nomenclature of Medicine Clinical Terms; ICD, International Classification of Diseases; LOINC, Logical Observation Identifier Names and Codes: ATC, Anatomical Therapeutic Chemical Classification System; RadLex, Radiological Lexicon; DOID, Disease Ontology; FMA, Foundational Model of Anatomy; OPS, Operationen- und Prozedurenschluessel (German coding system for procedures); SYMP, Symptom Ontology; LinkedCT, Linked Clinical Trials; DrugBank, Open Drug Data; SIDER, Side Effect Resource.
[7]See http://linkeddata.org/.

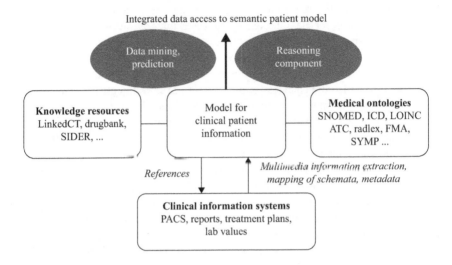

Figure 13.4   Semantic patient model and coding systems

13.4   Towards cognitive prostheses

A different direction concerns the integrated use of an MCPS for a cognitive pros-
thetic, thereby offering alternative interfaces to the MCPS for interaction, communication and control. Assistive devices are capable of restoring independence and
function to people suffering from musculoskeletal impairments. Traditional assistive
devices are exoskeletons [41]. New control techniques are needed to dynamically
shift the level of control between the human operator (the patient) and the intelligent
assistance tool using, for example, dynamical system modelling and stochastic
control. As with robot controllers that can utilise the strengths of industrial robot
arms (adept at handling lower-level control tasks) and humans (superior at handling
higher-level cognitive tasks) [1], a cognitive assistance tool will have to go through a
closed-loop perception and actuation cycle (closing the sensing → perception →
interpretation → action loop). A major issue here is learning to exploit knowledge
gained from observation of past actions and behaviours to predict likely human
responses and reactions. Towards cognitive assistance, a real-time depression monitoring system for the home has been developed [42]. It collects different behavioural
data, including sleep, weight, activities of daily living and speech prosody. The goal
was to potentially detect early signs of a depression episode. As this is an open-loop
system, it only generates reports for the caregivers.

Recent developments have in common that anomalies have to be detected in
medical condition-related activities and circumstances. Our first ambitious closed-
loop MCPS example implementation is a mixed reality intelligent user interface for
helping dementia patients. "Dementia" is a general term for a decline in mental
ability severe enough to interfere with daily life. Memory loss is an example.

Alzheimer's is the most common type of dementia and is projected to affect more than 107 million people worldwide by the year 2050 [43]. Cognitive models are approximations of a patient's mental abilities and limitations involving conscious mental activities (such as thinking, understanding, learning and remembering). One goal is coupling MCPS with augmented cognition for the patient. Some of the first results include using eye gaze and visualisation to augment memory [44], a smart wearable tool to enable people with severe speech and physical impairment (SSPI) to communicate by sentence fragments [45], attention engagement and cognitive state analysis for patient displays [46], and robot companions and smartpens for improved social communication of dementia patients [47].

Our reference architecture is a new kind of closed-loop humans-in-the-loop MCPS: effective shared control in cooperative MCPS tasks and aids need models of intelligent user interfaces such as mixed-initiative interaction [48, 49] and collaborative multimodality [50] to implement shared control in memory tasks in Kognit.[8] But the challenges of a high-confidence design and highly complex, interoperable medical device software and systems for intelligent prosthetics remain. Similar problems in the MCPS domain include minimally invasive surgical devices and the operating room of the future with robotically assisted surgery.[9]

13.5 Challenges and opportunities

Focusing on humans-in-the-loop applications, the following main challenges in feedback control systems have been identified (also see [5]):

1. understanding the complete spectrum of humans-in-the-loop control, since more sophisticated humans-in-the-loop applications are appearing,
2. modelling human behaviour of various types and identifying the best modelling schemes for each type and
3. incorporating these models into the formal feedback control methodology, which may require new results and theory to support formal performance guarantees.

Despite the worldwide scale and heterogeneous nature of the R&D challenges in MCPS, we believe that the following list of research directions will help us make significant progress in implementing an infrastructure for medical device integration and interoperation that accounts for those three feedback control challenges.

13.5.1 Clinical data intelligence models

Current data intelligence systems are restricted to only a few parameters, e.g., a few vital signs, for a clinical decision, limiting their scope. Over the past few years, concerning holistic patient data models, new mathematical methods have allowed the generation of so-called tensor models which can factor out many disadvantages of sparse data [51].

---

[8]See http://www.dfki.de/Kognit/.
[9]See http://www.davincisurgery.com/.

Together with the knowledge acquisition efforts for unstructured information repositories (e.g. medical image databases), the general need for large-scale digitisation of patient-related information comes to the fore again [52]. For example, patient records are often not captured electronically. In this sense, extended electronic health records, by new means of knowledge acquisition and digitisation, fuel data intelligence tasks.

The growing interest in capabilities such as home healthcare services, delivery of expert medical practice remotely (telemedicine) and online clinical laboratory analysis underscores the central role of advanced networking and distributed communication of medical information. Increased R&D focus on the specialised engineering of networked mobile medical device systems is needed. MCPS should respond to the demands for: (1) self-monitoring (quantified self) related to health and habits, while these solutions will also reduce the costly demand for secondary prevention, as well as cure and care for caregivers and insurers; and (2) big health data analysis and clinical data intelligence for individualised treatment [53]. On the side of design practice concerning verification and certification issues, especially for data intelligence-based controllers, it must be noted that MCPS are subject to regulatory oversight through an approval process (in the United States, the Food and Drug Administration (FDA) approves medical devices). Comprehensive background information can be found in the NITRD reports (e.g., [54]).

13.5.2   Semantic patient models and intelligent access

In the healthcare sector, more and more data about patients' health status as well as about medical background knowledge are becoming available. All these patients' data can be stored in different data repositories (e.g., in the various healthcare information systems),[10] in very heterogeneous formats (such as structured laboratory results or unstructured medical reports or images) and at distributed locations (such as the general physician, the hospital, some medical specialist, or the rehabilitation hospital). As of today, only a small percentage of the patient-related data can be and are (re)used in advanced clinical applications [55, 56]. This is mainly due to the fact that the available patient data are not semantically integrated, and thus a great deal of effort is needed to reuse the data sources in advanced clinical application. With the THESEUS MEDICO project, several clinical decision-support applications were able to demonstrate the high value of semantically integrated patient data. For instance, the reasoning-based application for lymphoma patient classification became possible through the semantic integration of medical image annotation with the related patient data [57, 58]. It has also been demonstrated [59] how semantically integrated patient data can be interpreted in a context-sensitive manner by integrating medical background knowledge to improve clinical decision support. Other applications of semantic patient models include pharmacokinetic models of drug absorption; these can be used to account for WCET requirements [60].

---

[10]PACS, Picture Archiving and Communication System; KIS, Kare Information Services; HIS, Hospital Information System; RIS, Radiology Information System; LIS, Laboratory Information System.

Tomorrow's physiological closed-loop systems, which will do more than simply alert caregivers if patients' states divert from the normal range, are, per definition, a replacement for physician reasoning, which was first discussed in 1979 [61]. It is only recently, with the advent of new computer-based models of medical decision making, that the question of when caregivers can concentrate on making more important clinical decision could be answered with a level of confidence. In order to provide trustworthy outcomes towards integrated decision support, a large amount of clinical computing effort must be combined with artificial intelligence techniques and knowledge engineering for a wide array of patient-treatment-related decisions.

13.5.3 Towards integrated decision support

Smart alarm systems are predecessors of future decision-support systems. Actuation-based MCPS are already quite complex and rarely incorporate user-centric designs. Apart from the ongoing discussion about human factors and ergonomics issues specific to a range of healthcare domains or applications, additional usability and utility studies have to be carried out in order to justify their integration into clinical workflows.

In general, integrated decision support has two components: first, to answer the question of how to get from guidelines to clinical decision support (for which a unified approach to translating and implementing medical knowledge [62] through semantic annotation is needed); and second, how to demonstrate that the support is relevant for clinical decisions (as in [63]). Towards this goal, mobile interaction and decision-support systems have been discussed recently [64]. Their potential extension to data analytics on a large scale has been tried by the following means:

- methods for knowledge extraction (leveraging social, population and clinical data) and personalisation via intelligent predictive analytics,
- design of personalised care systems to disseminate the discovered knowledge (and to enable patients to provide feedback to physicians about their ongoing care) and
- supporting personalised care delivery by interdisciplinary healthcare teams by modelling patient-focused workflows and supporting their adaptation.

One of the main problems in integrated decision support is also one of the main problems in MCPS in general: sensing medical parameters is quite easy, but perception and interpretation of what is sensed is hard when clinically relevant decisions are to be made. It remains unclear whether standard usability guidelines for industrial use case applications apply [50, 65]. However, we need empirical investigations so that we can better understand those specialised needs in MCPS. Specialised needs include an evidence-based and technology-aware certification process defined for approval of MCPS by regulating agencies, as their malfunctioning can do great harm to human health. But reliability is very challenging. Open problems of integrated decision-support MCPS include formal verification, validation and certification of MCPS.

13.6   Conclusion

In this chapter, we have explained the background of MCPS and presented an overview of some important related issues that have been investigated recently. In addition to outlining architectures and solutions, we have discussed some recent promising research and development directions. As an example, we highlighted the need to combine active and passive user input modes in clinical environments for knowledge acquisition and knowledge discovery towards enhanced real-time decision-making processes in MCPS. Furthermore, we believe that we need physiological closed-loop control systems with humans-in-the-loop extension. We argue that this would improve data acquisition and integration, and therefore would lead to better knowledge extraction for past and occurring events. Control theoretic methods, where the dialogue server proactively requests users for real-time expert input, may be helpful in this context.

MCPS systems share the following characteristics of cyber-physical systems: a tight integration of digital computation, responsible for control and communication in real time with a physical system, obeying the laws of physics and evolving in continuous time with actuators. These actuators control physical medical-related processes. But MCPS is unique in many respects and brings concerns about safety and dependability that have not been adequately researched in the field of human–computer interaction and intelligent user interfaces, in terms of usability and reliability in particular.

Acknowledgments

This work is partly supported by EIT Digital (http://www.eitdigital.eu/), in the MCPS activity (see http://www.dfki.de/MedicalCPS/). The author would like to thank industrial and academic project partners and software engineers.

References

[1]   Schirner G., Erdogmus D., Chowdhury K., Padir T. The future of human-in-the-loop cyber-physical systems. *Computer*, 46(1): 36–45, 2013.

[2]   Sonntag D., Weber M., Cavallaro A., Hammon M. Integrating digital pens in breast imaging for instant knowledge acquisition. *AI Magazine*, 35(1): 26–37, 2014.

[3]   Sonntag D., Zillner S., Schulz C., Weber M., Toyama T. Towards medical cyber-physical systems: Multimodal augmented reality for doctors and knowledge discovery about patients. In *Design, User Experience, and Usability. User Experience in Novel Technological Environments*, ed. Marcus A., Lecture Notes in Computer Science, vol. 8014, pp. 401–410. Springer, 2013.

[4]   Sonntag D. ERmed – towards medical multimodal cyber-physical environments. In *Foundations of Augmented Cognition. Advancing Human*

Performance and Decision-Making Through Adaptive Systems, eds Schmorrow D., Fidopiastis C., Lecture Notes in Computer Science, vol. 8534, pp. 359–370. Springer, 2014.

[5] Munir S., Stankovic J.A., Liang C.J.M., Lin S. Cyber physical system challenges for human-in-the-loop control. In *Proceedings of the 8th International Workshop on Feedback Computing*, Berkeley, CA. USENIX, 2013.

[6] Wood A.D., Stankovic J.A. Human in the loop: Distributed data streams for immersive cyber-physical systems. *ACM SIGBED Review*, 5(1): 20 (2pp.), 2008.

[7] Lee I., Sokolsky O., Chen S., *et al.* Challenges and research directions in medical cyber-physical systems. *Proceedings of the IEEE*, 100(1): 75–90, 2012.

[8] Kim B., Ayoub A., Sokolsky O., *et al.* Safety-assured development of the GPCA infusion pump software. In *Proceedings of the Ninth ACM International Conference on Embedded Software, EMSOFT'11*, New York, pp. 155–164. ACM, 2011.

[9] de Man F., Greuters S., Boer C., Veerman D., Loer S. Intra-operative monitoring – many alarms with minor impact. *Anaesthesia*, 68: 804–810, 2013.

[10] Carayon P. *Handbook of Human Factors and Ergonomics in Health Care and Patient Safety*, Second Edition. CRC Press, 2011.

[11] Massé F., Penders J., Serteyn A., van Bussel M., Arends J. Miniaturized wireless ECG-monitor for real-time detection of epileptic seizures. In *Wireless Health 2010, WH'10*, New York, pp. 111–117. ACM, 2010.

[12] Ramgopal S., Thome-Souza S., Jackson M., *et al.* Seizure detection, seizure prediction, and closed-loop warning systems in epilepsy. *Epilepsy & Behavior*, 37: 291–307, 2014.

[13] King A.L., Roederer A., Arney D., *et al.* GSA: A framework for rapid prototyping of smart alarm systems. In *Proceedings of the 1st ACM International Health Informatics Symposium, IHI'10*, New York, pp. 487–491. ACM, 2010.

[14] Wood A., Stankovic J.A., Virone G., *et al.* Context-aware wireless sensor networks for assisted living and residential monitoring. *Network, IEEE*, 22 (4): 26–33, 2008.

[15] Aranki D., Kurillo G., Yan P., Liebovitz D., Bajcsy R. Continuous, real-time, tele-monitoring of patients with chronic heart-failure – lessons learned from a pilot study. In *Proceedings of the 9th International Conference on Body Area Networks, BODYNETS 2014*, London, eds Fortino G., Suzuki J., Andreopoulos Y., Yuce M.R., Hao Y., Gravina R. ICST, 2014.

[16] Chen C., Favre J., Kurillo G., Andriacchi T.P., Bajcsy R., Chellappa R. Camera networks for healthcare, teleimmersion, and surveillance. *IEEE Computer*, 47(5): 26–36, 2014.

[17] Murugesan A., Sokolsky O., Rayadurgam S., Whalen M., Heimdahl M., Lee I. Linking abstract analysis to concrete design: A hierarchical approach to verify medical CPS safety. In *Proceedings of the ACM/IEEE 5th International Conference on Cyber-Physical Systems (with CPS Week 2014), ICCPS'14*, Washington, DC, pp. 139–150. IEEE Computer Society, 2014.

[18]   Wilhelm R., Engblom J., Ermedahl A., *et al.* The worst-case execution-time problem – overview of methods and survey of tools. *ACM Transactions on Embedded Computing Systems*, 7(3): 36 (53pp.), 2008.

[19]   Cucu-Grosjean L., Santinelli L., Houston M., *et al.* Measurement-based probabilistic timing analysis for multi-path programs. In *Proceedings of the 24th Euromicro Conference on Real-Time Systems (ECRTS)*, Pisa, pp. 91–101. IEEE, 2012.

[20]   Toser Z., Lorincz A. The cyber-physical system approach towards artificial general intelligence: The problem of verification. In *Proceedings of the 8th International Conference on Artificial General Intelligence, AGI*, Berlin, Lecture Notes in Computer Science, vol. 9205, pp. 373–383. Springer, 2015.

[21]   Chipalkatty R. Human-in-the-loop control for cooperative human–robot tasks. Dissertation. Georgia Institute of Technology, 2012.

[22]   Kozsik T., Lorincz A., Juhasz D., *et al.* Workflow description in cyber-physical systems. *Studia Universitatis Babes-Bolyai, Informatica*, 58(2): 20–30, 2013.

[23]   Agrawal A., Shiffman R.N. Using GEM-encoded guidelines to generate medical logic modules. In *American Medical Informatics Association Annual Symposium, AMIA 2001*, Washington, DC, pp. 7–11. AMIA, 2001.

[24]   Hussain T., Michel G., Shiffman R.N. The Yale Guideline Recommendation Corpus: A representative sample of the knowledge content of guidelines. *International Journal of Medical Informatics*, 78(5): 354–363, 2009.

[25]   Tuomainen M., Mykkänen J., Luostarinen H., Pöyhölä A., Paakkanen E. Model-centric approaches for the development of health information systems. *Studies in Health Technology and Informatics*, 129(1): 28–32, 2007.

[26]   Hewelt M., Kunde A., Weske M., Meinel C. Recommendations for medical treatment processes: the PIGS approach. *Business Process Management Workshops, BPM 2014*, Lecture Notes in Business Information Processing, vol. 202, pp. 16–27. Springer, 2015.

[27]   Ghidini C., Francescomarino C.D., Rospocher M., Tonella P., Serafini L. Semantics-based aspect-oriented management of exceptional flows in business processes. *IEEE Transactions on Systems, Man, and Cybernetics, Part C*, 42(1): 25–37, 2012.

[28]   Arney D., Plourde J., Schrenker R., Mattegunta P., Whitehead S.F., Goldman J.M. Design pillars for medical cyber-physical system middleware. In *Proceedings of the 5th Workshop on Medical Cyber-Physical Systems*, Dagstuhl, Germany, eds Turau V., Kwiatkowska M., Mangharam R., Weyer C., OpenAccess Series in Informatics (OASIcs), vol. 36, pp. 124–132. Schloss Dagstuhl–Leibniz-Zentrum fuer Informatik, 2014.

[29]   Sonntag D., Schulz C. A multimodal multi-device discourse and dialogue infrastructure for collaborative decision-making in medicine. In *Natural Interaction with Robots, Knowbots and Smartphones*, eds Mariani J., Rosset S., Garnier-Rizet M., Devillers L., pp. 37–47. Springer, 2014.

[30]   Sonntag D., Weber M., Hammon M., Cavallaro A. Integrating digital pens in breast imaging for instant knowledge acquisition. In *Innovative Applications of Artificial Intelligence, Twenty-Fifth IAAI Conference Proceedings*. AAAI, 2013.

[31] Sonntag D., Toyama T. On-body IE: A head-mounted multimodal augmented reality system for learning and recalling faces. In *Proceedings of the 9th International Conference on Intelligent Environments (IE)*, Athens, pp. 151–156. IEEE, 2013.

[32] Prange A., Sonntag D. Smartphone sketches for instant knowledge acquisition in breast imaging. In *Sketch: Pen and Touch Recognition, IUI 2014 Workshop, Haifa.* IUI, 2014.

[33] Chen J.J., Chakraborty S. Resource augmentation for uniprocessor and multiprocessor partitioned scheduling of sporadic real-time tasks. *Real-Time Systems*, 49(4): 475–516, 2013.

[34] Heikkila T., Strommer E., Kivikunnas S., *et al.* Low intrusive ehealth monitoring: Human posture and activity level detection with an intelligent furniture network. *Wireless Communications, IEEE*, 20(4): 57–63, 2013.

[35] Cleary J.G., Witten I. Data compression using adaptive coding and partial string matching. *IEEE Transactions on Communications*, 32(4): 396–402, 1984.

[36] Schürmann T., Grassberger P. Entropy estimation of symbol sequences. *Chaos*, 6(3): 414–427, 1996.

[37] Balcan M.F., Harvey N.J.A. Learning submodular functions. In *Machine Learning and Knowledge Discovery in Databases*, eds Flach P., Bie T., Cristianini N., Lecture Notes in Computer Science, vol. 7524, pp. 846–849. Springer, 2012.

[38] Zillner S., Sonntag D. Aligning medical ontologies by axiomatic models, corpus linguistic syntactic rules and context information. In *Proceedings of the 24th International Symposium on Computer-Based Medical Systems (CBMS)*, Bristol, pp. 1–6. IEEE, 2011.

[39] Oberkampf H., Zillner S., Bauer B., Hammon M. An OGMS-based model for clinical information (MCI). In *Proceedings of the 4th International Conference on Biomedical Ontology, ICBO 2013*, Montreal, eds Dumontier M., Hoehndorf R., Baker C.J.O., CEUR Workshop Proceedings, vol. 1060, pp. 97–100. CEUR, 2013.

[40] Oberkampf H., Gojayev T., Zillner S., Zühlke D., Auer S., Hammon M. From symptoms to diseases – creating the missing link. In *The Semantic Web. Latest Advances and New Domains – Proceedings of the 12th European Semantic Web Conference, ESWC 2015*, Portoroz, Slovenia, eds Gandon F., Sabou M., Sack H., d'Amato C., Cudré-Mauroux P., Zimmermann A., Lecture Notes in Computer Science, vol. 9088, pp. 652–667. Springer, 2015.

[41] Matthew R.P., Shia V., Tomizuka M., Bajcsy R. Optimal design for individualised passive assistance. In *Proceedings of the 6th Augmented Human International Conference, AH'15*, New York, pp. 69–76. ACM, 2015.

[42] Dickerson R.F., Gorlin E.I., Stankovic J.A. Empath: A continuous remote emotional health monitoring system for depressive illness. In *Proceedings of the 2nd Conference on Wireless Health, WH'11*, New York, pp. 5:1–5:10. ACM, 2011.

[43]   Brookmeyer R., Johnson E., Ziegler-Graham K., Arrighi H.M. Forecasting the global burden of Alzheimer's disease. *Alzheimer's & Dementia*, 3(3): 186–191, 2007.

[44]   Orlosky J., Toyama T., Sonntag D., Kiyokawa K. Using eye-gaze and visualization to augment memory. In *Distributed, Ambient, and Pervasive Interactions*, eds Streitz N., Markopoulos P., Lecture Notes in Computer Science, vol. 8530, pp. 282–291. Springer, 2014.

[45]   Voros G., Vero A., Pinter B., *et al.* Towards a smart wearable tool to enable people with SSPI to communicate by sentence fragments. In *Pervasive Computing Paradigms for Mental Health*, eds Cipresso P., Matic A., Lopez G., Lecture Notes of the Institute for Computer Sciences, Social Informatics and Telecommunications Engineering, vol. 100, pp. 90–99. Springer, 2014.

[46]   Toyama T., Sonntag D., Orlosky J., Kiyokawa K. Attention engagement and cognitive state analysis for augmented reality text display functions. In *Proceedings of the 20th International Conference on Intelligent User Interfaces, IUI'15*, New York, pp. 322–332. ACM, 2015.

[47]   Prange A., Sandrala I.P., Weber M., Sonntag D. Robot companions and smartpens for improved social communication of dementia patients. In *Proceedings of the 20th International Conference on Intelligent User Interfaces Companion, IUI Companion '15*, New York, pp. 65–68. ACM, 2015.

[48]   Hearst M.A. Trends & controversies: Mixed-initiative interaction. *IEEE Intelligent Systems*, 14(Sept./Oct.): 14, 1999.

[49]   Horvitz E. Trends & controversies: Uncertainty, action, and interaction: in pursuit of mixed-initiative computing. *IEEE Intelligent Systems*, 14(Sept./Oct.): 17–20, 1999.

[50]   Sonntag D. Collaborative multimodality. *KI – Künstliche Intelligenz*, 26(2): 161–168, 2012.

[51]   Nickel M., Jiang X., Tresp V. Reducing the rank in relational factorization models by including observable patterns. In *Advances in Neural Information Processing Systems 27: Annual Conference on Neural Information Processing Systems*, Montreal, eds Ghahramani Z., Welling M., Cortes C., Lawrence N.D., Weinberger K.Q., pp. 1179–1187. Neural Information Processing Systems Foundation, 2014.

[52]   Sonntag D., Tresp V., Zillner S., *et al.* The Clinical Data Intelligence Project (KDI). *Informatik Spektrum*, 38(3), 2015.

[53]   Gelissen J., Sonntag D. Special issue on health and wellbeing. *KI – Künstliche Intelligenz*, 29: 1–3, 2015.

[54]   NITRD. High Confidence Software and Systems Coordinating Group of the Networking and Information Technology Research and Development Program. *High-Confidence Medical Devices: Cyber-Physical Systems for 21st Century Health Care*. NITRD, 2009. See https://www.nitrd.gov/About/MedDevice-FINAL1-web.pdf/. Accessed: 20 December 2015.

[55]   Zillner S., Neururer S. Technology roadmap development for big data healthcare applications. *KI – Künstliche Intelligenz*, 29(2): 131–141, 2015.

[56] Kolias V.D., Stoitsis J., Golemati S., Nikita K.S. Utilizing semantic web technologies in healthcare. In *Concepts and Trends in Healthcare Information Systems*, eds Koutsouris D.D., Lazakidou A.A., Annals of Information Systems, vol. 16, pp. 9–19. Springer, 2014.

[57] Zillner S. Reasoning-based patient classification for enhanced medical image annotation. In *The Semantic Web: Research and Applications. Proceedings of the 7th Extended Semantic Web Conference, (ESWC 2010)*, Heraklion, Greece. Lecture Notes in Computer Science, vol. 6088, pp. 243–257. Springer, 2010.

[58] Zillner S., Seifert S., Erdt M., Daumke P., Kramer M. Semantic processing of medical data. In *Towards the Internet of Services: The THESEUS Research Program. Cognitive Technologies,* eds Wahlster W., Grallert H.J., Wess S., Friedrich H., Widenka T., pp. 343–356. Springer, 2014.

[59] Oberkampf H., Zillner S., Bauer B., Hammon M. Interpreting patient data using medical background knowledge. In *Proceedings of the 3rd International Conference on Biomedical Ontology (ICBO) 2012*, Graz, Austria. CEUR Workshop Proceedings, vol. 897 (5pp.). CEUR-WS, 2012.

[60] Mazoit J., Butscher K., Samii K. Morphine in postoperative patients: pharmacokinetics and pharmacodynamics of metabolites. *Journal of the American Society of Anesthesiologists* 105(1): 70–78, 2007.

[61] Shortliffe E.H., Buchanan B.G., Feigenbaum E.A. Knowledge engineering for medical decision making: A review of computer-based clinical decision aids. *Proceedings of the IEEE*, 67(9): 1207–1224, 1979.

[62] Middleton B., Kawamoto K., Reider J., Rosendale D., Shiffman R.N. From guidelines to clinical decision support: A unified approach to translating and implementing knowledge. In *American Medical Informatics Association Annual Symposium, AMIA 2012*, Chicago. AMIA, 2012.

[63] Chaney K., Shiffman R.N., Middleton B., White J., Reider J. Findings from a five-year clinical decision support demonstration project and the road ahead. In *American Medical Informatics Association Annual Symposium, AMIA 2013*, Washington, DC. AMIA, 2013.

[64] Sonntag D., Zillner S., Ernst P., Schulz C., Sintek M., Dankerl P. Mobile radiology interaction and decision support systems of the future. In *Towards the Internet of Services: The THESEUS Research Program. Cognitive Technologies*, eds Wahlster W., Grallert H.J., Wess S., Friedrich H., Widenka T., pp. 371–382. Springer, 2014.

[65] Sonntag D., Weihrauch C., Jacobs O., Porta D. *THESEUS Usability Guidelines for Use Case Applications*. DFKI and Federal Ministry of Education and Research Germany, 2010.

Index